PROBLEMS IN
COMPLEX VARIABLE THEORY

Modern Analytic *and* Computational Methods *in* Science *and* Mathematics

A GROUP OF MONOGRAPHS AND ADVANCED TEXTBOOKS

Richard Bellman, EDITOR
University of Southern California

Published

1. R. E. Bellman, R. E. Kalaba, and Marcia C. Prestrud, Invariant Imbedding and Radiative Transfer in Slabs of Finite Thickness, 1963
2. R. E. Bellman, Harriet H. Kagiwada, R. E. Kalaba, and Marcia C. Prestrud, Invariant Imbedding and Time-Dependent Transport Processes, 1964
3. R. E. Bellman and R. E. Kalaba, Quasilinearization and Nonlinear Boundary-Value Problems, 1965
4. R. E. Bellman, R. E. Kalaba, and Jo Ann Lockett, Numerical Inversion of the Laplace Transform: Applications to Biology, Economics, Engineering, and Physics, 1966
5. S. G. Mikhlin and K. L. Smolitskiy, Approximate Methods for Solution of Differential and Integral Equations, 1967
6. R. N. Adams and E. D. Denman, Wave Propagation and Turbulent Media, 1966
7. R. L. Stratonovich, Conditional Markov Processes and Their Application to the Theory of Optimal Control, 1968
8. A. G. Ivakhnenko and V. G. Lapa, Cybernetics and Forecasting Techniques, 1967
9. G. A. Chebotarev, Analytical and Numerical Methods of Celestial Mechanics, 1967
10. S. F. Feshchenko, N. I. Shkil', and L. D. Nikolenko, Asymptotic Methods in the Theory of Linear Differential Equations, 1967
11. A. G. Butkovskiy, Distributed Control Systems, 1969
12. R. E. Larson, State Increment Dynamic Programming, 1968
13. J. Kowalik and M. R. Osborne, Methods for Unconstrained Optimization Problems, 1968
14. S. J. Yakowitz, Mathematics of Adaptive Control Processes, 1969
15. S. K. Srinivasan, Stochastic Theory and Cascade Processes, 1969
16. D. U. von Rosenberg, Methods for the Numerical Solution of Partial Differential Equations, 1969
17. R. B. Banerji, Theory of Problem Solving: An Approach to Artificial Intelligence, 1969
18. R. Lattès and J.-L. Lions, The Method of Quasi-Reversibility: Applications to Partial Differential Equations. Translated from the French edition and edited by Richard Bellman, 1969
19. D. G. B. Edelen, Nonlocal Variations and Local Invariance of Fields, 1969
20. J. R. Radbill and G. A. McCue, Quasilinearization and Nonlinear Problems in Fluid and Orbital Mechanics, 1970
21. W. Squire, Integration for Engineers and Scientists, 1970
22. T. Parthasarathy and T. E. S. Raghavan, Some Topics in Two-Person Games, 1971
23. T. Hacker, Flight Stability and Control, 1970
24. D. H. Jacobson and D. Q. Mayne, Differential Dynamic Programming, 1970
25. H. Mine and S. Osaki, Markovian Decision Processes, 1970
26. W. Sierpiński, 250 Problems in Elementary Number Theory, 1970
27. E. D. Denman, Coupled Modes in Plasmas, Elastic Media, and Parametric Amplifiers, 1970
28. F. H. Northover, Applied Diffraction Theory, 1971
29. G. A. Phillipson, Identification of Distributed Systems, 1971
30. D. H. Moore, Heaviside Operational Calculus: An Elementary Foundation, 1971
32. V. F. Demyanov and A. M. Rubinov, Approximate Methods in Optimization Problems, 1970
33. S. K. Srinivasan and R. Vasudevan, Introduction to Random Differential Equations and Their Applications, 1971
34. C. J. Mode, Multitype Branching Processes: Theory and Applications, 1971
36. J. G. Krzyż, Problems in Complex Variable Theory, 1971
37. W. T. Tutte, Introduction to the Theory of Matroids, 1971

In Preparation

31. S. M. Roberts and J. S. Shipman, Two-Point Boundary Value Problems: Shooting Methods
35. R. Tomović and M. Vukobratović, General Sensitivity Theory
38. B. W. Rust and W. R. Burrus, Mathematical Programming and the Numerical Solution of Linear Equations

PROBLEMS IN COMPLEX VARIABLE THEORY

by

JAN G. KRZYŻ

Maria Curie-Skłodowska University, Lublin (Poland)
Institute of Mathematics, Polish Academy of Sciences

AMERICAN ELSEVIER PUBLISHING COMPANY, INC.
NEW YORK

PWN—POLISH SCIENTIFIC PUBLISHERS
WARSZAWA

1971

AMERICAN ELSEVIER PUBLISHING COMPANY, INC.
52 Vanderbilt Avenue, New York, N.Y. 10017

ELSEVIER PUBLISHING COMPANY, LTD.
Barking, Essex, England

ELSEVIER PUBLISHING COMPANY
335 Jan Van Galenstraat, P.O. Box 211
Amsterdam, The Netherlands

International Standard Book Number 70-153071

Library of Congress Catalog Card Number 0-444-00098-4

COPYRIGHT 1971 BY PAŃSTWOWE WYDAWNICTWO NAUKOWE
WARSZAWA (POLAND), MIODOWA 10

All rights reserved.
No part of this publication may be reproduced,
stored in a retrieval system, or transmitted
in any form or by any means, electronic,
mechanical, photocopying, recording,
or otherwise, without the prior
written permission of the publisher,
American Elsevier Publishing Company, Inc.
52 Vanderbilt Avenue, New York, N.Y. 10017.

PRINTED IN POLAND

To the memory of
Mieczysław Biernacki

Contents

Foreword   xiii
Notation   xv

PROBLEMS

1. Complex Numbers. Linear Transformations
   - 1.1. Sets and Sequences of Complex Numbers 3
   - 1.2. Spherical Representation 8
   - 1.3. Similarity Transformations 10
   - 1.4. Linear Transformations 11
   - 1.5. Symmetry 12
   - 1.6. Conformal Mappings Realized by Linear Transformations 13
   - 1.7. Invariant Points of Linear Transformations 14
   - 1.8. Hyperbolic Geometry [3], [21] 16

2. Regularity Conditions. Elementary Functions
   - 2.1. Continuity. Differentiability 19
   - 2.2. Harmonic Functions 21
   - 2.3. Geometrical Interpretation of the Derivative 23
   - 2.4. Conformal Mappings Connected with $w = z^2$ 24
   - 2.5. The Mapping $w = \frac{1}{2}(z+z^{-1})$ 25
   - 2.6. The Exponential Function and the Logarithm 27
   - 2.7. The Trigonometric and Hyperbolic Functions 28
   - 2.8. Inverse Trigonometric and Hyperbolic Functions 30
   - 2.9. Conformal Mapping of Circular Wedges 30

3. Complex Integration
   - 3.1. Line Integrals. The Index [1], [10] 33
   - 3.2. Cauchy's Theorem and Cauchy's Integral Formula 38
   - 3.3. Isolated Singularities 41
   - 3.4. Evaluation of Residues 43
   - 3.5. The Residue Theorem 45
   - 3.6. Evaluation of Definite Integrals Involving Trigonometric Functions 48
   - 3.7. Integrals over an Infinite Interval 49
   - 3.8. Integration of Many-Valued Functions [21] 52
   - 3.9. The Argument Principle. Rouché's Theorem 54

4. Sequences and Series of Analytic Functions
   - 4.1. Almost Uniform Convergence 57
   - 4.2. Power Series 58

4.3. Taylor Series	60
4.4. Boundary Behavior of Power Series	65
4.5. The Laurent Series	66
4.6. Summation of Series by Means of Contour Integration	69
4.7. Integrals Containing a Complex Parameter. The Gamma Function	72
4.8. Normal Families [1], [6], [16]	75

5. Meromorphic and Entire Functions
    - 5.1. Mittag–Leffler's Theorem [1] 77
    - 5.2. Partial Fractions Expansions of Meromorphic Functions [11] . . . 78
    - 5.3. Jensen's Formula. Nevanlinna's Characteristic [18] 80
    - 5.4. Infinite Products [1], [10] 82
    - 5.5. Factorization of an Entire Function [1], [6], [10] 84
    - 5.6. Factorization of Elementary Functions [11] 85
    - 5.7. Order of an Entire Function [6], [10], [13], [18] 88

6. The Maximum Principle
    - 6.1. The Maximum Principle for Analytic Functions 90
    - 6.2. Schwarz's Lemma [16] 91
    - 6.3. Subordination [6], [15], [16], [22] 92
    - 6.4. The Maximum Principle for Harmonic Functions 94

7. Analytic Continuation. Elliptic Functions
    - 7.1. Analytic Continuation [1], [2], [6], [10] 95
    - 7.2. The Reflection Principle [1] 96
    - 7.3. The Monodromy Theorem [1], [10] 98
    - 7.4. The Schwarz–Christoffel Formulae [1], [10] 100
    - 7.5. Jacobian Elliptic Functions sn, cn, dn [1], [6], [7], [10] 105
    - 7.6. The Functions σ, ζ, $\wp$ of Weierstrass [1], [6], [7], [10] 106
    - 7.7. Conformal Mappings Associated with Elliptic Functions [24] . . 108

8. The Dirichlet Problem
    - 8.1. Riemann's Mapping Theorem [1], [6], [10], [16] 110
    - 8.2. Poisson's Formula [1], [26], [27] 111
    - 8.3. The Dirichlet Problem [1], [14], [15], [19], [27] 113
    - 8.4. Harmonic Measure [1], [15], [16], [19], [25], [26], [27], ... 114
    - 8.5. Green's Function [15], [16], [17], [27] 116
    - 8.6. Bergman Kernel Function [6], [12], [24] 118

9. Two-Dimensional Vector Fields
    - 9.1. Stationary Two-Dimensional Flow of Incompressible Fluid [4], [9] 121
    - 9.2. Two-Dimensional Electrostatic Field [4] 124

10. Univalent Functions
    - 10.1. Functions of Positive Real Part [26], [27] 127
    - 10.2. Starshaped and Convex Functions [16], [17] 129
    - 10.3. Univalent Functions [16], [17], [20] 130
    - 10.4. The Inner Radius. Circular and Steiner Symmetrization [17], [20] 133
    - 10.5. The Method of Inner Radius Majorization [17] 136

SOLUTIONS

1. Complex Numbers. Linear Transformations 141
2. Regularity Conditions. Elementary Functions 155
3. Complex Integration 168
4. Sequences and Series of Analytic Functions 191
5. Meromorphic and Entire Functions 212
6. The Maximum Principle 224
7. Analytic Continuation. Elliptic Functions 229
8. The Dirichlet Problem 245
9. Two-Dimensional Vector Fields 256
10. Univalent Functions 260
Bibliography 277
Subject Index 279

Foreword

This collection of exercises in analytic functions is an enlarged and revised English edition of a Polish version first published in 1962. The book is mainly intended for mathematics students who are completing a first course in complex analysis, and its subject matter roughly corresponds to the material covered by Ahlfors's book [1]. Some chapters, for example, evaluation of residues, determination of conformal mappings, and applications in the two-dimensional field theory may be, however, of interest to engineering students. Most exercises are just examples illustrating basic concepts and theorems, some are standard theorems contained in most textbooks. However, the author does believe that the reconstruction of certain proofs could be instructive and is possible for an average mathematics student. When the subject matter of a particular chapter is not covered by standard textbooks, the numbers in parantheses given in the contents indicate a corresponding bibliography position which may be consulted for further information.

Some problems are due to the author, and some were adopted by the author from various sources. It was beyond the scope of author's possibility to trace the original sources and therefore the detailed references are omitted.

The second part of the book contains solutions of problems. In most cases a complete solution is given; in some cases, where no difficulties could be expected, or when an analogous problem has been already solved in a detailed manner, only a final solution is given. The author is well aware that it was extremely hard to avoid mistakes in a book of this kind. He did his best, however, to reduce their number to a minimum.

It is the author's pleasant duty to thank W. K. Hayman, Z. Lewandowski, and Q. I. Rahman, who suggested some problems included in this collection. Thanks are also due to Mrs. J. Zygmunt for her help in preparing the manuscript, as well as to M. Stark for his help and encouragement.

Lublin, July 1969   JAN G. KRZYŻ

Notation

1. Set theory

$a \in A$	a is an element of the set A
$a \notin A$	a is not an element of A
$B \subset A$	B is a subset of A
$A \cap B$	Intersection of sets A and B
$A \cup B$	Union of sets A and B
$A \setminus B$	The complement of B with respect to A
$A \setminus a$	The set A with the element a removed
$\{a : W(a)\}$	The set of all a having a property $W(a)$
$\overline{A}$	Closure of the set A
fr A	Boundary of the set A
∂A	The boundary cycle of a domain A taken with positive orientation
$\emptyset$	The empty set

2. Complex numbers

re z	The real part of a complex number $z = x+iy$, i.e. the real number x
im z	The imaginary part of a complex number $z = x+iy$, i.e. the real number y
$\bar{z}$	The conjugate of $z = x+iy$, i.e. the complex number $x-iy$
$|z|$	The absolute value of $z = x+iy$, i.e. $\sqrt{x^2+y^2}$
$\begin{cases}\arg z \\ \text{Arg } z\end{cases}$	The argument of $z \neq 0, \infty$, i.e. any angle θ satisfying the equations $|z|\cos\theta = \text{re } z$, $|z|\sin\theta = \text{im } z$. There exists a unique value of $\arg z$ which satisfies $-\pi < \arg z \leqslant \pi$. It is called the principal value of argument and is denoted Arg z

3. Sets of complex numbers

$[z_1, z_2]$	Closed line segment with end points z_1, z_2
(z_1, z_2)	Open line segment with end points z_1, z_2
$[z_1, z_2, \ldots, z_n]$	Polygonal line joining $z_1, z_2, \ldots, z_n$ in this order

$[z_1, z_2) = [z_1, z_2]\setminus z_2$, $(z_1, z_2] = [z_1, z_2]\setminus z_1$

$[0, +\infty)$, $(-\infty, 0]$ Positive and negative real axis

$[0, +i\infty)$, $(-i\infty, 0]$ Positive and negative imaginary axis

$(-\infty, +\infty)$ The set of all real numbers

$K(z_0; r)$ The open disk with center at z_0 and radius r

$K(\infty; 1)$ The set of all z with $|z| > 1$

$C(z_0; r)$ The circle with center at z_0 and radius r

H_+ (H_-) The upper (lower) half-plane

$\mathbf{C}$ ($\hat{\mathbf{C}}$) The finite (extended) plane

$(\mp; \mp)$ The open quadrants of $\mathbf{C}$, e.g. $(+; -) = \{z: \operatorname{re} z > 0, \operatorname{im} z < 0\}$

$\operatorname{conv} A$ The convex hull of a set A

$\operatorname{dist}(a; B) = \inf\{x: x = |a-b|, b \in B\}$

$\operatorname{dist}(A; B) = \inf\{x: x = |a-b|, a \in A, b \in B\}$

We use the same symbol $C(z_0; r)$ for $\operatorname{fr} K(z_0; r)$, as well as for $\partial K(z_0; r)$; and similarly, $[z_0, z_1]$ denotes either a set, or an oriented segment. We hope that this does not cause any misunderstanding.

4. Functions and mappings

$1:1$ One-to-one correspondence

$f(A)$ The image set of the set A under the mapping f

$\rho(a, b)$ The hyperbolic distance between $a, b \in K(0; 1)$, cf. Exercise 1.8.12

$\sigma(a, b)$ The spherical distance between $a, b \in \overline{\mathbf{C}}$

$\rho(a; f)$ The spherical derivative of f at the point a

$\omega(z; \gamma, G)$ Harmonic measure of an arc $\gamma \subset \operatorname{fr} G$ at a point $z \in G$ (G is a domain)

$g(z, z_0; G)$ Green's function of a domain G with the pole z_0

$r(z_0; G)$ The inner radius of a domain G at the point $z_0 \in G$

D_f The set of all values taken by f

$M(r, f)$ The l.u.b. of $|f(z)|$ on the circle $C(0; r)$

$n(\gamma, a)$ The index of the point a w.r.t. a closed, regular curve γ

$\rightrightarrows_G$ Almost uniform convergence of a sequence of functions in G

$\Gamma \sim 0 \pmod{G}$ The cycle Γ is homologous to zero w.r.t. the domain G

o, O If the quotient $f(t)/g(t)$ tends to 0 (or remains bounded) as $t \to t_0$, we write: $f(t) = o(g(t))$ (or $f(t) = O(g(t))$)

5. Families of function

$L_2(G)$ The class of functions analytic in a domain G and such that $\iint_G |f'|^2 < +\infty$

$\mathscr{P}$ The class of functions f analytic in $K(0; 1)$ such that $f(0) = 1$ and re $f(z) > 0$ in $K(0; 1)$

S The class of functions f analytic and univalent in $K(0; 1)$ such that $f(0) = 0$, $f'(0) = 1$

S^* The class of all $f \in S$ such that $f[K(0; 1)]$ is starshaped w.r.t. origin

S^c The class of all $f \in S$ such that $f[K(0; 1)]$ is a convex domain

Σ The class of all functions F analytic and univalent in $K(\infty; 1)$ whose Laurent expansion there has the form: $F(z) = z + b_0 + b_1 z^{-1} + \ldots$

Σ_0 The class of all $F \in \Sigma$ which do not take value 0

6. Abbreviations

a.u.   Almost uniform (convergence)

h-   Hyperbolic (or noneuclidean)

cont.   The problem is a continuation of the foregoing one

PROBLEMS

CHAPTER 1

Complex Numbers. Linear Transformations

1.1. SETS AND SEQUENCES OF COMPLEX NUMBERS

1.1.1. Find the real and imaginary parts of:

$$\frac{2}{1-3i}, \quad (1+i\sqrt{3})^6, \quad \left(\frac{1+i}{1-i}\right)^5, \quad \left(\frac{1+i\sqrt{3}}{1-i}\right)^4.$$

1.1.2. Find all complex z satisfying $\bar{z} = z^2$.

1.1.3. Show that for $|z| = r > 0$:

$$\mathrm{re}\, z = \frac{1}{2}\left(z + \frac{r^2}{z}\right), \quad \mathrm{im}\, z = \frac{1}{2i}\left(z - \frac{r^2}{z}\right).$$

1.1.4. Evaluate all complex z for which $(1+z)(1-z)^{-1}$ is (i) purely real; (ii) purely imaginary.

1.1.5. Show that unless $z = x+iy$ is real and negative, there exists a unique ζ with $\mathrm{re}\,\zeta > 0$ such that $\zeta^2 = z$.

1.1.6. Prove the identity

$$|z_1+z_2|^2 + |z_1-z_2|^2 = 2(|z_1|^2 + |z_2|^2)$$

and explain its geometrical meaning.

1.1.7. Show that

$$|a + \sqrt{a^2-b^2}| + |a - \sqrt{a^2-b^2}| = |a+b| + |a-b|.$$

1.1.8. Prove the identities:

(i) $|z_1(1+|z_2|^2) - z_2(1+|z_1|^2)|^2 = |z_1-z_2|^2|1-z_1\bar{z}_2|^2 - (z_1\bar{z}_2 - \bar{z}_1 z_2)^2$;
(ii) $|1+z_1\bar{z}_2|^2 + |z_1-z_2|^2 = (1+|z_1|^2)(1+|z_2|^2)$;
(iii) $|1-z_1\bar{z}_2|^2 - |z_1-z_2|^2 = (1-|z_1|^2)(1-|z_2|^2)$.

1.1.9. Show that for $\theta \neq 0, \mp 2\pi, \mp 4\pi, \ldots$
$$1+e^{i\theta}+ \ldots +e^{in\theta} = \sin\tfrac{1}{2}(n+1)\theta(\sin\tfrac{1}{2}\theta)^{-1}\,e^{in\theta/2}$$

1.1.10. Suppose $|z_1| = |z_2| = |z_3| = 1$. Show that z_1, z_2, z_3 are vertices of an equilateral triangle, iff $z_1+z_2+z_3 = 0$.

1.1.11. Suppose $|z_k| = 1$, $k = 1$ to 4. Show that z_k are vertices of a rectangle, iff $\sum_{k=1}^{4} z_k = 0$.

1.1.12. Let $z_1 * z_2$ be the inner (scalar) product of vectors $[0, z_1]$, $[0, z_2]$ and $z_1 \times z_2$ the component of the outer (vector) product perpendicular to the z-plane. Show that
$$\bar{z}_1 z_2 = z_1 * z_2 + z_1 \times z_2 \cdot i.$$

1.1.13. Prove that the area enclosed by the triangle $[z_1, z_2, z_3, z_1]$ whose orientation is positive, is equal $\tfrac{1}{2}\mathrm{im}(\bar{z}_1 z_2 + \bar{z}_2 z_3 + \bar{z}_3 z_1)$.

1.1.14. Prove that the area enclosed by the positively oriented polygon $[z_1, z_2, \ldots, z_n, z_1]$ is equal $\tfrac{1}{2}\mathrm{im}(\bar{z}_1 z_2 + \bar{z}_2 z_3 + \ldots + \bar{z}_n z_1)$.

1.1.15. Find the angle α between two vectors $[z_1, z_2]$, $[z_3, z_4]$. Also find the distance of z_3 from the straight line through z_1, z_2.

1.1.16. Show that, if the centers of gravity and of circum-circle of a triangle coincide, the triangle must be equilateral.

1.1.17. Discuss the curves defined by the following complex functions of a real variable (unless stated otherwise, a, b, ω are real and positive and $-\infty < t < +\infty$).
 (i) $z = \exp(a+bit)$;
 (ii) $z = (1+it)^{-1}$;
 (iii) $z = ae^{it} + a^{-1}e^{-it}$, $0 \leqslant t \leqslant 2\pi$;
 (iv) $z = at + be^{i\omega t}$;
 (v) $z = a(1-it)e^{it}$;
 (vi) $z = (1+e^{it})^2$, $0 \leqslant t \leqslant 2\pi$;
 (vii) $z = t^2 + it^4$, $0 \leqslant t < +\infty$;
 (viii) $z = t + it^{-1}$, $t > 0$.

1.1.18. Suppose $z(t) = x(t) + iy(t)$ and x, y are real differentiable functions of $t \in (a, b)$. Explain the geometrical meaning of $z'(t) = x'(t) + iy'(t)$. Show that $(e^{it})' = ie^{it}$.

1.1. SETS AND SEQUENCES

1.1.19. Given the curve $z(\theta) = r(\theta)e^{i\theta}$, where $r(\theta)$ is a positive, differentiable function of θ. Find the angle α between the tangent and the radius vector.

1.1.20. Suppose $z(t) = x(t)+iy(t)$ is a complex, differentiable function of a real parameter $t \in (a, b)$ that does not vanish in (a, b). Show that

(i) $\dfrac{d}{dt}\arg z(t) = (xy'-x'y)|z(t)|^{-2}$;

(ii) $\dfrac{d}{dt}|z(t)| = (xx'+yy')|z(t)|^{-1}$.

1.1.21. Explain the geometrical meaning of the following sets of complex numbers:

(i) $\{z: |z-a| = |z-b|\}$, $a \neq b$;
(ii) $\{z: |z+c|+|z-c| \leqslant 2a\}$, $a > 0$, $|c| < a$;
(iii) $\left\{z: 0 < \arg\dfrac{z+i}{z-i} < \dfrac{\pi}{4}\right\}$;
(iv) $\{z: 0 \leqslant \operatorname{re} iz < 1\}$;
(v) $\{z: \operatorname{re} z^2 > \alpha\}$, $\alpha > 0$;
(vi) $\left\{z: \left|\dfrac{z+1}{z-1}\right| < 1\right\}$;
(vii) $\{z: |z|+\operatorname{re} z \leqslant 1\}$;
(viii) $\{z: |z^2-1| < 1\}$;
(ix) $\{z: \operatorname{re}[z(z+i)(z-i)^{-1}] > 0\}$.

1.1.22. Show that the set $\{z: \arg(z-a)(z-b) = \operatorname{const}\}$ is an arc of an equilateral hyperbola whose center is located at $\tfrac{1}{2}(a+b)$.

1.1.23. Evaluate all $R > 0$ for which the set
$$\{z: |z^2+az+b| < R\}$$
is connected.

1.1.24. Explain the geometrical meaning of the set
$$\{z: A|z|^2-\bar{B}z-B\bar{z}+C = 0\},$$
where A, C are real, $A \neq 0$, $|B|^2 > AC$.

1.1.25. Find the radius and the center of Apollonius circle
$$|z-a|\,|z-b|^{-1} = k \quad (k \neq 1, k > 0).$$

1.1.26. Find the equation of the circle through three not collinear points z_1, z_2, z_3 (cf. Ex. 1.1.24).

1.1.27. Suppose $|z_1| \neq 0, 1$ and $0, z_1, z_2$ are not collinear. Show that the circle through $z_1, z_2, \bar{z}_1^{-1}$ has $[z_1(1+|z_2|^2)-z_2(1+|z_1|^2)](z_1\bar{z}_2-\bar{z}_1 z_2)^{-1}$ as center and its radius is $|z_1-z_2|\,|1-z_1\bar{z}_2|\,|z_1\bar{z}_2-\bar{z}_1 z_2|^{-1}$. Also show that $\bar{z}_2^{-1}$ is situated on this circle.

1.1.28. Suppose m_1, m_2, m_3 are nonnegative, $m_1+m_2+m_3 = 1$, and z_1, z_2, z_3 are not collinear. Show that

(i) the point $z_0 = m_1 z_1 + m_2 z_2 + m_3 z_3$ belongs to the closed triangle T with vertices z_1, z_2, z_3;

(ii) conversely, for any $z_0 \in T$ there exists a unique system of nonnegative m_1, m_2, m_3 with $m_1+m_2+m_3 = 1$ such that $z_0 = m_1 z_1 + m_2 z_2 + m_3 z_3$. The numbers m_j are called *barycentric coordinates* of z_0 w.r.t. T.

1.1.29. The intersection of all closed and convex sets containing a given set A is called the *convex hull* of A and is denoted $\text{conv}\, A$. Show that

$$\text{conv}\{z_1, z_2, \ldots, z_n\} = \left\{ \zeta : \zeta = \sum_{k=1}^n m_k z_k;\ m_k \geq 0,\ k=1,\ldots,n;\ \sum_{k=1}^n m_k = 1 \right\}.$$

1.1.30. Show that $\text{conv}\{z_1, z_2, \ldots, z_n\} = \bigcup T_{klm}$, where T_{klm} is the closed triangle with vertices z_k, z_l, z_m and the summation ranges over all triples $\{k, l, m\}$ of positive integers $\leq n$.

1.1.31. Prove that the equality $\sum_{k=1}^n (\zeta - z_k)^{-1} = 0$ implies:

$$\zeta \in \text{conv}\{z_1, z_2, \ldots, z_n\}.$$

1.1.32. Prove following theorem (due to Gauss and Lucas): all zeros of the derivative of a polynomial are contained in the convex hull of zeros of the given polynomial.

1.1.33. Show that

$$\lim_{n\to\infty} \left(1 + \frac{x+iy}{n}\right)^n = e^x(\cos y + i \sin y).$$

1.1.34. Discuss the behavior of the sequence $\{z_n\}$,

$$z_n = (1+i)\left(1+\frac{i}{2}\right)\cdots\left(1+\frac{i}{n}\right).$$

1.1.35. Show that, if $\{\zeta_n\}$ and $\sum |b_n|$ both converge, then the series $\sum \zeta_n b$ is convergent, too.

1.1.36. Suppose $\operatorname{re} z_i \geq 0$ ($n = 1, 2, \ldots$) and both $\sum z_n$, $\sum z_n^2$ converge. Show that also $\sum |z_n|^2$ is convergent.

1.1.37. Prove Toeplitz's theorem: Suppose (a_{nk}) is an infinite matrix of complex numbers ($n, k = 1, 2, \ldots$) which satisfies:

(i) $\sum_{k=1}^{\infty} |a_{nk}| \leq A$ for $n = 1, 2, \ldots$;

(ii) $\lim_{k \to \infty} a_{nk} = 0$ for $n = 1, 2, \ldots$;

(iii) $\lim_{n \to \infty} \left(\sum_{k=1}^{\infty} a_{nk} \right) = 1$.

Then for any positive integer n and any convergent $\{\zeta_n\}$ the series $\sum_{k=1}^{\infty} a_{nk} \zeta_k$ is convergent.

Moreover, if $z_n = \sum_{k=1}^{\infty} a_{nk} \zeta_k$, then $\lim_{n \to \infty} z_n$ exists and is equal $\lim_{n \to \infty} \zeta_n$.

1.1.38. Suppose $\dfrac{p_1 + p_2 + \ldots p_n}{|p_1| + |p_2| + \ldots + |p_n|} \geq M > 0$ for $n = 1, 2, \ldots$ and $\lim_{n \to \infty}(|p_1| + |p_2| + \ldots + |p_n|) = +\infty$. Show that for any convergent $\{z_n\}$

$$\lim_{n \to \infty} \frac{p_1 z_1 + p_2 z_2 + \ldots + p_n z_n}{p_1 + p_2 + \ldots + p_n} = \lim_{n \to \infty} z_n.$$

1.1.39. Suppose $z_n = u_0 + u_1 + \ldots + u_{n-1} + c u_n$ is a convergent sequence and $\operatorname{re} c > \frac{1}{2}$. Show that also $w_n = u_0 + u_1 + \ldots + u_{n-1} + u_n$ is convergent and has the same limit.

1.1.40. Suppose $\{p_n\}$ is a sequence of positive reals monotonically increasing to infinity. Show that for any convergent series $\sum z_n$ with complex terms we have:

$$w_n = p_n^{-1}(p_1 z_1 + p_2 z_2 + \ldots p_n z_n) \to 0.$$

1.1.41. Show that f $\sum \mu_n z_n$ converges and $\mu_n \to 0$ then

$$\lim_{n \to \infty} \mu_n(z_1 + z_2 + \ldots + z_n) = 0.$$

1.1.42. Suppose $\{u_n\}$, $\{v_n\}$ converge to u and v resp. Show that

$$w_n = \frac{1}{n}(u_1 v_n + u_2 v_{n-1} + \ldots + u_n v_1)$$

converges to uv.

1.2. SPHERICAL REPRESENTATION

1.2.1. Suppose $Ox_1x_2x_3$ is the system of rectangular coordinates whose axes Ox_1, Ox_2 coincide with real and imaginary axes Ox, Oy of the complex plane $\mathbf{C}$. Suppose, moreover, that the ray emanating from the north pole $N(0; 0; 1)$ of the unit sphere S: $x_1^2+x_2^2+x_3^2 = 1$ and intersecting S at $A(x_1; x_2; x_3)$ intersects $\mathbf{C}$ at the point z. The point $z = x+iy$ is called the stereographic projection of $A(x_1; x_2; x_3)$ whereas A is called the spherical image of z. Show that

$$x_1 = \frac{z+\bar{z}}{1+|z|^2}, \quad x_2 = \frac{z-\bar{z}}{i(1+|z|^2)}, \quad x_3 = \frac{|z|^2-1}{1+|z|^2},$$

and

$$z = \frac{x_1+ix_2}{1-x_3}.$$

Hence the points of the sphere S (also called Riemann sphere) can be used for geometrical representation of complex numbers.

1.2.2. Find the spherical images of $e^{i\alpha}$, $-1+i$, $3-4i$.

1.2.3. Describe spherical images of northern and southern hemisphere.

1.2.4. Show that any straight line in $\mathbf{C}$ has a circle through N as its spherical image.

1.2.5. Show that the stereographic projection of any circle on S not containing N is also a circle.

1.2.6. Show that the spherical images of z, $\bar{z}^{-1}$ are points symmetric w.r.t. the plane Ox_1x_2.

1.2.7. Find the relation between the spherical images of following points: (i) z, $-z$; (ii) z, $\bar{z}$; (iii) z, z^{-1}.

1.2.8. If φ, θ denote the geographical latitude and longitude of A respectively, show that the stereographic projection of A has the representation: $z = e^{i\theta}\tan(\frac{1}{4}\pi+\frac{1}{2}\varphi)$.

1.2.9. Show that z, ζ correspond to antipodal points of S, iff $z\bar{\zeta} = -1$.

1.2.10. Prove that the circle $A|z|^2+Bz+\bar{B}\bar{z}+C = 0$ (A, C real) has a great circle on S as its spherical image, iff $A+C = 0$.

1.2.11. Show that $C(z_0; R)$ is the stereographic projection of a great circle on S, iff $R^2 = 1+|z_0|^2$.

1.2.12. Find the stereographic projection of the great circle joining the points $(\frac{3}{13}; -\frac{4}{13}; \frac{12}{13})$, $(-\frac{2}{3}; \frac{2}{3}; \frac{1}{3})$.

1.2. SPHERICAL REPRESENTATION

1.2.13. The distance $\sigma(z_1, z_2)$ between two points on S whose stereographic projections are z_1, z_2 is called the spherical distance between z_1 and z_2. Show that

$$\sigma(z_1, z_2) = 2|z_1 - z_2|[(1+|z_1|^2)(1+|z_2|^2)]^{-1/2}.$$

1.2.14. Suppose $d\sigma$, ds are lengths of infinitesimal arcs on S and C resp. corresponding to each other under stereographic projection. Suppose, moreover, the arc of length ds emanates from the point $z \in C$. Show that

$$\frac{d\sigma}{ds} = 2(1+|z|^2)^{-1}.$$

Show, moreover, that the angle between any two regular arcs in C and the angle between their spherical images are equal.

1.2.15. Suppose the sphere S is rotated by the angle φ round the diameter whose end points have $a, -\bar{a}^{-1}$ (cf. 1.2.9) as stereographic projections. Suppose, moreover, z, ζ are stereographic projections of points corresponding to each other under this rotation. Show that

$$\frac{\zeta - a}{1 + \bar{a}\zeta} = e^{i\varphi} \frac{z-a}{1+\bar{a}z}.$$

1.2.16. Suppose $A_1, A_2 \in S$ and a_1, a_2 are stereographic projections of A_1 and A_2, resp. Find the set of all points $a \in C$ such that a is the stereographic projection of a point $A \in S$ equidistant from A_1 and A_2.

1.2.17. Find the radius of the circle on S whose stereographic projection is $C(a; r)$.

1.2.18. Suppose Γ is a regular arc on S and γ is its stereographic projection. Show that the length $l(\Gamma)$ of Γ is equal to

$$\int_\gamma \frac{2}{1+|z|^2} ds.$$

1.2.19. Find the stereographic projection of a rhumb line on S, i.e. of a line on S which cuts all meridians at the same angle.

1.2.20. Find the length $l(\Gamma)$ of the rhumb line Γ joining the points whose stereographic projection are $z_1 = r_1$, $z_2 = r_2 e^{i\alpha}$, $0 < \alpha < 2\pi$. Evaluate $l(\Gamma)$ for $z_1 = \frac{1}{\sqrt{3}}$, $z_2 = \frac{1}{2}(3+i\sqrt{3})$.

1.2.21. Evaluate the area of a spherical domain D being the spherical image of a regular domain Δ in C.

1.2.22. Show that the area $|T|$ of a spherical triangle T with angles α, β, γ is equal $\alpha+\beta+\gamma-\pi$.

1.2.23. Evaluate the area of the spherical triangle T whose vertices are $(0; 2^{-1/2}; 2^{-1/2})$, $(2^{-1/2}; 2^{-1/2}; 0)$, $(0; 1; 0)$.

1.3. SIMILARITY TRANSFORMATIONS

1.3.1. Show that each similarity transformation $w = az+b$ ($a \neq 0$) can be composed of a translation, a rotation and a homothety with center at the origin.

1.3.2. Prove that a similarity transformation
 (i) carries circles into circles and
 (ii) parallel straight lines into parallel straight lines;
 (iii) leaves the ratio $(z_3-z_1)/(z_3-z_2)$ unchanged;
 (iv) leaves the angle between two curves unchanged.

1.3.3. Find a similarity transformation mapping the strip
$$\{z: 0 < \operatorname{re} z < 1\}$$
onto the strip
$$\{w: |\operatorname{im} w| < \tfrac{1}{2}\pi\}$$
so that $(z = \tfrac{1}{2}) \leftrightarrow (w = 0)$.

1.3.4. Find the most general similarity transformation mapping
 (i) the upper half-plane onto itself;
 (ii) the upper half-plane onto the lower one;
 (iii) the upper half-plane onto the right half-plane.

1.3.5. Find the similarity transformation mapping the segment $[a, b]$ onto $[A, B]$ so that $a \leftrightarrow A$, $b \leftrightarrow B$.

1.3.6. Find the similarity transformation mapping the triangle with vertices $0, 1, i$ onto the triangle with vertices $0, 2, 1+i$.

1.3.7. Find the similarity transformation mapping the strip
$$\{x+iy: kx+b_1 \leqslant y \leqslant kx+b_2\}, \quad b_1 < b_2,$$
onto the strip
$$\{w: 0 \leqslant \operatorname{re} w \leqslant 1\}$$
so that $(z = ib_2) \leftrightarrow (w = 0)$.

1.4. LINEAR TRANSFORMATIONS

1.3.8. Show that for any similarity transformation $w = az+b$ ($a \neq 0, 1$) there exists a unique invariant point z_0; show that the transformation can be composed of rotation and a homothety center at z_0.

1.3.9. Find the invariant point z_0, the angle of rotation and the ratio of homothety for the transformations in (i) Exercise 1.3.6; (ii) Exercise 1.3.7.

1.3.10. Show that for any similarity transformation $w = az+b$ ($a \neq 0, 1$) there exists a family of logarithmic spirals invariant under the transformation.

1.4. LINEAR TRANSFORMATIONS

1.4.1. Show that any linear transformation $w = (az+b)/(cz+d)$ (a, b, c, d are complex constants, $ad-bc \neq 0$) is composed of a translation: $z_1 = z+\alpha$, an inversion: $z_2 = 1/z_1$ and a similarity transformation: $w = Az_2+B$ (some of these transformations may, however, fall out).

1.4.2. Prove that any regular arc γ and its image arc under inversion $w = 1/z$ intersect the radius vector at the same angles.

1.4.3. Show that the angle between any regular arcs γ_1, γ_2, and the angle between their image arcs under any linear transformation are equal.

1.4.4. Show that the inversion $w = 1/z$ carries the circle $C(a; r)$ into the circle with radius $r||a|^2-r^2|^{-1}$ and center $\bar{a}(|a|^2-r^2)^{-1}$ whenever $|a| \neq r$. Find the image line in case $|a| = r$.

1.4.5. Show that under inversion any circle through $-1, 1$ is carried into itself.

1.4.6. Show that under linear transformations circles are mapped onto circles or straight lines.

1.4.7. Prove that under inversion, and also under linear transformations the cross-ratio

$$(z_1, z_2, z_3, z_4) = \frac{z_1-z_3}{z_1-z_4} : \frac{z_2-z_3}{z_2-z_4}$$

remains unchanged.

1.4.8. Find the linear transformation carrying a, b, c into $0, 1, \infty$, respectively.

1.4.9. Find the images of the following lines under inversion:
 (i) the family of circles $C(a; |a|)$;
 (ii) the family of parallel straight lines $y = x+b$;

(iii) the family of straight lines through the origin;
(iv) the family of straight lines through z_0 $(z_0 \neq 0)$;
(v) the parabola $y = x^2$.

1.4.10. Show that the cross-ratio (z_1, z_2, z_3, z_4) is real, iff all the four points z_k lie on circle or on a straight line.

1.4.11. Show that all linear transformations form a group.

1.4.12. Prove that all linear transformations $w = (mz+n)/(pz+q)$ where m, n, p, q are integers satisfying $mq-np = 1$, also form a group.

1.5. SYMMETRY

Two points z, z^* $(z \neq z^*)$ are called *symmetric* w.r.t. L where L is a circle, or a straight line, if every circle through z, z^* intersects L at a right angle. In particular, if $L = C(a; r)$ and $z \neq a, \infty$ then z^* lies on the ray through z and with origin at a and $|z-a||z^*-a| = r^2$. The point z^* is called *reflection* of z with respect to L.

1.5.1. If z^* is the reflection of z with respect to $C(a; r)$, then $z^* = a + r^2/(\bar{z}-\bar{a})$.

1.5.2. Show that $C(a; r)$ is an Apollonius circle for points z, z^* (i.e. $|(\zeta-z)/(\zeta-z^*)| = $ const for $\zeta \in C(a; r)$).

1.5.3. Find the reflection of $2+i$ w.r.t. $C(i; 3)$.

1.5.4. Find the reflections w.r.t. $C(0; 1)$ of (i) the circle $C(1; 1)$; (ii) the hyperbola $x^2-y^2 = 1$.

1.5.5. Show that any circle orthogonal to $C(0; 1)$ remains unchanged after a reflection w.r.t. $C(0; 1)$.

1.5.6. Show that the symmetry w.r.t. L remains preserved under linear transformations: if $z \leftrightarrow w$, $z^* \leftrightarrow w^*$, $L \leftrightarrow L_1$ and z, z^* are symmetric w.r.t. L, then also w, w^* are symmetric w.r.t. L_1.

1.5.7. Given the point z, find its reflection z^* w.r.t. the straight line through a_1, a_2.

1.5.8. Show that the reflection w.r.t. L_1 followed by the reflection w.r.t. L_2 is a linear transformation. When is the resulting transformation independent of the order of reflections?

1.5.9. Prove that the points symmetric w.r.t. both circles $C(a_1; r_1)$, $C(a_2; r_2)$ are roots of the equation $\bar{a}_2 - \bar{a}_1 = r_1^2(z-a_1)^{-1} - r_2^2(z-a_2)^{-1}$.

1.5.10. Given the straight line L through a_1, a_2 and the circle $C(a;r)$, find the pair of points symmetric w.r.t. both lines.

1.5.11. Find all circles orthogonal to both circles $C(0;1)$, $C(1;4)$.

1.6. CONFORMAL MAPPINGS REALIZED BY LINEAR TRANSFORMATIONS

Given three points z_k and their image points w_k ($k = 1, 2, 3$) under a linear transformation w, we can determine w by solving the equation $(w, w_1, w_2, w_3) = (z, z_1, z_2, z_3)$.

We can also use the symmetry invariance in order to determine the linear transformation: if we know that C_w is the image of C_z and w is the image z, then also the point w^* symmetric to w w.r.t. C_w must be the image of z^*, z^* being the reflection of z w.r.t. C_z.

1.6.1. Find the image domain of the given domain in z-plane under the given linear transformation:
  (i) $\{z:\ \mathrm{re}\,z > 0,\ \mathrm{im}\,z > 0\}$, $\quad w = (z-i)(z+i)^{-1}$;
  (ii) $\{z:\ |z| < 1,\ \mathrm{im}\,z > 0\}$, $\quad w = (2z-i)(2+iz)^{-1}$;
  (iii) $\{z:\ 0 < \arg z < \tfrac{1}{4}\pi\}$, $\quad w = z(z-1)^{-1}$;
  (iv) $\{z:\ 0 < \mathrm{re}\,z < 1\}$, $\quad w = (z-1)(z-2)^{-1}$;
  (v) $\{z:\ 1 < |z| < 2\}$, $\quad w = z(z-1)^{-1}$.

1.6.2. Find the linear transformation carrying the circle $C(0;1)$ into a straight line parallel to the imaginary axis, the point $z = 4$ into the point $w = 0$ and leaving the circle $C(0;2)$ invariant.

1.6.3. Find the linear transformation carrying the points a, b, c, d on the real axis ($a < b < c < d$) into $-k^{-1}$, -1, 1, k^{-1} ($0 < k < 1$). Evaluate k.

1.6.4. Find all linear transformations carrying the upper half-plane and the points 0, -1 into itself.

1.6.5. Find the image domains of the unit disk and its upper half under the linear transformation $w = (5-4z)(4z-2)^{-1}$.

1.6.6. Find the linear transformation carrying $C(0;1)$ into $C(1;1)$ so that the points 0, 1 correspond to $\tfrac{1}{2}$, 0 respectively.

1.6.7. Find the linear transformation carrying the outside of $C(0;1)$ into the right half-plane and the points $z = 1$, $-i$, -1 into $w = i$, 0, $-i$. What are the images of concentric circles center at the origin?

1.6.8. Using the property of symmetry invariance find all the linear transformations carrying the unit disk onto itself.

1.6.9. Find the linear transformation carrying $C(0;1)$ into $C(1;2)$ and the points $z = -1, 0$ into $w = -1, i$.

1.6.10. Find the linear transformation mapping the unbounded doubly connected domain in the extended plane whose boundary consists of $C(5;4)$, $C(-5;4)$ onto $\{w: 1 < |w| < R\}$. Evaluate R.

1.6.11. Find the linear transformation mapping the ring domain with boundary $C(1;1) \cup C(i;\sqrt{6})$ onto $\{w: 1 < |w| < R\}$. Evaluate R.

1.6.12. Find the linear transformation mapping the right half-plane with removed closed disk $\overline{K}(h;R)$, $h > R$, onto $\{w: \rho < |w| < 1\}$. Show that

$$\rho = \frac{h}{R} - \sqrt{\left(\frac{h}{R}\right)^2 - 1}.$$

1.6.13. Find a linear transformation mapping the bounded domain whose boundary consists of $C(0;2)$, $C(1;1)$ onto a strip bounded by two straight lines parallel to the imaginary axis.

1.7. INVARIANT POINTS OF LINEAR TRANSFORMATIONS

1.7.1. If $w = (az+b)/(cz+d)$, $ad-bc \neq 0$, and $\alpha = (a\alpha+b)/(c\alpha+d)$, then α is called an *invariant point of the given linear transformation*. Find the general linear transformation with two different and finite invariant points α, β.

1.7.2. Show that the general linear transformation with invariant points 0, ∞ is a similarity $w = Az$ $(A \neq 0)$.

1.7.3. A transformation T whose inverse T^{-1} is identical with T is called an *involution*. Show that a linear transformation $(az+b)/(cz+d)$ different from identity is an involution, iff $a+d = 0$.

1.7.4. Show that an involution different from identity has always two different invariant points.

1.7.5. Prove that any linear transformation with two different invariant points can be written in the standard form:

$$(w-\alpha)/(w-\beta) = A(z-\alpha)/(z-\beta).$$

1.7.6. Show that if $\Delta = (d-a)^2 + 4bc$ and the sign of $\sqrt{\Delta}$ is suitably chosen, then

$$A = (a+d+\sqrt{\Delta})/(a+d-\sqrt{\Delta}).$$

1.7. INVARIANT POINTS

1.7.7. Bring the linear transformation $w = (z+i)/(z-i)$ to the standard form of Exercise 1.7.5.

1.7.8. Prove that a linear transformation with only one invariant point ∞ is a translation. Also prove that a linear transformation with only one finite invariant point α (or the *parabolic transformation*) has the form

$$(w-\alpha)^{-1} = (z-\alpha)^{-1} + h \quad (h \neq 0).$$

1.7.9. Find the parabolic transformation mapping $C(0; R)$ onto itself whose only invariant point is $z = R$.

1.7.10. If α, β are invariant points and $A = |A|e^{i\theta}$ (cf. Ex. 1.7.5) then the circle C_z with diameter $[\alpha, \beta]$ is carried into a circle C_w with radius $R = \frac{1}{2}|\alpha-\beta| |\cos\theta|^{-1}$.

1.7.11. The sequence $\{z_n\}$ is defined by the recurrence formula: $z_{n+1} = f(z_n)$, where f is a linear transformation with at most 2 invariant points and z_0 is given. Discuss the convergence of $\{z_n\}$.

1.7.12. Find the points of accumulation of the sequence $\{z_n\}$:

$$z_0 = 0, \quad z_{n+1} = \frac{z_n+i}{z_n-i}.$$

1.7.13. If A in Exercise 1.7.5 is real, the corresponding linear transformation is called *hyperbolic*, if $A = e^{i\varphi}$ (with real φ) it is called *elliptic*. Prove that in both cases there exists a family of circles such that any circle of the family is mapped onto itself under the transformation.

1.7.14. Prove that for any parabolic transformation with an invariant point α there exists a family of circles tangent to each other at α and such that each circle of the family is mapped onto itself under the given transformation.

1.7.15. Suppose α, β are invariant points of a linear transformation which is not an identity and carries a circle C into itself. Show that either α, β are situated on C, or are symmetric w.r.t. C.

1.7.16. Suppose a linear transformation which is neither elliptic, nor hyperbolic, has two finite and different invariant points. Show that no circle can be mapped onto itself by this transformation.

1.7.17. Suppose $w = (az+b)/(cz+d)$, $ad-bc = 1$ and $a+d$ is real. Prove that the transformation is elliptic if $|a+d| < 2$, hyperbolic if $|a+d| > 2$ and parabolic if $|a+d| = 2$.

1.7.18. Show that the rotations of the Riemann sphere correspond to elliptic transformations in the plane after stereographic projection.

1.7.19. Suppose α, β are invariant points of an elliptic transformation and $|\alpha-\beta| \geqslant 2$. Prove that this transformation corresponds to a rotation of the Riemann sphere followed by a translation.

1.7.20. Show that the linear transformation
$$w = (az+b)/(-\bar{b}z+\bar{a}), \quad |a|+|b| > 0,$$
corresponds to a rotation of the Riemann sphere. Evaluate stereographic projections of the end-points of the diameter being the axis of rotation, as well as the angle of rotation in terms of a and b.

1.7.21. Find the linear transformation representing the rotation of the Riemann sphere by an angle $\frac{1}{3}\pi$ round the diameter with end-points $(\frac{2}{3}; \frac{2}{3}; \frac{1}{3})$, $(-\frac{2}{3}; -\frac{2}{3}; -\frac{1}{3})$.

1.7.22. Find the general linear transformation representing a rotation of the Riemann sphere by an angle π. Show that this is an involution.

1.7.23. Show that for any involution with two finite invariant points α, β which is different from identity the factor A in Exercise 1.7.5 equals -1. Prove that the straight line through α, β is mapped onto itself.

1.7.24. Find all straight lines remaining invariant under the involution
$$2wz+i(w+z)-2 = 0.$$

1.8. HYPERBOLIC GEOMETRY

In hyperbolic (Lobachevski–Bolyai) geometry the axioms of Euclid are valid except for the parallel axiom: *there are at least two different straight lines in the plane through a given point not on the straight line L which do not meet L*. There is a very simple and elegant way essentially due to Poincaré of satisfying the axioms of non-Euclidean geometry by a suitable choice of certain configurations in Euclidean space. The *points* in the hyperbolic plane or h-points are the points of the unit disk $K(0; 1)$. The *straight lines* (hyperbolic straight lines, or h-lines) are the arcs of circles, or straight line segments orthogonal to the unit circle and interior to it. Hyperbolic *motions* are linear transformations mapping $K(0; 1)$ onto itself. Two sets of h-points are congruent if there exists an h-motion carrying one set intoanother one.

1.8. HYPERBOLIC GEOMETRY

We can also introduce in a natural way h-distance in the hyperbolic plane which is invariant under h-motion. Complex numbers and linear transformations are very convenient tools in analytic treatment of hyperbolic geometry.

1.8.1. Prove that there exists a unique h-line through any two h-points represented by z_1, z_2 ($z_1 \neq z_2$, $|z_1| < 1$, $|z_2| < 1$).

1.8.2. Prove that there exists a unique h-line through a given point z_1 in a given direction $e^{i\alpha}$ (i.e. meeting $C(0; 1)$ at $e^{i\alpha}$).

1.8.3. The unit circle $C(0; 1)$ is called the *h-line at infinity*. Two h-lines meeting at infinity (i.e. two circular arcs orthogonal to $C(0; 1)$ intersecting each other at a point on $C(0; 1)$ are called *h-parallels*. Show that there are two h-parallels to a given h-line L through a given h-point z_1 not on L, as well as infinitely many h-lines through z_1 not meeting L.

1.8.4. Find the general form of an h-rotation (i.e. an h-motion with a unique invariant h-point z_0).

1.8.5. Find a general h-translation, i.e. an h-motion with two invariant points on the h-line at infinity.

1.8.6. Find a general h-boundary rotation, i.e. an h-motion with a unique invariant point on $C(0; 1)$.

1.8.7. Find a general h-motion. Verify the group property for h-motions.

1.8.8. Write parametric equation of an h-segment $[z_1, z_2]_h$, i.e. a subarc of h-line with end-points z_1, z_2.

1.8.9. Suppose the h-segments $[a, z]_h$, $[b, w]_h$ are congruent in the sense of hyperbolic geometry. Verify that

$$|(z-a)/(1-\bar{a}z)| = |(w-b)/(1-\bar{b}w)|.$$

1.8.10. Suppose C and Γ are two regular curves situated in the unit disk and carried into each other under an h-motion. Show that

$$\int_C \frac{|dz|}{1-|z|^2} = \int_\Gamma \frac{|d\zeta|}{1-|\zeta|^2}.$$

1.8.11. Consider all regular curves situated inside $K(0; 1)$ and joining two fixed points 0, R ($0 < R < 1$) of $K(0; 1)$. Show that

$$\int_\gamma \frac{|dz|}{1-|z|^2}$$

has a minimum for γ being the segment $[0, R]$, the minimum being equal

$$\frac{1}{2}\log\frac{1+R}{1-R} = \operatorname{artanh} R.$$

Hint: Verify first that we can restrict ourselves to regular curves with parametric representation $\theta = \theta(r)$, where θ, r are polar coordinates.

1.8.12. If C is a regular curve situated inside $K(0; 1)$ then

$$\int_C \frac{|dz|}{1-|z|^2}$$

is called *hyperbolic* (or *h-*) *length* of C. Show that the h-segment with end points z_1, z_2 is the curve with shortest h-length among all regular curves in $K(0; 1)$ joining z_1 to z_2. The h-length of $[z_1, z_2]_h$ is called *hyperbolic* (or *h-*) *distance* $\rho(z_1, z_2)$ of points z_1, z_2. Also show that

$$\rho(z_1, z_2) = \operatorname{artanh}|z_1-z_2||1-z_1\bar{z}_2|^{-1}.$$

1.8.13. Find the h-circle with h-radius R, i.e.

$$\{z:\ \rho(z_0, z) = R\}.$$

Also find its h-length l_h.

1.8.14. Verify the usual properties of a metric for $\rho(z_1, z_2)$.

1.8.15. Show that

$$\rho(z_1, z_2) = \tfrac{1}{2}\log(z_1, z_2, e^{i\beta}, e^{i\alpha}),$$

where $e^{i\alpha}$, z_1, z_2, $e^{i\beta}$ are successive points of a circle orthogonal to $C(0; 1)$.

1.8.16. Suppose a regular domain D, $\bar{D} \subset K(0; 1)$, is carried under h-motion into Ω. Prove that

$$\iint_D \frac{dx\,dy}{(1-x^2-y^2)^2} = \iint_\Omega \frac{d\xi\,d\eta}{(1-\xi^2-\eta^2)^2}.$$

The integral on the left is called *hyperbolic* (or *h-*) *area* of D and will be denoted $|D|_h$.

1.8.17. Find $|\Omega|_h$ for $\Omega = \{z:\ |z| < R\}$.

1.8.18. Consider an h-triangle T, i.e. a domain bounded by three h-segments with angles α, β, γ. Show that

$$|T|_h = \tfrac{1}{4}[\pi-(\alpha+\beta+\gamma)].$$

Hint: Take the origin as one of the vertices.

1.8.19. Evaluate the h-area of an h-triangle with vertices z_1, z_2, z_3.

CHAPTER 2

Regularity Conditions. Elementary Functions

A complex function $w = f(z)$ defined on a set of complex numbers A is actually defined by a pair of real-valued functions $u(x, y)$, $v(x, y)$ of two real variables x, y ($x+iy = z$) with a common domain A. In a formally identical manner as in real analysis we can introduce the notions of limit, continuity and differentiability. If $f(z) = u(x, y)+iv(x, y)$ is differentiable at $z_0 = x_0+iy_0$, then u, v have partial derivatives at z_0 satisfying Cauchy–Riemann equations at this point: $u_x = v_y$, $u_y = -v_x$. On the other hand, if all the four partials of first order of u, v exist in some neighborhood of z_0, are continuous and satisfy Cauchy–Riemann equations at z_0, then $f = u+iv$ is *differentiable* at z_0, i.e.

$$f(z_0+h) = f(z_0)+ah+h\eta(h), \quad \text{where} \quad \lim_{h \to 0} \eta(h) = 0;$$

the constant a is called the *derivative* of f at z_0. The most interesting and most important case occurs when f is defined and has a derivative at every point of some domain (or open, connected set) D in the plane. Then f is called *analytic*, *holomorphic* or *regular* in D. Regularity has far-reaching consequences that go much beyond what one can obtain from differentiability in the real case. The theory of analytic functions has as its central theme just the investigation of these consequences. So, for example, regularity in a domain D implies the existence of derivatives of all orders at all points of D. Since the definitions of the derivative in real and complex domain are formally identical, the usual rules of differentiation as the formulas concerning the derivative of a sum, a product or a quotient, as well as the chain rule, remain the same in complex case.

2.1. CONTINUITY. DIFFERENTIABILITY

2.1.1. Discuss the continuity at $z = 0$ of functions defined at $z \neq 0$ as follows:

(i) $\dfrac{\operatorname{re} z}{1+|z|}$; (ii) $z^{-1} \operatorname{re} z$; (iii) $z^{-2} \operatorname{re} z^2$; (iv) $z^{-2} (\operatorname{re} z^2)^2$ and equal 0 at $z = 0$.

2.1.2. Suppose f is defined and uniformly continuous in $K(0;1)$. Prove that for any sequence $\{z_n\}$, $z_n \in K(0;1)$, convergent to ζ ($|\zeta| = 1$) there exists a limit $\varphi(\zeta)$ depending only on ζ, and not on a particular choice of $\{z_n\}$. Also prove that F: $F(z) = f(z)$ for $z \in K(0;1)$, $F(\zeta) = \varphi(\zeta)$ for $\zeta \in C(0;1)$, is continuous in $\overline{K}(0;1)$.

2.1.3. Verify that the function

$$f: \quad f(0) = 0, \quad f(z) = |z|^{-2}(1+i)\operatorname{im} z^2 \quad \text{for} \quad z \neq 0,$$

satisfies Cauchy–Riemann equations at $z = 0$. Is f differentiable at $z = 0$?

2.1.4. Verify that $f(z) = z \operatorname{re} z$ is differentiable at $z = 0$ only.

2.1.5. Suppose $f(z) = u(x,y) + iv(x,y)$ and the limit

$$\lim_{h \to 0} \operatorname{re} h^{-1}[f(z_0+h) - f(z_0)]$$

exists. Show that the partials u_x, v_y at z_0 both exist and are equal.

2.1.6. Suppose $u(x,y)$, $v(x,y)$ are continuous and have continuous partials of first order at $z_0 = x_0 + iy_0$.
If $f = u + iv$ and the limit

$$\lim_{h \to 0} |h|^{-1} |f(z_0+h) - f(z_0)|$$

exists, then either f, or $\bar{f} = u - iv$ has a derivative at z_0.

2.1.7. Verify that the following functions fulfill the Cauchy–Riemann equations in the whole plane: (i) $f(z) = z^3$; (ii) $f(z) = e^x \cos y + i e^x \sin y$.

2.1.8. Verify that $f(z) = x(x^2+y^2)^{-1} - iy(x^2+y^2)^{-1}$ is analytic in $\mathbf{C} \setminus 0$.

2.1.9. If $f = u + iv$ is analytic and satisfies $u^2 = v$ in a domain D, then f is a constant.

2.1.10. Suppose

$$\Delta u = \frac{\partial^2 u}{\partial x^2} + \frac{\partial^2 u}{\partial y^2}.$$

If f is analytic and does not vanish in a domain D, then

$$\Delta |f(z)| = |f(z)|^{-1} |f'(z)|^2, \quad z \in D.$$

2.1.11. Prove that for an analytic function f:

$$\Delta(|f|^2) = 4|f'|^2.$$

2.1.12. Write Cauchy–Riemann equations for
$$f(z) = U(r, \theta) + iV(r, \theta), \quad \text{where} \quad z = re^{i\theta}.$$
Express f' in terms of partials of U, V.

2.1.13. Prove that $f(z) = z^n$ (n is a positive integer) satisfies Cauchy–Riemann equations and $f'(z) = nz^{n-1}$.

2.2. HARMONIC FUNCTIONS

A real-valued function u of two real variables x, y (resp. of one complex variable $z = x + iy$) defined in a domain D is said to be *harmonic* in D, if it has continuous partial derivatives of second order that satisfy in D Laplace's equation:
$$\Delta u = u_{xx} + u_{yy} = 0.$$
Notice that continuity of partial derivatives of second order implies continuity u_x and u_y, as well as continuity of u. Two functions u, v harmonic in a domain D and satisfying Cauchy–Riemann equations in D: $u_x = v_y$, $u_y = -v_x$ are called *conjugate harmonic* functions. Any pair of conjugate harmonic functions u, v determines an analytic function $u + iv$.

2.2.1. Find all the functions harmonic in $\mathbf{C} \setminus (-\infty, 0]$ which are constant one the rays $\arg z = \text{const}$.

2.2.2. Find all the functions harmonic in $\mathbf{C} \setminus 0$ which are constant on the circles $C(0; r)$.

2.2.3. Verify that the functions $u = \log|z|$, $v = \arg z$ are conjugate harmonic functions in $\mathbf{C} \setminus (-\infty, 0]$ and $\text{Log } z = \log|z| + i \text{Arg } z$, where $\text{Arg } z$ is the principal value of argument: $-\pi < \text{Arg } z < \pi$, is analytic in $\mathbf{C} \setminus (-\infty, 0]$.

2.2.4. Verify that $e^x \cos y$, $e^x \sin y$ are conjugate harmonic functions in $\mathbf{C}$. Also verify that the analytic function
$$\exp z = e^x \cos y + i e^x \sin y$$
fulfills the identity:
$$\exp \text{Log } z = \text{Log} \exp z = z$$
in $\mathbf{C} \setminus (\infty, 0]$.

2.2.5. Show that $\dfrac{d}{dz} \text{Log } z = z^{-1}$ in $\mathbf{C} \setminus (\infty, 0]$ and $\dfrac{d}{dz} \exp z = \exp z$ in $\mathbf{C}$.

2.2.6. Suppose $w = f(z)$ is analytic in a domain D and $f(D) \cap (-\infty, 0] = \varnothing$. Show that $F(z) = \log|f(z)| + i \text{Arg } f(z)$ is analytic in D. Evaluate F'.

2.2.7. Suppose u is harmonic in a domain D. Verify that $f = u_x - iu_y$ is analytic in D.

2.2.8. Suppose v_1, v_2 are harmonic and conjugate with u in a domain D. Verify that $v_1 - v_2$ is a constant in D.

2.2.9. Suppose u is harmonic in $\{z\colon |z| > 0\}$ and homogeneous of degree m, $m \neq 0$, i.e. for any $t > 0$, $u(tz) = t^m u(z)$. Verify that $v = m^{-1}(yu_x - xu_y)$ is a conjugate harmonic function.

2.2.10. Find a conjugate harmonic function v for u equal:
(i) $x^2 - y^2 + xy$; (ii) $x^3 + 6x^2y - 3xy^2 - 2y^3$; (iii) $x(x^2+y^2)^{-1}$; (iv) $(x^2-y^2) \times (x^2+y^2)^{-2}$. Evaluate in each case a corresponding analytic function $u+iv$ as depending on $z = x+iy$.

2.2.11. Find a conjugate harmonic function v for
$$u(x, y) = \frac{(1+x^2+y^2)x}{1+2(x^2-y^2)+(x^2+y^2)^2}.$$
Write $u+iv$ as depending on $z = x+iy$.

2.2.12. Show that a function u harmonic in a domain D has a conjugate harmonic function v in D, iff $f = u_x - iu_y$ admits a primitive in D.

2.2.13. Find a conjugate harmonic function v for
$$u(x, y) = e^x(x\cos y - y\sin x).$$

2.2.14. Write Laplace's equation in polar coordinates r, θ. Verify that $r^n\cos n\theta$, $r^n\sin n\theta$ are harmonic for any positive integer n.

2.2.15. Discuss the existence of nonconstant harmonic functions having the form: (i) $\varphi(xy)$; (ii) $\varphi(x+\sqrt{x^2+y^2})$; (iii) $\varphi(x^2+y)$, where φ is a suitable, real-valued function of one real variable. Find a corresponding conjugate harmonic function in case it does exist.

2.2.16. Given a real-valued function F with continuous partial derivatives of second order in a domain D such that $F_x^2 + F_y^2 > 0$ in D. Suppose $a < F(x, y) < b$ for $x+iy \in D$ and ψ is a real-valued, continuous function of $t \in (a, b)$ such that $(F_{xx}+F_{yy})(F_x^2+F_y^2)^{-1} = \psi \circ F$. Then there exists a real-valued function φ defined in (a, b) and such that $\varphi \circ F$ is harmonic in D. Evaluate φ as depending on ψ.

2.2.17. Find an analogue of Exercise 2.2.16 in case F is given in polar coordinates r, θ.

2.2.18. Verify the existence of functions u harmonic in $\mathbf{C}\setminus(-\infty, 0]$ and constant on confocal parabolas with foci at the origin and vertices on $(0, +\infty)$. Find all these functions.

2.2.19. Find all the functions f analytic in $\mathbf{C}\setminus 0$ such that $|f|$ has a constant value on circles $x^2+y^2-ax = 0$.

2.2.20. Find all the functions f analytic in $\mathbf{C}\setminus(-\infty, 0]$ such that $\arg f$ has a constant value on circles $C(0; r)$.

2.2.21. Verify the existence of functions $u(r, \theta)$ harmonic in $\mathbf{C}\setminus(-\infty, 0]$ having a constant value on arcs of logarithmic spirals $r = ke^{\lambda\theta}$, where r, θ are polar coordinates, λ is fixed for all the spirals and k is a parameter determining the individual arcs.

2.2.22. Find all the functions regular in $\mathbf{C}\setminus 0$ whose absolute value is constant on lemniscates $r^2 = a^2\sin 2\theta$.

2.3. GEOMETRICAL INTERPRETATION OF THE DERIVATIVE

If f is analytic in a domain D, $z_0 \in D$ and $f'(z_0) \neq 0$, then

$$f(z_0+h) = f(z_0)+hf'(z_0)+O(h^2) \quad \text{as} \quad h \to 0.$$

This means that locally f is a similarity transformation composed of a rotation by the angle $\arg f'(z_0)$, a homothety with the ratio $|f'(z_0)|$ followed by a translation $f(z_0)$. The angle between any regular arcs intersecting at z_0 and the angle between the image arcs are equal, therefore the mapping realized by an analytic function f with $f'(z) \neq 0$ is said to be *conformal*. An analytic function realizing a conformal and homeomorphic mapping of a domain D is said to be *univalent* in D.

2.3.1. The linear transformation $w = (z+1)/(z-1)$ carries the boundary of the upper half-disk of $K(0; 1)$ into two rays emanating from the origin (why?). Find the angle α between the image ray of $(-1, 1)$ and the positive real axis as well as the local length distortion λ at $z = -1$.

2.3.2. Given a linear transformation $w = (az+b)/(cz+d)$ with $c \neq 0$, find the sets of all z for which
  (i) the length of infinitesimal segments is preserved;
  (ii) the direction of infinitesimal segments is preserved.

2.3.3. Suppose $z = z(t)$ is a differentiable, complex-valued function of a real variable $t \in (a, b)$ such that $z'(t) \neq 0$ and $w = f(z)$ is a conformal mapping defined in a domain D containing all the points $z(t)$, $t \in (a, b)$. Show that if $\arg[f'(z(t))z'(t)]$ is constant in (a, b), then the image of the arc $z = z(t)$, $a < t < b$, is a straight line segment (not necessarily bounded).

2.3.4. Show that for any linear mapping $w = (az+b)/(cz+d)$, $ad-bc \neq 0$, $c \neq 0$, some straight line has as its image a parallel straight line.

2.3.5. Find the sets of all z where an infinitesimal segment is expanded, contracted or preserved under the given transformation:
(i) $w = z^2$; (ii) $w = z^2 + 2z$; (iii) $w = z^{-1}$.

2.3.6. Find local magnification and the angle of local rotation at $z_0 = -3+4$ under the mapping $w = z^3$.

2.3.7. Show that the Jacobian $\dfrac{\partial(u, v)}{\partial(x, y)}$ of the mapping $f = u + iv$, where f is analytic in a domain D, is equal $|f'|^2$. Give a geometrical interpretation.

2.3.8. Verify that if $f = u + iv$ is analytic and $f'(z_0) \neq 0$, then the lines $u = \text{const}$, $v = \text{const}$, intersect at z_0 at the right angle.

2.3.9. Find the lines $u = \text{const}$, $v = \text{const}$ for the mappings
(i) $w = z^2$; (ii) $w = \text{Log}\, z$.

2.3.10. Find the length of the image arc under the univalent mapping $f \colon D \to \mathbf{C}$ of the arc given by the equation $z = z(t)$, $a \leqslant t \leqslant b$. Also find the area of the image domain of Ω, $\Omega \subset D$.

2.3.11. Show that under the mapping $w = z^2$, the image curve of $C(1; 1)$ is the cardioid $w(\varphi) = 2(1+\cos\varphi)e^{i\varphi}$. Find its length and the area enclosed.

2.3.12. Evaluate the integral $\iint_D (x^2+y^2)\,dx\,dy$, where D is a domain situated in $\{z \colon \operatorname{re} z > 0,\ \operatorname{im} z > 0\}$ whose boundary consists of the segment $[1, 2]$ and three arcs of hyperbolas $x^2 - y^2 = 1$, $x^2 - y^2 = 4$, $xy = 1$.

Hint: Cf. Ex. 2.3.9 (i).

2.3.13. Evaluate the length of the image arc of the segment $[0, i]$ under the mapping $w = z(1-z)^{-2}$.

2.4. CONFORMAL MAPPINGS CONNECTED WITH $w = z^2$

2.4.1. Evaluate the maximal error in $K(i; \tfrac{1}{10})$ if the mapping $w = z^2$ of this disk is replaced by its differential at $z = i$.

2.4.2. Find the image domain of the square: $0 < x < 1$, $0 < y < 1$ and the length of the boundary of the image domain under the mapping $w = z^2$, $z = x+iy$.

2.4.3. Find the univalent mapping of the domain $\{z: \operatorname{re} z > 0, \operatorname{im} z > 0\}$ onto $K(0; 1)$ such that $z_0 = 1+i$ corresponds to the center.

2.4.4. Find the univalent mapping of $K(0; 1)$ onto the inside of
$$w(\theta) = 2(1+\cos\theta)e^{i\theta}, \quad 0 \leqslant \theta \leqslant 2\pi.$$

2.4.5. Find the univalent mapping of the domain situated on the right-hand side branch of the hyperbola $x^2 - y^2 = a^2$ onto $K(0; 1)$ which carries the focus of the hyperbola into $w = 0$ and the vertex into $w = -1$.

2.4.6. Find the univalent conformal mapping of the domain
$$\{z = x+iy: -\infty < y < +\infty, 2px < y^2\}, \quad p > 0$$
onto the unit disk such that the points $z = -\frac{1}{2}$ and $z = 0$ correspond to $w = 0$ and $w = 1$ respectively.

2.4.7. Find the univalent conformal mapping of the domain bounded by the branch of hyperbola: $x^2 - y^2 = 1$, $x > 0$, and the rays $\arg z = \mp\frac{1}{4}\pi$ onto the strip $\{w: |\operatorname{im} z| < 1\}$.

2.4.8. Show that the mapping $z = a\sqrt{2w}(1+w^2)^{-1/2}$ carries the unit disk $\{w: |w| < 1\}$ into the domain bounded by the branches of the hyperbola $x^2 - y^2 = a^2$.

2.4.9. Map conformally the inside of the right half of the lemniscate $|w^2 - a^2| = \rho^2$, $0 < \rho \leqslant a$, onto the unit disk.

2.4.10. Map conformally the inside of lemniscate $|w^2 - a^2| = \rho^2$, $\rho > a$, onto the unit disk.

2.4.11. Map conformally the strip domain between the parabolas: $y^2 = 4(x+1)$, $y^2 = 8(x+2)$ onto the strip $\{w: |\operatorname{im} w| < 1\}$.

2.5. THE MAPPING $w = \frac{1}{2}(z+z^{-1})$

2.5.1. Suppose C is an arbitrary circle through $-1, 1$ and z_1, z_2 do not lie on C and satisfy $z_1 z_2 = 1$. Show that one of these points lies inside C and another one outside C.

2.5.2. Show that the mapping $w = \frac{1}{2}(z+z^{-1})$ carries both the inside and the outside of any circle C through the points $z = \mp 1$ in a 1:1 manner onto the same domain in the w-plane. Find the image domain.
Hint: Show that $(w-1)/(w+1) = [(z-1)/(z+1)]^2$.

2.5.3. Show that the image domain of the upper half-plane under the mapping $w = \frac{1}{2}(z+z^{-1})$ is $\mathbf{C} \setminus \{(-\infty, -1] \cup [1, +\infty)\}$.

2.5.4. Show that the image domain of the unit disk under the mapping $w = \frac{1}{2}(z+z^{-1})$ is $\mathbf{C}\setminus[-1, 1]$.

2.5.5. Suppose C is a circle through $z = \mp 1$ and Γ is a circle having a common tangent with C at $z = 1$ and situated in the outside of C. Describe the image curve of Γ under the mapping $w = \frac{1}{2}(z+z^{-1})$.

2.5.6. Find the image curves of:
(i) circles $C(0; R)$;
(ii) rays $\arg z = \theta$
under the mapping $w = \frac{1}{2}(z+z^{-1})$.

2.5.7. Map conformally the ellipse $\{w\colon |w-2|+|w+2| < 100/7\}$ slit along $[-2, 2]$ onto the annulus $\{z\colon 1 < |z| < R\}$. Evaluate R.

2.5.8. Map conformally the outside of the unit disk onto the outside of the ellipse: $\{w\colon |w-c|+|w+c| > 2a\}$ $(c^2 = a^2-b^2,\ a, b, c > 0)$.

2.5.9. Map conformally the domain $\mathbf{C}\setminus\{\overline{K}(0; 1) \cup [-a, -1] \cup [1, b]\}$ $(a > 1, b > 1)$ onto the outside of the unit disk.

2.5.10. Map conformally the outside of the unit disk slit along $(-\infty, -1)$ onto:
(i) w-plane slit along the negative real axis;
(ii) the right half-plane.

2.5.11. Map conformally the domain whose boundary consists of three rays: $(-\infty, -1]$, $[1, +\infty)$, $[2i, +i\infty)$ and of the upper half of $C(0; 1)$ onto a half-plane.

2.5.12. Map conformally the domain bounded by two confocal ellipses:
$$\{w\colon 2\sqrt{5} < |w-2|+|w+2| < 6\}$$
onto an annulus $\{z\colon R_1 < |z| < R_2\}$. Evaluate R_1/R_2.

2.5.13. Map conformally the domain bounded by the right-hand branches of the hyperbolas $u^2\cos^{-2}\alpha - v^2\sin^{-2}\alpha = 1$, $u^2\cos^{-2}\beta - v^2\sin^{-2}\beta = 1$ $(0 < \alpha < \beta < \frac{1}{2}\pi)$ onto the angle $\{z\colon \alpha < \arg z < \beta\}$.

Hint: Cf. Excercise 2.5.6 (ii).

2.5.14. Show that the image W of the point w under the mapping $W = w^3 - 3w$ describes three times an ellipse with foci $-2, 2$, when w describes once a confocal ellipse.

2.5.15. Show that under the mapping $W = w^3 - 3w$ the quadrant $\{w: \operatorname{re} w > 0, \operatorname{im} w < 0\}$ is carried 1:1 onto the complementary domain of the set $\{W: \operatorname{re} W \geqslant 0, \operatorname{im} W \geqslant 0\} \cup [-2, 0]$.

2.5.16. Map conformally the part of the z-plane to the left of the right-hand branch of the hyperbola $x^2 - y^2 = 1$ on a half-plane.
Hint: Map the upper half of the given domain by the mapping $W = z^2$.

2.6. THE EXPONENTIAL FUNCTION AND THE LOGARITHM

2.6.1. Use the identity:
$$\exp z = e^z = e^x \cos y + i e^x \sin y, \quad z = x + iy,$$
to verify that
(i) $|e^z| = e^x$;
(ii) $\exp(z + 2\pi i) = \exp z$;
(iii) $\exp(z_1 + z_2) = (\exp z_1)(\exp z_2)$.

2.6.2. Show that for any complex $w \neq 0$ and any real α the equation $e^z = w$ has exactly one solution z satisfying $\alpha < \operatorname{im} z \leqslant \alpha + 2\pi$.

2.6.3. Find the image domain of the strip $-\pi < \operatorname{im} z < \pi$ under the mapping $w = e^z$. Also find the images of segments $(x_0 - \pi i, x_0 + \pi i)$ and of straight lines $y = y_0$.

2.6.4. Find the image of the straight line $y = mx + n$ under the mapping $w = e^z$ ($m \neq 0$).

2.6.5. Find the image domain of the strip $mx - \pi < y < mx + \pi$ under the mapping $w = e^z$.

2.6.6. For which z is the exponential function (i) real; (ii) purely imaginary? Evaluate the real and imaginary parts of $\exp(2+i)$ up to 4 decimals.

2.6.7. Find the image domain of the square $|x-a| < \varepsilon$, $|y| < \varepsilon$ under the mapping $w = e^z$ (a, ε are real, $0 < \varepsilon < \pi$, $z = x + iy$). Evaluate the limit of the ratio of the areas of both domains as $\varepsilon \to 0$.

2.6.8. Show that the principal branch of the logarithm maps conformally $K(1; r)$, $0 < r < 1$, onto a convex domain symmetric w.r.t. the straight lines: $\operatorname{im} w = 0$, $\operatorname{re} w = \frac{1}{2} \log(1 - r^2)$.

2.6.9. Show that the function $w = \operatorname{Log}[(z-\alpha)/(z-\beta)]$, where Log denotes the principal branch of logarithm corresponding to $|\arg z| < \pi$, is univalent in $\mathbf{C} \setminus [\alpha, \beta]$. Find the image domain of $\mathbf{C} \setminus [\alpha, \beta]$ and also the images of:

(i) circular arcs with end-points α, β;
(ii) Apollonius circles for the points α, β;
(iii) the point $z = \infty$.

2.6.10. Show that the function
$$W = (\beta-\alpha)\{\text{Log}[(z-\alpha)/(z-\beta)]\}^{-1}$$
is univalent in $\mathbf{C}\setminus[\alpha, \beta]$. Find the image domain.

2.6.11. Suppose f is analytic in a domain D and does not take real, non-positive values in D. Show that
$$|f(z)| = \varphi(x)\psi(y), \quad x+iy = z \in D,$$
implies: $f(z) = \exp(az^2+bz+c)$ where a is a real constant and b, c are complex constants.

2.6.12. Find a conformal mapping of the sphere into the w-plane such that the straight lines $v = $ const are image lines of parallels and the straight lines $u = $ const are the image lines of meridians ($u+iv = w$). Express u, v in terms of geographical coordinates θ, φ on the sphere.

Hint: If z is the stereographic projection of a point on the sphere, then $w = f(z)$ is analytic.

2.7. THE TRIGONOMETRIC AND HYPERBOLIC FUNCTIONS

2.7.1. Starting from the definition of e^z (Ex. 2.6.1) verify that
$$\frac{1}{2}(e^{iz}+e^{-iz}), \quad \frac{1}{2i}(e^{iz}-e^{-iz})$$
are analytic in $\mathbf{C}$ and coincide with $\cos z$ and $\sin z$ resp. on the real axis. This defines the functions sin and cos for complex z.

2.7.2. Write $\cos z$, $\sin z$, $\tan z = \sin z/\cos z$ in the form $u(x, y)+iv(x, y)$ where u, v are real-valued functions of real variables x, y ($x+iy = z$).

Verify that u, v are conjugate harmonic functions.

2.7.3. Show that
$$|\sin z|^2 = \sin^2 x + \sinh^2 y, \quad |\cos z|^2 = \cos^2 x + \sinh^2 y.$$
Find all zeros of sine and cosine.

2.7.4. Verify that $|\sin z| \geqslant 1$ on the boundary of any square with vertices $\pi(m+\tfrac{1}{2})(\mp 1 \mp i)$, $m = 0, 1, 2, \ldots$

2.7.5. Verify that $|\cos z| \geqslant 1$ on the boundary of any square with vertices $\pi m(\mp 1 \mp i)$, $m = 1, 2, \ldots$

2.7. TRIGONOMETRIC AND HYPERBOLIC FUNCTIONS

2.7.6. Show that $|\sin z| \leqslant \cosh R$, $|\cos z| \leqslant \cosh R$ for any $z \in \bar{K}(0; R)$.

2.7.7. Show that for any z with $|\operatorname{im} z| \geqslant \delta > 0$
$$|\tan z| \leqslant [1+(\sinh \delta)^{-2}]^{1/2}, \quad |\cot z| \leqslant [1+(\sinh \delta)^{-2}]^{1/2}.$$

2.7.8. Verify the identity $\sin^2 z + \cos^2 z = 1$ for all complex z.

2.7.9. Verify the identity $\sin \bar{z} = \overline{\sin z}$ and its analogues for cos, tan, cot.

2.7.10. Find all z for which $\sin z$, $\cos z$, $\tan z$ are
(i) real; (ii) purely imaginary.

2.7.11. Evaluate $\cos(5-i)$, $\sin(1-5i)$ up to 4 decimals. Show that if $z_0 = \frac{1}{2}\pi + i\log(4+\sqrt{15})$ then $\sin z_0 = 4$.

2.7.12. Verify for complex z_1, z_2 the addition formulas:
$$\cos(z_1+z_2) = \cos z_1 \cos z_2 - \sin z_1 \sin z_2,$$
$$\sin(z_1+z_2) = \sin z_1 \cos z_2 + \cos z_1 \sin z_2.$$

2.7.13. Write $\cosh z = \frac{1}{2}(e^z+e^{-z})$, $\sinh z = \frac{1}{2}(e^z-e^{-z})$ in the form $u(x,y) + iv(x,y)$, $x+iy = z$.

2.7.14. Express $|\sinh z|^2$, $|\cosh z|^2$ as functions of x, y.

2.7.15. Find the relation between corresponding trigonometric and hyperbolic functions and give a geometric interpretation.

2.7.16. Verify the identity:
$$(1+i)\cot(\alpha+i\beta)+(1-i)\cot(\alpha-i\beta) = 2\frac{\sin 2\alpha + \sinh 2\beta}{\cosh 2\alpha - \cos 2\beta}.$$

2.7.17. Find the image domain of the strip $|\operatorname{re} z| < \frac{1}{2}\pi$ under the mapping $w = \sin z$. Find the image arcs of segments $(-\frac{1}{2}\pi+iy_0, \frac{1}{2}\pi+iy_0)$ and of straight lines $x = x_0$ and verify the univalence of sine in the strip considered.

2.7.18. Find the image domain of the rectangle: $0 < \operatorname{re} z < \frac{1}{2}\pi$, $0 < \operatorname{im} z < a$, under the mapping $w = \sin z$. Are the angles at all vertices preserved?

2.7.19. Show that cosine is a univalent function in the strip $0 < \operatorname{re} z < \frac{1}{2}\pi$, the image domain being the right half-plane with they ray $[1, +\infty)$ removed.

2.7.20. Map 1:1 conformally the strip $0 < \operatorname{re} z < \frac{1}{2}\pi$ onto the unit disk slit along a radius.

2.7.21. Map 1:1 conformally the domain D to the left of the parabola $y^2 = 4(1-x)$ onto the unit disk.
Hint: Consider the image domain of $D \setminus (-\infty, 0]$ under the mapping $t = \sqrt{z}$.

2.8. INVERSE TRIGONOMETRIC AND HYPERBOLIC FUNCTIONS

2.8.1. Find all solutions of the equation $z = \cos w$ in terms of the logarithm.

2.8.2. Show that $w = \text{Arccos}\, z$ maps 1:1 conformally the z-plane slit along the rays $(-\infty, -1]$, $[1, +\infty)$ onto the strip $0 < \text{re}\, w < \pi$.

2.8.3. Find $\text{Arccos}\tfrac{1}{4}(3+i)$.

2.8.4. Show that $w = \text{Arctan}\, z$ maps 1:1 conformally the z-plane slit along the rays $[i, +i\infty)$, $[-i, -i\infty)$ onto the strip $|\text{re}\, w| < \tfrac{1}{2}\pi$. Also show that

$$w = \frac{1}{2i} \text{Log} \frac{1+iz}{1-iz}.$$

2.8.5. Discuss the image arcs of Apollonius circles with limit points $-i, i$ and the image arcs of circles through these points under the mapping $w = \text{Arctan}\, z$.

2.8.6. Evaluate

(i) $\text{Arctan}(1+2i)$; (ii) $\text{Arctan}\, e^{i\theta}$, $-\tfrac{1}{2}\pi < \theta < \tfrac{1}{2}\pi$.

2.8.7. Find the image domain of the unit disk under the mapping $w = \text{Arctan}\, z$.

2.8.8. Find the principal branch Artanh of the inverse of $\tanh$ (i.e. the branch whose restriction to the real axis coincides with the real function artanh) in terms of Log.

Find the image domain of the unit disk under Artanh.

2.8.9. Find the image domain of the quadrant $\text{re}\, z > 0$, $\text{im}\, z > 0$ under $w = \text{Arsinh}\, z$.

2.8.10. Show that $w = \arcsin e^z$ maps the z-plane slit along the rays $y = k\pi$, $-\infty < x \leqslant 0$ ($k = 0, \mp 1, \mp 2, \ldots$) onto the upper half-plane.

2.9. CONFORMAL MAPPING OF CIRCULAR WEDGES

Elementary functions such as logarithm, exponential function, linear functions, power $w = z^\alpha = \exp(\alpha \log z)$, carry some families of rays, circular arcs, or straight lines, into a similar family, e.g., $\text{Log}\, z$ carries the family of rays $\arg z = \text{const}$ into the family of parallel straight lines. A suitable superposition of such transformations enables us to map any circular wedge, i.e. a simply connected domain whose boundary consists of two circular arcs (not necessarily different, either arc can be replaced by a straight line segment), onto a disk, or half-plane. In particular the mapping $w = z^\alpha$ for α suitably chosen carries an angular sector with vertex at the origin into a half-plane. On the other hand the exponential

2.9. CONFORMAL MAPPING OF CIRCULAR WEDGES

function transforms zero angle into a half-plane. It may even happen that a particular circular triangle with two right angles can be mapped onto a circular wedge and hence onto a half-plane.

2.9.1. Map 1:1 conformally a circular wedge in the z-plane whose sides intersect at a, b and make an angle α onto a half-plane.

Hint: Consider first a linear mapping carrying a, b into 0, ∞ respectively.

2.9.2. Map 1:1 conformally the angle

$$D = \{z\colon \alpha < \arg z < \beta\}, \quad 0 < \beta - \alpha < 2\pi,$$

onto the right half-plane.

2.9.3. Map 1:1 conformally the upper half of the unit disk onto the upper half-plane.

2.9.4. Map 1:1 conformally the circular sector: $0 < \arg z < \tfrac{1}{3}\pi$, $0 < |z| < 1$, onto the unit disk.

2.9.5. Map the circular wedge $\{z\colon |z| < 1\} \cap \{z\colon |z+i\sqrt{3}| > 2\}$ onto the upper half-plane.

2.9.6. Map the wedge $K(-1; \sqrt{2}) \cap K(1; \sqrt{2})$ onto $K(0; 1)$ so that $0 \leftrightarrow 0$.

2.9.7. Map the strip $\{z\colon |\mathrm{im}\, z| < \tfrac{1}{2}\}$ onto $K(0; 1)$ so that $0 \leftrightarrow 0$. Find the image arcs of segments $\mathrm{re}\, z = x_0$ and straight lines $\mathrm{im}\, z = y_0$.

2.9.8. Map the domain $K(0; 1) \cap (\mathbf{C} \setminus \overline{K}(\tfrac{1}{2}; \tfrac{1}{2}))$ onto $K(0; 1)$.

2.9.9. Map the complementary domain of the set $\overline{K}(i; 1) \cap \overline{K}(-i; 1)$ onto the outside of $K(0; 1)$ so that $\infty \leftrightarrow \infty$.

2.9.10. Map the circular triangle $Oa_1 a_2$:

$$K(1; 1) \cap [\mathbf{C} \setminus \overline{K}(1-i; \sqrt{2})] \cap \{z\colon |z/(z-2)| < 2\}$$

onto the upper half-plane.

2.9.11. Map the w-plane slit along the ray $(-\infty, -\tfrac{1}{4}]$ onto the unit disk so that $0 \leftrightarrow 0$.

2.9.12. Map the w-plane slit along the rays $(-\infty, -\tfrac{1}{2}]$, $[\tfrac{1}{2}, +\infty)$ onto the unit disk so that $0 \leftrightarrow 0$.

2.9.13. Map the upper half-plane with the points of the upper half of the unit disk removed onto the upper half-plane.

2.9.14. Map the upper half-plane slit along the segment $(0, ih]$, $h > 0$, onto the upper half-plane.

2.9.15. Map the upper half-plane slit along the ray $[ih, +i\infty)$, $h > 0$, onto the upper half-plane.

2.9.16. Show that the mapping

$$w = \frac{(1-\rho)^2}{\rho} \cdot \frac{z}{(1-z)^2}, \quad 0 < \rho < 1,$$

carries the unit disk $|z| < 1$ slit along the radius $(-1, 0]$ onto the w-plane slit along the negative real axis.

2.9.17. Map the unit disk $|z| < 1$ onto the w-plane slit along the negative real axis so that $(z = 0) \leftrightarrow (w = 1)$.

2.9.18. Map the unit disk $|z| < 1$ slit along the radius $(-1, 0]$ onto the unit disk $|t| < 1$ so that the points $t = 0$, $z = \rho$ $(0 < \rho < 1)$ correspond to each other.

2.9.19. Map the domain $K(0; 1) \setminus (-1, -\rho]$, $0 < \rho < 1$, onto the unit disk $|t| < 1$ so that $0 \leftrightarrow 0$.

2.9.20. Assume that $|a| < 1$, $|b| < 1$, $a \neq b$, and γ is the circle through a, b orthogonal to $C(0; 1)$. Let $D(a, b)$ be the unit disk slit along the arc of γ joining b to $C(0; 1)$, situated inside $C(0; 1)$ and not containing a. Map conformally $D(a, b)$ onto the unit disk $|t| < 1$ so that $(\zeta = a) \leftrightarrow (t = 0)$.

2.9.21. Evaluate $\left(\dfrac{d\zeta}{dt}\right)_{t=0}$ for the mapping of Exercise 2.9.20.

2.9.22. Show that the mapping $w = \rho z(z+\rho)(1+\rho z)^{-1}$, $\rho > 1$, carries 1:1 the outside of the unit disk in the z-plane into the outside of a circular arc on the circle $C(0; \rho)$ symmetric w.r.t. the real axis. Verify that the angle subtended by the arc is equal to $4\arcsin\rho^{-1}$.

2.9.23. Find the 1:1 conformal mapping of the domain $\hat{\mathbf{C}} \setminus \overline{K}(0; 1)$ onto the w-plane slit along the arc: $|w| = 1$, $|\arg w| \leqslant \frac{1}{2}\pi$ so that the points at ∞ correspond to each other.

2.9.24. Find the 1:1 conformal mapping of the domain $\hat{\mathbf{C}} \setminus \overline{K}(0; 1)$ onto the w-plane slit along the arc: $|w| = R$, $|\arg w| \leqslant \alpha$ so that the points at ∞ correspond to each other.

CHAPTER 3

Complex Integration

3.1. LINE INTEGRALS. THE INDEX

Suppose $F = u+iv$ is a complex-valued function of a real variable $t \in [a, b]$; F is said to be *integrable* (e.g. in Riemann's sense), if both real-valued functions u, v are integrable; then the integral $\int_a^b F(t)dt$ is defined as $\int_a^b u(t)dt + i\int_a^b v(t)dt$.
Most of the properties of the real integral also hold for the integral of a complex-valued function. The line integrals of complex-valued functions which are a very important tool in complex analysis, can be reduced to the integrals over an interval.

Suppose γ is a regular curve, i.e. a curve which is represented by the equation $z = z(t)$, $a \leqslant t \leqslant b$, with continuous $z(t)$ having a piecewise continuous non-vanishing derivative $z'(t)$. If f is a complex-valued function defined for all z on γ, then the *line integral* $\int_\gamma f(z)dz$ may be defined as $\int_a^b G(t)dt$, where $G(t) = f(z(t))z'(t) = f(z(t))[x'(t)+iy'(t)]$. The line integral does not depend on the choice of parameter; if $-\gamma$ is the arc with the opposite orientation, i.e. the arc defined by the equation $z = z(-\tau)$, $-b \leqslant \tau \leqslant -a$, then $\int_{-\gamma} f(z)dz = -\int_\gamma f(z)dz$.

If f is defined in a domain D and has a primitive F in D (i.e. F is analytic in D and $F' = f$) then for any regular arc γ contained in D with end points z_1, z_2 we have $\int_\gamma f(z)dz = F(z_2)-F(z_1)$. In particular, for any closed curve C in D we have $\int_C f(z)dz = 0$. Conversely, if the line integral of f over any regular, closed curve in D vanishes, then f has a primitive in D.

The notion of line integral can be extended on *chains* Γ, which are linear forms $\sum_{m=1}^n k_m \gamma_m$, where γ_m are regular curves and k_m are integers, the integral $\int_\Gamma$ being defined as $\sum_{m=1}^n k_m \int_{\gamma_m}$. If all curves γ_m are closed, the corresponding chain is said to be a *cycle*.

Similarly as in real analysis, we also consider unoriented line integrals. If $f = u+iv$ is defined for all points of a regular curve γ and s is the arc-length on γ, then $\int_\gamma u\,ds + i\int_\gamma v\,ds$ is called *unoriented line integral* and is denoted $\int_\gamma f\,ds$, or $\int_\gamma f|dz|$.

3.1.1. Evaluate the line integral $\int_\gamma \operatorname{re} z\,dz$ for: (i) $\gamma = [0, 1+i]$; (ii) $\gamma = C(0; r)$.

3.1.2. Show that, for a regular curve γ with the equation $z = z(t)$, $a \leqslant t \leqslant b$, we have
$$\int_\gamma f(z)|dz| = \int_a^b f(z(t))|z'(t)|\,dt.$$

3.1.3. Evaluate the integral $\int_{C(0;1)} |z-1|\,|dz|$.

3.1.4. Evaluate the integral $\int_\gamma |z|\,dz$ for:
(i) $\gamma = [-i, i]$;
(ii) γ being the left-hand half of $C(0; 1)$ joining $-i$ to i;
(iii) γ being the right-hand half of $C(0; 1)$ joining $-i$ to i.

3.1.5. Verify the inequalities:
(i) $\left|\int_a^b f(t)\,dt\right| \leqslant \int_a^b |f(t)|\,dt$, $a < b$;
(ii) $\left|\int_\gamma f(z)\,dz\right| \leqslant \int_\gamma |f(z)|\,|dz|$.

3.1.6. Show that $\left|\int_\Gamma f(z)\,dz\right| \leqslant ML$, where $M = \sup|f(z)|$ on γ and L is the length of γ.

3.1.7. Show that $\int_{\partial Q} \dfrac{dz}{z-z_0} = 2\pi i$, where ∂Q is the closed polygonal line $[z_0-a-ia, z_0+a-ia, z_0+a+ia, z_0-a+ia, z_0-a-ia]$.

3.1.8. Prove that the function $(z-z_0)^{-1}$ analytic in $\mathbf{C}\setminus z_0$ has no primitive in this domain.

3.1.9. Suppose Γ is a contour, i.e. a regular, closed positively oriented simple (or Jordan) curve and A is the area enclosed by Γ. Prove that
$$\int_\Gamma x\,dz = -i\int_\Gamma y\,dz = \tfrac{1}{2}\int_\Gamma \bar{z}\,dz = iA.$$

3.1. LINE INTEGRALS. THE INDEX

3.1.10. Evaluate $\int_\Gamma z\bar{z}^{-1}dz$, where Γ is the boundary of the upper half of the annulus $\{z: 1 < |z| < 2\}$ with positive orientation.

3.1.11. Evaluate $\int_\Gamma |z|\bar{z}\,dz$, where Γ is the boundary of the upper half of $K(0; 1)$ with positive orientation.

3.1.12. Assume u, v are real-valued functions having continuous partials of first order in some neighborhood of $z_0 = x_0 + iy_0$. Prove that $f = u + iv$ is differentiable w.r.t. z at z_0, iff

$$\lim_{r \to 0} \frac{1}{\pi r^2} \int_{C(z_0;r)} f(z)\,dz = 0.$$

3.1.13. Prove the Weierstrass mean value theorem:

If f is a complex-valued function continuous in a domain D which has a primitive F in D, then for any $a, b \in D$ such that $[a, b] \subset D$ we have: $F(b) - F(a) = (b-a)\zeta$ with $\zeta \in \text{conv}\,\Gamma$, where $\text{conv}\,\Gamma$ is the convex hull of the curve Γ with the equation $w = f(a + (b-a)t)$, $0 \leqslant t \leqslant 1$.

Hint: Consider the integral sums of $\int_{[a,b]} f(z)\,dz$ after parametrization and interpret Δt_k as masses.

3.1.14. Suppose F is analytic in a convex domain G and $\text{re}\,F'(z) > 0$ for any $z \in G$. Prove that F is univalent in G (i.e. $F(z_1) \neq F(z_2)$ for any $z_1 \neq z_2$).

3.1.15. Prove that the polynomial $P(z) = a + nz + z^n$ is univalent in $K(0; 1)$.

3.1.16. Prove that the function $F(z) = z + e^z$ is univalent in the left half-plane.

3.1.17. Prove that the function $F(z) = z^3 + 3z$ is univalent in the domain to the right of the right-hand branch of the hyperbola $x^2 - y^2 = 1$.

3.1.18. Suppose φ is a complex-valued function continuous on a regular curve γ and

$$F(z) = \int_\gamma (\zeta - z)^{-1} \varphi(\zeta)\,d\zeta.$$

Show that F is continuous and analytic in any domain not containing any points of γ and verify that

$$F'(z) = \int_\gamma (\zeta - z)^{-2} \varphi(\zeta)\,d\zeta.$$

3.1.19. Suppose γ is a closed, regular curve and a is a point not on γ. Show that

$$n(\gamma, a) = \frac{1}{2\pi i} \int_\gamma \frac{dz}{z-a}$$

is an integer. The function $n(\gamma, a)$ is called the index of the point a with respect to γ, or the winding number of γ with respect to a.

Hint: Consider the function

$$h(t) = \int_\alpha^t \frac{z'(\tau)}{z(\tau)-a} d\tau$$

where $z = z(t)$, $\alpha \leqslant t \leqslant \beta$, is the equation of γ. Verify that $u(t) = [z(t) - a] \exp(-h(t))$ is a constant.

3.1.20. Show that $n(\gamma, a)$ has a constant value for all $a \in D$, if the domain D does not contain any points of γ.

3.1.21. Suppose γ is situated in a disk Δ. Verify that $n(\gamma, a) = 0$ for any $a \in \mathbf{C} \setminus \Delta$.

3.1.22. Suppose γ is a regular, closed curve not meeting the negative real axis. Show that for any $a \in (-\infty, 0)$ we have $n(\gamma, a) = 0$.

3.1.23. Prove Jordan's theorem for curves starlike w.r.t. the origin: If γ has the equation $z(\theta) = r(\theta)e^{i\theta}$, $0 \leqslant \theta \leqslant 2\pi$, $r(0) = r(2\pi)$ and $r(\theta)$ is a positive, continuously differentiable function of θ, then $\hat{\mathbf{C}} \setminus \{\gamma\} = D_0 \cup D_\infty$, where D_0, D_∞ are disjoint domains with a common boundary γ. Moreover, $n(\gamma, z) = 1$ for all $z \in D_0$ and $n(\gamma, z) = 0$ for all $z \in D_\infty$.

3.1.24. Suppose $\{\gamma_k\}$ is a sequence of closed, regular curves such that $n(\gamma_k, 0) = 1$ for all k and $\delta_k = \inf_{\zeta \in \gamma_k} |\zeta| \to +\infty$ as $k \to +\infty$. Show that for any fixed z there exists k_0 such that $n(\gamma_k, z) = 1$ for all $k > k_0$.

3.1.25. The notion of index can be extended on cycles: If $\Gamma = \sum_{m=1}^n k_m \gamma_m$ and a does not lie on any γ_m, then $n(\Gamma, a)$ is defined as $\sum_{m=1}^n k_m n(\gamma_m, a)$.

Suppose $C_1 = C(0; r_1)$, $C_2 = C(0; r_2)$, $r_1 < r_2$. Evaluate $n(\Gamma, a)$ for $|a| < r_1$, $r_1 < |a| < r_2$, $r_2 < |a|$ and (i) $\Gamma = C_1 + C_2$; (ii) $\Gamma = C_1 - C_2$.

3.1.26. Suppose γ is a cycle, a, b are points not on curves of γ and $n(\gamma, a) = n(\gamma, b)$. Show that

$$\int_\gamma (z-a)^{-1}(z-b)^{-1} dz = 0.$$

3.1.27. Prove that for any positive integers m, n and a, b such that $n(\gamma, a) = n(\gamma, b)$ we have

$$\int_\gamma (z-a)^{-m}(z-b)^{-n} = 0.$$

3.1. LINE INTEGRALS. THE INDEX

3.1.28. Suppose W_n is a polynomial of degree at most n and $a \in K(0; r)$. Prove that

$$\int_{C(0;r)} \frac{W_n(z)}{z^{n+1}(z-a)}\, dz = 0.$$

3.1.29. If $|a| < r < |b|$, show that

$$\int_{C(0;r)} (z-a)^{-1}(z-b)^{-1}\, dz = 2\pi i (a-b)^{-1}.$$

3.1.30. Evaluate the integral

$$\int_{C(0;r)} (z-a)^{-m}(z-b)^{-n}\, dz,$$

where m, n are positive integers and $|a| < r < |b|$.

3.1.31. Verify that

$$\int_{C(0;2)} (1+z^2)^{-1}\, dz = 0.$$

3.1.32. If $\varGamma$ is a cycle and W is a polynomial, show that

$$\int_{\varGamma} W(z)\, dz = 0.$$

3.1.33. Suppose $\varGamma$ is a cycle and a is a point not on any curve of $\varGamma$. Show that

$$\int_{\varGamma} (z-a)^{-n}\, dz = 0 \quad \text{for any integer } n \geq 2.$$

3.1.34. Suppose $R(z)$ is a rational function, $a_1, \ldots, a_r$ are the zeros of the denominator, $A_1, \ldots, A_r$ are coefficients of $(z-a_1)^{-1}, \ldots, (z-a_r)^{-1}$ in the development of R in partial fractions and $\varGamma$ is a cycle such that all its curves omit the points $a_1, \ldots, a_r$. Show that

$$\int_{\varGamma} R(z)\, dz = 2\pi i \sum_{j=1}^{r} A_j n(\varGamma, a_j).$$

3.1.35. If $|a| \neq r$, show that

$$\int_C |z-a|^{-2} |dz| = 2\pi r \big||a|^2 - r^2\big|^{-1}, \quad C = C(0; r).$$

Hint: Transform the unoriented integral into an oriented one; put $|dz| = -ir\, dz/z$.

3.1.36. If $f(z) = z(1-z)^{-2}$ and $0 < r < 1$, show that

$$\frac{1}{2\pi} \int_0^{2\pi} |f(re^{i\theta})|\, d\theta = \frac{r}{1-r^2}.$$

3.1.37. If Γ is a regular arc joining the points $z = 1$, $z = re^{i\varphi}$ ($0 \leqslant \varphi \leqslant 2\pi$) and omitting the origin, show that

$$\int_\Gamma \frac{dz}{z} = \log r + i\varphi + 2k\pi i,$$

where k is an integer depending on Γ.

3.2. CAUCHY'S THEOREM AND CAUCHY'S INTEGRAL FORMULA

Cauchy's theorem is a fundamental theorem in the theory of analytic functions. There are several forms of Cauchy's theorem, the simplest one being Goursat lemma, or Cauchy's theorem for a rectangle:

If f is analytic in a domain D and the closed rectangle R: $a \leqslant \operatorname{re} z \leqslant b$, $c \leqslant \operatorname{im} z \leqslant d$ is contained in D, then the line integral of f over the boundary ∂R of R vanishes: $\int_{\partial R} f(z)\, dz = 0$.

Cauchy's theorem for a rectangle is on the one hand quite useful in proving local properties of analytic functions via Cauchy's integral formula and on the other hand it gives rise to more general statements, e.g. to the following homological version of Cauchy's theorem.

Suppose f is analytic in a domain D and Γ is a cycle homologous to zero with respect to D (in notation: $\Gamma \sim 0 \pmod{D}$), *i.e. each closed curve in Γ is contained in D and the index $n(\Gamma, a) = 0$ for any $a \in \mathbf{C} \setminus D$. Then $\int_\Gamma f(z)\, dz = 0$.*

In particular, any closed curve in a simply connected domain is a cycle homologous to zero, hence for any function analytic in a simply connected domain D and any closed, regular curve γ in D we have: $\int_\gamma f(z)\, dz = 0$ which is Cauchy's theorem for simply connected domains.

From Cauchy's theorem Cauchy's integral formula can be derived. Suppose f is analytic in a domain D and $\Gamma \sim 0 \pmod{D}$ is such that the curves in Γ omit some point $a \in D$. Then

$$n(\Gamma, a) f(a) = \frac{1}{2\pi i} \int_\Gamma \frac{f(z)}{z-a}\, dz.$$

3.2. CAUCHY'S THEOREM AND INTEGRAL FORMULA

In particular we may assume that Γ is a contour γ situated together with its inside in D and a is a point inside γ. Then $n(\gamma, a) = 1$ for positive orientation of γ and under these assumptions

$$f(a) = \frac{1}{2\pi i} \int_\gamma \frac{f(z)}{z-a} dz.$$

Hence the existence of derivatives of all orders, as well as the representation formula for nth derivative follow:

$$f^{(n)}(a) = \frac{n!}{2\pi i} \int_\gamma \frac{f(z)}{(z-a)^{n+1}} dz.$$

3.2.1. Evaluate $\int_\Gamma (1+z^2)^{-1} dz$, where Γ is the ellipse $x^2 + 4y^2 = 1$.

3.2.2. Evaluate $\int_C \frac{e^z \cos z}{(1+z^2)\sin z} dz$, where $C = C(2+i; \sqrt{2})$.

3.2.3. If f is analytic in $K(0; R) \setminus 0$, show that the value of $\int_0^{2\pi} f(re^{i\theta}) d\theta$, $0 < r < R$, does not depend on r.

3.2.4. Evaluate $\int_0^{2\pi} f(re^{i\theta}) d\theta$ in case is analytic in $K(0; R)$ and $0 < r < R$.

3.2.5. If u is harmonic in $K(0; R)$ and $0 < r < R$, show that

$$u(0) = \frac{1}{2\pi} \int_0^{2\pi} u(re^{i\theta}) d\theta.$$

3.2.6. Evaluate $\int_0^{2\pi} \log|re^{i\theta} - a| d\theta$, $r < |a|$.

3.2.7. If f is analytic in $K(a; R) \setminus 0$ and $\lim_{z \to a} (z-a)f(z) = A$, show that

$$\int_{C(a;r)} f(z) dz = 2\pi i A \quad \text{for any} \quad r \in (0, R).$$

3.2.8. Suppose $D = \mathbf{C} \setminus H$, H is an enumerable set of points $a_1, \ldots, a_n, \ldots$ with $\lim a_n = \infty$ and f is analytic in D. If γ is a closed, regular curve in D and $A_k = \lim_{z \to a_k} (z-a_k)f(z)$ exists for all k, show that $\int_\gamma f(z) dz = 2\pi i \sum_k A_k n(\gamma, a_k)$. Verify that the sum contains a finite number of non-zero terms.

3.2.9. Suppose C is a closed, regular curve omitting the points $0, 1, -1$. Find all possible values of $\int_C \frac{dz}{z(z^2-1)}$.

3.2.10. Evaluate $\int_{C(a;a)} \frac{z}{z^4-1} dz$, $a > 1$.

3.2.11. Evaluate $\int_{C(0;2a)} \frac{e^z}{z^2+a^2} dz$, $a > 0$.

3.2.12. Using Cauchy's formula for $f''(a)$ evaluate $\frac{1}{2\pi i} \int_C \frac{ze^z}{(z-a)^3} dz$, where C is a contour containing a inside.

3.3.13. Evaluate in a similar way the integral

$$\frac{1}{2\pi i} \int_C \frac{e^z}{z(1-z)^3} dz$$

in case: (i) $C = C(0; \frac{1}{2})$; (ii) $C = C(1; \frac{1}{2})$.

3.2.14. Evaluate $\int_{C(0;R)} \frac{f(z)}{(z-a)(z-b)} dz$ in case $a, b \in K(0; R)$ for f analytic in some domain containing $\bar{K}(0; R)$.

Prove Liouville's theorem: Any function analytic in $\mathbf{C}$ and bounded is a constant.

3.2.15. Using Cauchy's formula for the derivative evaluate

$$\int_{C(0;r)} \frac{dz}{(z-b)(z-a)^m}, \quad |a| < r < |b|.$$

3.2.16. Suppose C is a contour containing 0 inside and leaving z outside and f is analytic in $\mathbf{C} \setminus 0$ and bounded outside C. Prove that

$$f(z) = \frac{1}{2\pi i} \int_C \frac{zf(\zeta)}{(z-\zeta)} d\zeta.$$

Evaluate the right-hand side term for z situated inside C and $\neq 0$.

3.2.17. Let P_n be the polynomial $(z-z_1)(z-z_2) \ldots (z-z_n)$ with $z_j \neq z_k$ for $j \neq k$. How many values can take the integral $\int_C \frac{dz}{P_n(z)}$ for C being contours with positive orientation omitting all z_k?

3.3. ISOLATED SINGULARITIES

If f is analytic in $K(a;r)\setminus a$, then the point a is called an *isolated singularity* of f. If the limit $\lim_{z\to a} f(z)$ exists, then the singularity is called *removable*: we can extend the domain of f putting $f(a) = \lim_{z\to a} f(z)$ and obtain in this way always a function analytic in the whole disk $K(a;r)$. If $\lim_{z\to a} f(z) = \infty$, then the point a is said to be a *pole*. In this case there exists a finite $A\neq 0$ and a positive integer n such that $\lim_{z\to a}(z-a)^n f(z) = A$. The number n is called the *order of the pole*. Otherwise a is called an *essential isolated singularity*. If f is analytic in $\{z: |z|>R\}$, then $f(1/z)$ has an isolated singularity at 0 and the character of singularity of f at ∞ is defined to be the same as that of $f(1/z)$ at $z = 0$.

3.3.1. Suppose f is analytic in $K(a;R)\setminus a$ and $\lim_{z\to a}(z-a)f(z) = 0$. Prove that for any closed, regular curve in $K(a;R)\setminus a$ we have $\int_\gamma f(z)\,dz = 0$.

3.3.2. Show that under the assumptions of Excercise 3.3.1 we have

$$f(\zeta) = \frac{1}{2\pi i}\int_{C(a;r)}\frac{f(z)}{z-\zeta}\,dz, \quad \text{where} \quad 0 < |\zeta - a| < r < R.$$

3.3.3. Under assumptions of Excercise 3.3.1 prove that a finite limit $\lim_{r\to a} f(z) = b$ exists and the function: $\varphi(z) = f(z)$, $z \in K(a;R)\setminus a$, $\varphi(a) = b$, is analytic in $K(a;R)$.

3.3.4. Show that if f is analytic in $K(0;R)$ and $f(0) = 0$ then the point $z = 0$ is a removable singularity of $f(z)/z$.

3.3.5. Show that $\exp(1/z)$ has $z = 0$ as an essential isolated singularity.

3.3.6. Show that $\sin(1/z)$ has 0 as an essential isolated singularity.

3.3.7. Prove that $z = k\pi$, $k = 0, \mp 1, \mp 2, \ldots$, are poles of order 2 of $(\sin z)^{-2}$.

3.3.8. State the character of isolated singularities (also possibly at infinity) for the following functions:

(i) $\dfrac{1}{z-z^3}$; (ii) $\dfrac{z^5}{(1-z)^2}$; (iii) $\dfrac{e^z}{1+z^2}$;

(iv) $\dfrac{1-e^z}{1+e^z}$; (v) $\exp\dfrac{z}{1-z}$; (vi) $(e^z-1)^{-1}\exp\dfrac{1}{1-z}$;

(vii) $\exp\tan\dfrac{1}{z}$; (viii) $\sin\left(\cos\dfrac{1}{z}\right)^{-1}$.

3.3.9. Verify that a function analytic in the open plane $\mathbf{C}$ and having $z = \infty$ as a removable singularity, is a constant.

3.3.10. Show that a function analytic in the extended plane except for a finite number of poles, is rational.

3.3.11. Verify Taylor's formula: if f is analytic in $K(a; R)$ then for any positive integer n there exists a function f_n analytic in $K(a; R)$ such that for any $z \in K(a; R)$ we have:

$$f(z) = f(a) + \frac{f'(a)}{1!}(z-a) + \ldots + \frac{f^{(n-1)}(a)}{(n-1)!}(z-a)^{n-1} + (z-a)^n f_n(z).$$

Hint: $f(z) = f(a) + (z-a)f_1(z)$, where f_1 is analytic in $K(a; R)$.

3.3.12. If $f(a) = f'(a) = \ldots = f^{(m-1)}(a) = 0$ and $f^{(m)}(a) \neq 0$, the point a is said to be a zero of f of order m. Show that $(z-a)^{-m}f(z)$ has a removable singularity at a zero a of order m.

3.3.13. Prove that $[f(z)]^{-1}$ has a pole of order m at a, iff f has a zero of order m at this point.

3.3.14. Suppose f is analytic in the extended plane except for a finite number of poles and p, q are the numbers of zeros and poles counted with due multiplicity. Prove that $p = q$.

3.3.15. If f is analytic in $K(a; R)$ and $f^{(n)}(a) = 0$ for all positive integers n, show that f is a constant.

Hint: Use 3.3.11 and estimate $|f(z) - f(a)|$ by means of Cauchy's formula.

3.3.16. Suppose f is a nonconstant analytic function in $K(a; r)$. Show that there exists a positive integer m such that $f^{(m)}(a) \neq 0$.

3.3.17. Suppose f is analytic and not identically 0 in $K(a; r)$. Show that $1/f$ is either analytic in some neighborhood of a, or has a pole at a.

3.3.18. Suppose f is analytic in $K(a; R) \setminus a$ and does not take in this domain any value from $K(w_0; \delta)$. Prove that a is either a pole, or a removable singularity.

3.3.19. Suppose a is an essential isolated singularity of f, P is a nonconstant polynomial. Prove that a is also an essential isolated singularity of $P \circ f$.

3.3.20. Suppose A is an enumerable set of points contained in the disk $K(a; R)$ having a as its only point of accumulation. Suppose, moreover, that f is analytic in $K(a; R) \setminus (A \cup \{a\})$ and each point of $A \setminus a$ is a pole of f. Prove that the set of values taken by f is dense in the plane.

3.4. EVALUATION OF RESIDUES

3.4.1. Suppose f is analytic in $K(a;r) \setminus a$. Show that there exists a unique complex number A such that for any closed, regular curve γ in $K(a;r) \setminus a$:
$$\int_\gamma [f(z) - A(z-a)^{-1}]dz = 0.$$
The constant A is called the *residue* of f at a and is denoted $\operatorname{res}(a;f)$.

Hint: Consider a cycle $\Gamma = \gamma - mC(a;\delta)$ which is $\sim 0 \pmod{K(a;r) \setminus a}$.

3.4.2. If φ is analytic in $K(a;r)$ and $f(z) = \sum_{k=1}^n A_k(z-a)^{-k} + \varphi(z)$, show that $\operatorname{res}(a;f) = A_1$.

3.4.3. Express $\dfrac{z^3+z^2+2}{z(z^2-1)^2}$ as a sum of partial fractions and evaluate the residues at $0, 1, -1$.

3.4.4. If a is a pole of first order of f, prove that $\operatorname{res}(a;f) = \lim_{h \to 0} hf(a+h)$.

3.4.5. Suppose Γ is analytic in $\mathbf{C}$ except for the set of nonpositive integers, where it has poles of first order. Suppose, moreover, $\Gamma(1) = 1$ and $z\Gamma(z) = \Gamma(z+1)$ for all regular points. Prove that $\operatorname{res}(-n; \Gamma) = \dfrac{(-1)^n}{n!}$.

3.4.6. Evaluate the residues of: (i) $\dfrac{e^z}{z(z-1)}$ at $z = 0, 1$; (ii) $\dfrac{e^{\pi z}}{1+z^2}$ at $z = \mp i$.

3.4.7. If f is odd (i.e. $f(z) \equiv -f(-z)$), show that
$$\operatorname{res}(a;f) = \operatorname{res}(-a;f).$$
If f is even (i.e. $f(z) \equiv f(-z)$), show that
$$\operatorname{res}(a;f) = -\operatorname{res}(-a;f).$$

3.4.8. Prove de l'Hospital's rule for analytic functions: if f, g are analytic in $K(a;r) \setminus a$, $\lim_{z \to a} f(z) = \lim_{z \to a} g(z) = 0$ (or $\lim_{z \to a} g(z) = \infty$) and the finite, or infinite limit $\lim_{z \to a} \dfrac{f'(z)}{g'(z)}$ exists, then also the limit $\lim_{z \to a} \dfrac{f(z)}{g(z)}$ exists and both limits are equal.

3.4.9. If z_ν is the pole of $f(z) = (z^4+a^4)^{-1}$, show that $\operatorname{res}(z_\nu;f) = -\tfrac{1}{4}z_\nu a^{-4}$.

3.4.10. Show that the residues of $f(z) = z^{n-1}/(z^n+a^n)$ at all its poles are equal $1/n$.

3.4.11. Suppose $k \geq 2$ is an integer and f is analytic in some neighborhood of a. Prove that
$$\operatorname{res}(a;\, (z-a)^{-k}f(z)) = \frac{f^{(k-1)}(a)}{(k-1)!}.$$

3.4.12. Suppose $z_k = (k+\tfrac{1}{2})\pi$ and f is analytic in some domain containing the real axis. Prove that
$$\operatorname{res}(z_k;\, f(z)(\cos z)^{-2}) = f'(z_k).$$

3.4.13. Show that
$$\operatorname{res}\left(0;\, \frac{\sin \alpha z}{z^3 \sin \beta z}\right) = \frac{\alpha}{6\beta}(\beta^2 - \alpha^2), \quad \beta \neq 0.$$

3.4.14. Prove that:
 (i) $\operatorname{res}(1;\, e^z(z-1)^{-4}) = \tfrac{1}{6}e$;
 (ii) $\operatorname{res}\left(i;\, \dfrac{\exp(a \operatorname{Log} z)}{(1+z^2)^2}\right) = \dfrac{1}{4i}(1-a)\exp(\tfrac{1}{2}a\pi i)$ (a real);
 (iii) $\operatorname{res}(ai;\, (z^2+a^2)^{-2}e^{miz}) = \dfrac{-ie^{-am}}{4a^3}(am+1),\ a \neq 0$;
 (iv) $\operatorname{res}(i;\, [(1+z^2)\cosh\tfrac{1}{2}\pi z]^{-1}) = \dfrac{1}{2\pi i}$;
 (v) $\operatorname{res}(z_1;\, (z-z_1)^{-n}(z-z_2)^{-1}) = (-1)^{n-1}(z_1-z_2)^{-n},\ n = 1, 2, \ldots$;
 (vi) $\operatorname{res}(i;\, (1+z^2)^{-n}) = -i \cdot 2^{-2n+1}\binom{2n-2}{n-1},\ n = 1, 2, \ldots$;
 (vii) $\operatorname{res}(-1;\, z^{2n}(1+z)^{-n}) = (-1)^{n+1}\dfrac{(2n)!}{(n-1)!(n+1)!},\ n = 1, 2, \ldots$;
 (viii) $\operatorname{res}(n^2\pi^2;\, \sqrt{z}/\sin\sqrt{z}) = (-1)^n 2\pi^2 n^2,\ n = 1, 2, \ldots$

3.4.15. If f is analytic in $\mathbf{C}\setminus \overline{K}(0; R)$ and $r > R$, then $\operatorname{res}(\infty; f)$ is defined as
$$-\frac{1}{2\pi i}\int_{C(0;\, r)} f(z)\, dz.$$

If $f(z) = P(z) + Az^{-1} + \varphi(z)z^{-2}$ for $|z| > R$, where P is a polynomial and $\varphi(z) = O(1)$ as $z \to \infty$, show that $\operatorname{res}(\infty; f) = -A$.

3.4.16. Evaluate:
 (i) $\operatorname{res}\left(\infty;\, \operatorname{Log}\dfrac{z-a}{z-b}\right)$;
 (ii) $\operatorname{res}\left(\infty;\, \sqrt{(z-a)(z-b)}\right)$.

3.4.17. Suppose f is analytic in $\mathbf{C}$ except for a finite number of points. Show that the sum of residues at all singularities (the point at infinity included) equals to zero.

3.4.18. Evaluate:
(i) $\operatorname{res}(\infty;\, (z+1)^{-3}\sin 2z)$;
(ii) $\operatorname{res}\left(\infty;\, z^3 \cos\dfrac{1}{z-2}\right)$.

3.4.19. Evaluate the residues of the following functions at all isolated singularities: (i) $(z^3-z^5)^{-1}$; (ii) $e^z/z^2(z^2+4)$; (iii) $\cot^2 z$; (iv) $\cot^3 z$; (v) $\sin\dfrac{z}{z+1}$
(vi) $[z(1-e^{-hz})]^{-1}$; (vii) $z^n \sin\dfrac{1}{z}$, $n = 0, \mp 1, \mp 2, \ldots$

3.4.20. Suppose φ is analytic in a neighborhood of a and $\varphi'(a) \neq 0$. If f has a simple pole at the point $\varphi(a)$ and $\operatorname{res}(\varphi(a);f) = A$, show that $\operatorname{res}(a;\, f \circ \varphi) = A/\varphi'(a)$.

3.4.21. If f is analytic in $K(a;R) \setminus a$, then f has in $K(a;R) \setminus a$ the following Laurent expansion:

$$f(z) = \sum_{n=-\infty}^{\infty} A_n (z-a)^n \quad \text{and} \quad \operatorname{res}(a;f) = A_{-1}.$$

Evaluate in this way:
(i) $\operatorname{res}\left(0;\, z^{-2} e^{1/z} \operatorname{Log}\dfrac{1-\alpha z}{1-\beta z}\right)$;
(ii) $\operatorname{res}\left(1;\, \operatorname{Log} z \cos\dfrac{1}{z-1}\right)$.

3.5. THE RESIDUE THEOREM

If f is analytic in a domain D except for a set of isolated singularities a_k, then for any cycle γ homologous to zero w.r.t. D and contained in $D \setminus \bigcup_k \{a_k\}$ we have

$$\int_\gamma f(z)\,dz = 2\pi i \sum_k n(\gamma, a_k) \operatorname{res}(a_k; f).$$

The sum on the right is finite since $n(\gamma, a_k) \neq 0$ only for a finite number of a_k. In most applications γ is a contour with positive orientation. Then $n(\gamma, a_k)$ is either 0 in case a_k is situated in the unbounded component of $\mathbf{C} \setminus \{\gamma\}$, or 1 in case a_k is in the bounded component.

The residue theorem enables us to evaluate line integrals of analytic functions over closed curves by means of residues. In what follows all the contours considered as paths of integration are taken with positive orientation unless stated otherwise.

3.5.1. Evaluate the integral $\int_C \dfrac{dz}{1+z^4}$, where C is the ellipse $x^2-xy+y^2+x+y=0$.

3.5.2. Evaluate the integrals:

(i) $\int_\Gamma \dfrac{dz}{1+z^4}$, $\Gamma = C(1; 1)$;

(ii) $\int_\Gamma \dfrac{dz}{(z-1)^2(1+z^2)}$, $\Gamma = C(1+i; \sqrt{2})$;

(iii) $\int_\Gamma \dfrac{dz}{(z^2-1)^2(z-3)^2}$, Γ is the asteroid $x^{2/3}+y^{2/3} = 2^{2/3}$;

(iv) $\int_\Gamma \dfrac{dz}{z^3(z^{10}-2)}$, $\Gamma = C(0; 2)$.

3.5.3. Show that:

(i) $\int_C \dfrac{dz}{1+z^3} = -\dfrac{2}{3}\pi i$, C is the ellipse $2x^2+y^2-\dfrac{3}{2} = 0$;

(ii) $\int_{C(0;R)} \dfrac{z^2}{\exp(2\pi i z^3)-1} dz = 2n+1$, $n < R^3 < n+1$, n being a positive integer;

(iii) $\int_\gamma \dfrac{e^{\pi z}}{2z^2-i} dz = \dfrac{1}{2}\pi(i-1)e^{\pi/2}$, γ is the boundary of $K(0; 1) \cap (+;+)$.

3.5.4. Verify that

$$\int_{C(0;2)} \dfrac{z^3 \exp(1/z)}{1+z} dz = -\dfrac{2}{3}\pi i.$$

3.5.5. Let ∂Q be the boundary of the square Q with vertices $\pi(\mp 1 \mp i)$ and let w be a point inside Q. Prove that

$$\mathrm{Log}\, w = \dfrac{1}{2\pi i} \int_{\partial Q} \dfrac{ze^z}{e^z-w} dz.$$

3.5. THE RESIDUE THEOREM

3.5.6. If $a, b \in K(0; r)$ and n is a positive integer, show that
$$\int_{C(0;r)} z^n \operatorname{Log} \frac{z-a}{z-b} \, dz = \frac{2\pi i}{n+1} (b^{n+1} - a^{n+1}).$$

3.5.7. If $a, b \in K(0; r)$ show that
$$\int_{C(0;r)} \sqrt{(z-a)(z-b)} \, dz = -\tfrac{1}{4} \pi i (a-b)^2$$
for the branch of square root equal to $z + O(1)$ near infinity.

3.5.8. Under the assumptions of Excercise 3.5.7 show that:

(i) $\displaystyle \int_{C(0;r)} \frac{z}{\sqrt{(z-a)(z-b)}} \, dz = \pi i (a+b);$

(ii) $\displaystyle \int_{C(0;r)} \frac{dz}{\sqrt{(z-a)(z-b)}} = 2\pi i.$

3.5.9. Evaluate the integrals:

(i) $\displaystyle \int_{C(0;2)} \sqrt{\frac{z}{z+1}} \, dz;$

(ii) $\displaystyle \int_{C(0;1)} \frac{dz}{\sqrt{4z^2+4z+3}}$

for the branch of square root equal to $1+o(1)$ and $\dfrac{1}{2z}(1+o(1))$ respectively near infinity.

3.5.10. Verify that
$$\int_{C(0;2)} \frac{z^n}{\sqrt{1+z^2}} \, dz = \begin{cases} 0 & \text{for } n \text{ odd}, \\ (-1)^k \dfrac{1 \cdot 3 \cdot \ldots \cdot (2k-1)}{2 \cdot 4 \cdot \ldots \cdot 2k} 2\pi i \end{cases}$$
$$\text{for } n = 2k, \; k = 1, 2, \ldots$$

3.5.11. Let Γ be the parabola $x = y^2$ with the orientation corresponding to increasing y and let $\sqrt{1+z^2}$ be the branch of square root with positive real part in the right half-plane and in a neighborhood of the origin. Show that the improper line integral
$$\int_\Gamma \frac{dz}{(1+z^4)\sqrt{1+z^2}} = \frac{1}{2} \pi i \sqrt{1+\sqrt{2}} \,.$$

3.6. EVALUATION OF DEFINITE INTEGRALS INVOLVING TRIGONOMETRIC FUNCTIONS

3.6.1. Let $R(u, v)$ be a rational function of two variables u, v. Show that the definite integral $\int_0^{2\pi} R(\cos\theta, \sin\theta)\,d\theta$ may be considered as a parametrized line integral $\int_{C(0;1)} R_1(z)\,dz$, where R_1 is a rational function of z.

3.6.2. Prove that

$$\int_0^{2\pi} \frac{\sin^2\theta}{a+b\cos\theta}\,d\theta = \frac{2\pi}{b^2}(a-\sqrt{a^2-b^2}), \quad 0 < b < a.$$

3.6.3. Prove that

(i) $\dfrac{1}{2\pi}\displaystyle\int_0^{2\pi} \frac{\cos^2\theta}{1+a^2-2a\cos(\theta-\varphi)}\,d\theta = \frac{1+a^2\cos 2\varphi}{2(1-a^2)}$;

(ii) $\displaystyle\int_0^{2\pi} \frac{\cos^2 3\theta}{1+a^2-2a\cos 2\theta}\,d\theta = \pi\,\frac{1-a+a^2}{1-a},\ 0<|a|<1.$

3.6.4. If n is a positive integer, show that

$$\int_0^{2\pi} \frac{(1+2\cos\theta)^n \cos n\theta}{3+2\cos\theta}\,d\theta = \frac{2\pi}{\sqrt{5}}(3-\sqrt{5})^n.$$

3.6.5. Prove that:

(i) $\displaystyle\int_0^{2\pi} \frac{d\theta}{(1+a\cos\theta)^2} = \frac{2\pi}{(1-a^2)^{3/2}},\ |a|<1$;

(ii) $\displaystyle\int_0^{2\pi} \frac{d\theta}{1-2a\cos\theta+a^2} = \frac{2\pi}{|a^2-1|},\ |a|\neq 1.$

3.6.6. Prove that

$$\int_{-\pi}^{\pi} \frac{x\sin x}{1+a^2-2a\cos x}\,dx = \frac{2\pi}{a}\log\frac{1+a}{a},\quad a>1.$$

Also evaluate the integral for $0<a<1$.

Hint: Integrate $z(a-e^{-iz})^{-1}$ over the boundary ∂R_n of the rectangle R_n with vertices $\mp\pi$, $\mp\pi+in$.

3.6.7. By integrating $e^z z^{-n-1}$ round $C(0; 1)$ show that

(i) $\int_0^{2\pi} \exp(\cos\theta)\cos(n\theta-\sin\theta)\,d\theta = 2\pi/n!$;

(ii) $\int_0^{2\pi} \exp(\cos\theta)\sin(n\theta-\sin\theta)\,d\theta = 0$.

3.7. INTEGRALS OVER AN INFINITE INTERVAL

The theorem of residues is quite useful in evaluating a large number of real integrals. Let e.g. f be a function analytic on the real axis and in the upper half-plane except for a finite number of isolated singularities a_k with $\mathrm{im}\,a_k > 0$, $k = 1, 2, \ldots, n$. The residue theorem as applied to f and the upper semicircle gives us

(3.7A) $$\int_{-R}^{R} f(x)\,dx + \int_{\Gamma(R)} f(z)\,dz = 2\pi i \sum_{k=1}^{n} \mathrm{res}(a_k; f)$$

where $R > \max\{|a_1|, |a_2|, \ldots, |a_n|\}$ and $\Gamma(R)$ is the upper semicircle of $C(0; R)$. If $\lim_{R\to\infty} \int_{\Gamma(R)} f(z)\,dz = 0$ then we obtain at once from (3.7A):

$$\int_{-\infty}^{+\infty} f(x)\,dx = 2\pi i \sum_{k=1}^{n} \mathrm{res}(a_k; f)$$

which gives a convenient method of evaluating certain integrals over an infinite interval.

3.7.1. Show that

(i) $\int_0^{+\infty} \dfrac{x^2}{x^4+6x^2+13}\,dx = \dfrac{\pi}{8}$;

(ii) $\int_{-\infty}^{+\infty} \dfrac{x^2-x+2}{x^4+10x^2+9}\,dx = \dfrac{5}{12}\pi$;

(iii) $\int_0^{+\infty} \dfrac{x^2}{(x^2+a^2)^3}\,dx = \dfrac{1}{16}\pi a^{-3}$, $a > 0$;

(iv) $\int_{-\infty}^{+\infty} \dfrac{dx}{(x^2+a^2)(x^2+b^2)^2} = \dfrac{\pi(a+2b)}{2ab^3(a+b)^2}$, $a, b > 0$;

(v) $\int_{-\infty}^{+\infty} \dfrac{x^6}{(x^4+a^4)^2}\,dx = \dfrac{3\sqrt{2}\pi}{16a}$, $a > 0$.

3.7.2. Verify the inequality:
$$\int_0^{\pi} \exp(-A\sin\theta) < \pi/A, \quad A > 0.$$

Hint: Show that $\sin\theta > 2\theta/\pi$ for $\theta \in (0, \tfrac{1}{2}\pi)$.

3.7.3. Use Exercise 3.7.2 and the equality $e^{ix} = \cos x + i\sin x$ to show that

(i) $\displaystyle\int_0^{+\infty} \frac{\cos x}{x^2+a^2} dx = \frac{\pi e^{-a}}{2a}, \quad a > 0;$

(ii) $\displaystyle\int_0^{+\infty} \frac{x\sin x}{(x^2+a^2)^2} dx = \frac{\pi e^{-a}}{4a}, \quad a > 0;$

(iii) $\displaystyle\int_0^{+\infty} \frac{\cos mx}{(x^2+a^2)^2} dx = \frac{\pi}{4a^3}(1+am)e^{-am}, \quad a, m > 0;$

(iv) $\displaystyle\int_0^{+\infty} \frac{\cos x}{(x^2+a^2)(x^2+b^2)} dx = \frac{\pi}{b^2-a^2}\left(\frac{e^{-a}}{a} - \frac{e^{-b}}{b}\right), \quad a \neq b, \ a, b > 0;$

(v) $\displaystyle\int_0^{+\infty} \frac{\cos x}{(1+x^2)^3} dx = \frac{7\pi}{16e};$

(vi) $\displaystyle\int_0^{+\infty} \frac{\cos ax}{1+x^2+x^4} dx = \frac{\pi}{\sqrt{3}}\sin\left(\frac{\pi}{6}+\frac{a}{2}\right)\exp\left(-\frac{a\sqrt{3}}{2}\right), \quad a > 0.$

3.7.4. If f is a continuous function in the sector $\{z: 0 < |z-a| \leqslant \delta, \theta_1 \leqslant \arg(z-a) \leqslant \theta_2\}$ and $\lim_{z \to a}(z-a)f(z) = b$, show that

$$\lim_{r \to 0} \int_{\gamma_r} f(z)\,dz = ib(\theta_2 - \theta_1).$$

where $\gamma_r = \{z: z = a+re^{i\theta}, \theta_1 \leqslant \theta \leqslant \theta_2\}$ for $0 < r \leqslant \delta$.

3.7.5. By using the identity: $2\sin^2 z = \mathrm{re}(1-e^{2iz})$, z is real, and by integrating $f(z) = z^{-2}(1-e^{2iz})$ over the contour consisting of $[-R, -r]$, $-\Gamma(r)$, $[r, R]$, $\Gamma(R)$ ($0 < r < R$) show that

$$\int_0^{+\infty} \left(\frac{\sin x}{x}\right)^2 dx = \frac{\pi}{2}.$$

3.7. INTEGRALS OVER AN INFINITE INTERVAL

3.7.6. Prove in a similar way that

$$\int_0^{+\infty} \frac{\sin^2 mx}{x^2(x^2+a^2)} dx = \frac{\pi}{4a^3}(2am + e^{-2am} - 1) \quad (a, m > 0).$$

3.7.7. Verify the identity

$$\sin^3 x = \text{im}[\tfrac{1}{4}(1-e^{3ix}) - \tfrac{3}{4}(1-e^{ix})]$$

for real x and prove that

$$\int_0^{+\infty} \left(\frac{\sin x}{x}\right)^3 dx = \frac{3}{8}\pi.$$

3.7.8. By integrating $f(z) = z^{-3}(z^2+a^2)^{-1}[z+i(e^{iz}-1)]$ $(a > 0)$ over a suitable contour show that

$$\int_0^{+\infty} \frac{x - \sin x}{x^3(x^2+a^2)} dx = \frac{\pi}{2a^4}(1 - a + \tfrac{1}{2}a^2 - e^{-a}).$$

3.7.9. By integrating $f(z) = e^{az}(1+e^z)^{-1}$ $(0 < a < 1)$ round the rectangle with vertices $\mp R$, $\mp R + 2\pi i$ show that

$$\int_{-\infty}^{+\infty} \frac{e^{ax}}{1+e^x} dx = \frac{\pi}{\sin a\pi}.$$

Also prove that

(i) $\displaystyle\int_0^{+\infty} \frac{t^{a-1}}{1+t} dt = \frac{\pi}{\sin a\pi}$;

(ii) $\displaystyle\int_0^{+\infty} \frac{x^{m-1}}{1+x^n} dx = \frac{\pi}{n\sin(m\pi/n)}$ $(0 < m < n).$

3.7.10. Verify in a similar manner that

$$I = \int_{-\infty}^{+\infty} \frac{e^{ax}}{1+e^x+e^{2x}} dx = \int_0^{+\infty} \frac{t^{a-1}}{1+t+t^2} dt = \frac{2\pi}{\sqrt{3}} \cdot \frac{\sin \tfrac{1}{3}\pi(1-a)}{\sin a\pi} \quad (0 < a < 2).$$

3.7.11. By integrating $f(z) = [(1+z^2)\cosh\tfrac{1}{2}\pi z]^{-1}$ round the square Q_N with vertices $\mp N$, $\mp N + 2iN$, where N is an integer, show that

$$\int_0^{+\infty} [(1+x^2)\cosh\tfrac{1}{2}\pi x]^{-1} dx = \log 2.$$

3.7.12. By integrating $f(z) = e^{az}(e^{-2iz}-1)^{-1}$ $(a > 0)$ round the rectangle with vertices 0, π, $\pi+iR$, iR suitably indented show that

$$\int_0^{+\infty} \frac{\sin ay}{e^{2y}-1} dy = \frac{\pi}{4}\coth\frac{\pi a}{2} - \frac{1}{2a}.$$

3.7.13. By integrating $f(z) = e^{aiz}/\sinh z$ (a real) round the rectangle with vertices $\mp R$, $\mp R+2\pi i$ suitably indented show that

$$\int_0^{+\infty} \frac{\sin ax}{\sinh x} dx = \frac{\pi}{2}\tanh\frac{\pi a}{2}.$$

3.7.14. By integrating $e^{az}/\cosh \pi z$ $(-\pi < a < \pi)$ round the rectangle with corners at $\mp R$, $\mp R+i$ and suitable indentations, show that

$$\int_0^{+\infty} \frac{\cosh ax}{\cosh \pi x} dx = \frac{1}{2}\left(\cos\frac{a}{2}\right)^{-1}.$$

3.7.15. Prove that

(i) $\displaystyle\int_0^{+\infty} \left(\frac{1}{x} - \frac{1}{\sinh x}\right)\frac{dx}{x} = \log 2;$

(ii) $\displaystyle\int_0^{+\infty} \frac{1-\cos ax}{x\sinh x} dx = \log\cosh\tfrac{1}{2}a\pi$ (a real);

(iii) $\displaystyle\int_0^{+\infty} \frac{x^{a-1}}{(1+x)^2} dx = \frac{\pi(1-a)}{\sin \pi a}$ $(0 < a < 2).$

3.8. INTEGRATION OF MANY-VALUED FUNCTIONS

3.8.1. By integrating $f(z) = (\text{Log}\,z)^2(z^2+a^2)^{-1}$ round the upper half of the annulus $\{z: r < |z| < R\}$, $0 < r < a < R$, prove that

$$\int_0^{+\infty} \frac{(\log x)^2}{x^2+a^2} dx = \frac{\pi}{8a}[\pi^2+4(\log a)^2],$$

$$\int_0^{+\infty} \frac{\log x}{x^2+a^2} dx = \frac{\pi}{2a}\log a.$$

3.8.2. Prove in a similar way that

$$\int_0^{+\infty} \frac{\log x}{(1+x^2)^2} dx = -\frac{\pi}{4}.$$

3.8. INTEGRATION OF MANY-VALUED FUNCTIONS

3.8.3. By integrating $(1+z^2)^{-2} \exp(a \operatorname{Log} z)$ round the contour of Exercise 3.8.1, show that

$$\int_0^{+\infty} \frac{x^a}{(1+x^2)^2}\,dx = \frac{\pi}{4}(1-a)\left(\cos\frac{\pi a}{2}\right)^{-1}, \quad -1 < a < 3.$$

3.8.4. By integrating $(1+z^2)^{-1} \operatorname{Log} z$ round the boundary of

$$\{z: r < |z| < R\} \cap \{z: 0 < \arg z < \tfrac{1}{4}\pi\},$$

show that

$$\int_0^{+\infty} \frac{\log t}{1+t^4}\,dt = -\frac{\pi^2}{8\sqrt{2}}, \quad \int_0^{+\infty} \frac{t^2 \log t}{1+t^4}\,dt = \frac{\pi^2}{8\sqrt{2}}.$$

3.8.5. Show that

(i) $\displaystyle\int_0^{+\infty} \frac{\log x}{\sqrt{x}(x^2+a^2)^2}\,dx = 2^{-3/2}\pi a^{-5/2}(\tfrac{3}{2}\log a - 1 - \tfrac{3}{4}\pi) \quad (a > 0);$

(ii) $\displaystyle\int_0^{+\infty} \frac{\log(1+x^2)}{1+x^2}\,dx = \pi \log 2;$

Hint: Integrate $(1+z^2)^{-1} \operatorname{Log}(z+i)$ round the upper half of $K(0; R)$.

(iii) $\displaystyle\int_0^1 \frac{\log(x+x^{-1})}{1+x^2}\,dx = \frac{\pi}{2}\log 2.$

3.8.6. If $a > 0$, prove that

$$\int_0^{+\infty} \frac{dx}{(x^2+a^2)(\log^2 x + \pi^2)} = \frac{\pi}{2a}\left(\log^2 a + \frac{\pi^2}{4}\right)^{-1} - (1+a^2)^{-1}.$$

Hint: Integrate $[(z^2+a^2)\operatorname{Log} z]^{-1}$ round the annulus $\{z: r < |z| < R\}$ slit along $[-R, -r]$.

3.8.7. By integrating $\log(z-a)$ $(0 < a < 1)$ round the cycle consisting of $C(0; 1)$, $C(a; r)$ and $[a+r, 1]$, $a+r < 1$, show that

$$\int_0^{2\pi} \log|e^{i\theta} - a|\,d\theta = 0.$$

3.8.8. By integrating $(z^2-1)^{-1}\log z$ round the boundary of

$$\{z: r < |z| < R\} \cap (+; +),$$

show that
$$\int_0^{+\infty} \frac{\log x}{x^2-1}\,dx = \frac{\pi^2}{4}.$$

3.8.9. Prove that

(i) $\displaystyle\int_0^1 \frac{dx}{(x+1)\sqrt[3]{x^2(1-x)}} = \frac{\pi\sqrt[3]{4}}{\sqrt{3}};$

(ii) $\displaystyle\int_0^1 \frac{x^{2n}}{\sqrt[3]{x(1-x^2)}}\,dx = \frac{\pi}{\sqrt{3}} \cdot \frac{1\cdot 4\cdot \ldots \cdot(3n-2)}{3\cdot 6 \cdot \ldots \cdot 3n}.$

3.8.10. By integrating $z^{1-p}(1-z)^p(1+z)^3$ round the boundary of
$$K(0;R)\setminus\{\bar{K}(0;r)\cup\bar{K}(1;r)\cup[0,1]\}, \quad 0<r<\tfrac{1}{2},\ R>\tfrac{3}{2},$$
show that
$$\int_0^1 x^{1-p}(1-x)^p(1+x)^{-3}\,dx = 2^{p-3}\pi p(1-p)(\sin p\pi)^{-1} \quad (-1<p<2).$$

3.8.11. Prove that
$$\int_0^1 (1-x^n)^{-1/n}\,dx = \pi[\sin(\pi/n)]^{-1} \quad (n=2,3,\ldots).$$

3.9. THE ARGUMENT PRINCIPLE. ROUCHÉ'S THEOREM

Suppose f is meromorphic in a domain D and $\varDelta$, $\bar{\varDelta}\subset D$, is a subdomain whose boundary consists of a finite system γ of closed, regular curves $\gamma_1,\ldots,\gamma_n$ with a positive orientation w.r.t. $\varDelta$ (i.e. the index $n(\gamma,z)=1$ for each $z\in\varDelta$). Suppose, moreover, that f is continuous and does not vanish on γ and maps γ onto a cycle $\varGamma$. According to the argument principle, *the index $n(\varGamma,0)$ is equal to the difference between the number of zeros and the number of poles of f in $\varDelta$, zeros and poles being counted as many times as their order indicates.*

In particular, if f is analytic in D, we can determine the number of zeros of f. The number of zeros can be also evaluated by using Rouché's theorem:

If f, g are analytic in D and continuous in its closure $\bar{D}$ and $|g(z)|<|f(z)|$ on the boundary of D, then $f+g$ and f have the same number of zeros in D.

Again the zeros are counted with due multiplicity.

3.9.1. If F is analytic in $K(0;1)$, continuous in $\bar{K}(0;1)$ and $|F(z)|<1$ for all $z\in\bar{K}(0;1)$, show that the equation $F(z)-z=0$ has a unique root in $K(0;1)$.

3.9.2. Show that the polynomial $P(z) = z^8 + 3z^3 + 7z + 5$ has exactly two roots in the first quadrant $(+;+)$.

3.9.3. Show that the polynomial $P(z) = z^4 + 2z^3 - 2z + 10$ has exactly one root in each quadrant.

3.9.4. If a, b are real and different from zero, show that the polynomial $P(z) = z^{2n} + a^2 z^{2n-1} + b^2$ has for even n exactly n roots of positive real part and $n-1$ roots of positive real part for odd n.

3.9.5. Show that $P(z) = z^5 + 2z^3 + 2z + 3$ has exactly one root in the first quadrant.

3.9.6. Verify that all zeros of $P(z) = z^5 - z + 16$ are contained in the annulus $1 < |z| < 2$ and two of them have a positive real part.

3.9.7. Show that exactly two roots of $P(z) = z^5 + 5z^3 + 2z^2 + 4z + 1$ are situated in the right half-plane.

3.9.8. Show that the polynomial $P(z) = z^4 + 3z + 3$ has a unique root in the strip $D = \{z: 0 < \operatorname{im} z < 1\}$.

3.9.9. Show that the polynomial $P(z) = z^4 + iz^3 + 1$ has a unique root in the first quadrant and four roots in the disk $K(0; \tfrac{3}{2})$.

3.9.10. Evaluate the number q of roots situated in the unit disk for the polynomials: (i) $z^4 - 5z + 1$; (ii) $z^8 - 4z^5 + z^2 - 1$.

3.9.11. If $a > e$, show that the equation $az^n = e^z$ has exactly n solutions in $K(0; 1)$.

3.9.12. Show that no roots of the equation $(z+1)e^{-z} = z+2$ are contained in the right half-plane.

3.9.13. If $\lambda > 1$, show that the equation $z + e^{-z} = \lambda$ has one solution with positive real part.

3.9.14. Prove that the polynomial $1 + z + az^n$, $n \geq 2$, has for any complex a at least one root in $\overline{K}(0; 2)$.

3.9.15. Suppose $0 < |a| < 1$ and p is a positive integer. Show that the equation $(z-1)^p = ae^{-z}$ has exactly p simple roots with positive real part and all of these are located inside $K(1; 1)$.

3.9.16. Prove that all roots of the equation $\tan z - z = 0$ are real and each interval $((n-\tfrac{1}{2})\pi, (n+\tfrac{1}{2})\pi)$, $n \neq 0$, contains exactly one root λ_n.

3.9.17. If $|a_m| < 1$ ($m = 1, 2, \ldots, n$), $|b| < 1$ and
$$F(z) = \prod_{m=1}^{n} \frac{z - a_m}{1 - \bar{a}_m z},$$
show that the equation $F(z) = b$ has exactly n roots in the unit disk.

3.9.18. If F is analytic in $K(0; a)$ and continuous in $\bar{K}(0; a)$, $|F(z)| > m$ on $C(0; a)$ and $|F(0)| < m$, show that F has at least one zero in $K(0; a)$.

3.9.19. If f is analytic in the annulus $A = \{z : r < |z| < R\}$, $f(z) \neq a$ for all $z \in A$ and $\gamma_t = f(C(0; t))$, show that for any $t \in (r, R)$ the index $n(\gamma_t, a)$ has the same value.

3.9.20. If f is analytic in a domain D except for one simple pole z_0 and continuous in $\bar{D} \setminus z_0$, $|f(z)| = 1$ on $\bar{D} \setminus D$, show that f takes in D exactly once every value a with $|a| > 1$.

CHAPTER 4

Sequences and Series of Analytic Functions

4.1. ALMOST UNIFORM CONVERGENCE

Following Saks and Zygmund [10] we shall call a sequence of functions $\{f_n\}$ defined in an open set G *almost uniformly* (a.u.) *convergent on G to a function f*, if $\{f_n\}$ tends to f uniformly on each compact subset of G. In what follows we use the notation: $f_n \rightrightarrows_G f$.

If all functions f_n are analytic in a domain D and $f_n \rightrightarrows_D f$, then also f is analytic in D. Moreover, for any fixed, positive integer k, $f_n^{(k)} \rightrightarrows_D f^{(k)}$.

We can also consider a.u. convergent series of functions. The so-called M-test of Weierstrass yields a quite convenient, sufficient condition for a.u. convergence of functional series. Let $\sum u_n$ be a series of functions defined on an open set G. If for each compact subset F of G there exists a convergent series $\sum M_n$ with positive terms such that $|u_n(z)| \leqslant M_n$ for all $z \in F$, then $\sum u_n$ is a.u. convergent on G.

4.1.1. Show that the sequence $\{z\exp(-\tfrac{1}{2}n^2 z^2)\}$ is uniformly convergent on the real axis and at the same time it is not a.u. convergent in any disk $K(0; r)$.

4.1.2. Prove that the series

$$\sum_{n=1}^{\infty} u_n(z) = \sum_{n=1}^{\infty} z^n[(1-z^n)(1-z^{n+1})]^{-1}$$

is a.u. convergent in $K(0; 1)$ to $z(1-z)^{-2}$ and also a.u. convergent in $\mathbf{C} \setminus \overline{K}(0; 1)$ to $(1-z)^{-2}$.

4.1.3. Prove that the series $\sum_{n=0}^{\infty} 3^{-n}\sin nz$ is a.u. convergent in the strip $|\operatorname{im} z| < \log 3$, its sum f being analytic in this domain. Evaluate $f'(0)$.

4.1.4. If f is analytic in $K(0; 1)$ and $f(0) = 0$, show that the series $\sum_{n=1}^{\infty} f(z^n)$ is a.u. convergent in $K(0; 1)$ and its sum is analytic.

4.1.5. Prove that in any closed disk $\bar{K}(z_0; r)$ leaving outside all negative integers the series $\sum_{n=1}^{\infty} (-1)^{n+1}(z+n)^{-1}$ is uniformly convergent and its sum is analytic in $\mathbf{C}\setminus\{-1; -2; -3; \ldots\}$.

4.1.6. Suppose γ is a closed, regular curve not meeting any negative integer and f is the analytic function of Exercise 4.1.5. Evaluate $\int_\gamma f(z)\,dz$.

4.1.7. If $|z| < 1$ and $\tau(n)$ is the number of divisors of the integer n, show that

$$\sum_{n=1}^{\infty} z^n(1-z^n)^{-1} = \sum_{n=1}^{\infty} \tau(n) z^n.$$

Prove the a.u. convergence of both series in $K(0; 1)$.

4.1.8. Prove the identity

$$1 + \binom{k+1}{1} z + \binom{k+2}{2} z^2 + \ldots = (1-z)^{-k-1},$$

$k = 0, 1, 2, \ldots$, $|z| < 1$.

4.1.9. Find the set of points of convergence of the functional series $\sum u_n(z)$ in case $u_n(z)$ equals to: (i) $z^n(1+z^{2n})^{-1}$; (ii) $n^{-2}\cos nz$; (iii) $(z+n)^{-2}$; (iv) $(q^n z + q^{-n} z^{-1} - 2)^{-1}$ $(0 < q < 1)$.

4.1.10. Prove that the series

$$1 + \sum_{n=0}^{\infty} \frac{z^2(z^2+1^2)\ldots(z^2+n^2)}{[(n+1)!]^2}$$

s convergent for any z, its sum being analytic in the finite plane.

4.1.11. By using the identity

$$\frac{\pi^2}{\sin^2 \pi z} = \sum_{n=-\infty}^{\infty} (z-n)^{-2},$$

$z \neq 0, \mp 1, \mp 2, \ldots$ (cf. e.g. Ex. 4.5.15) find the sum $\sum_{n=-\infty}^{\infty} (z-n)^{-3}$.

4.1.12. If the sequence of functions $\{f_n\}$ analytic and univalent in a domain G is a.u. convergent in G and $f = \lim f_n$, show that f is univalent, unless it is a constant.

4.2. POWER SERIES

A series of functions of the form $\sum_{n=0}^{\infty} a_n(z-a)^n$ is called a *power series*. The least upper bound R of nonnegative r for which the sequence $\{|a_n| r^n\}$ is bounded is called the *radius of convergence of the given power series* and the disk $K(a; R)$ is called the *disk of convergence*. In case $R > 0$ the power series is a.u. convergent in its disk of convergence, its sum being analytic in this disk. Outside its disk of convergence, i.e. in $\mathbf{C}\setminus \overline{K}(a; R)$, the power series is divergent. The radius of convergence can be evaluated by means of the Cauchy–Hadamard formula: $R = (\overline{\lim} \sqrt[n]{|a_n|})^{-1}$.

4.2.1. If $a_n \neq 0$ ($n = 1, 2, \ldots$) and the limit $\lim a_n/a_{n+1} = q$ exists, show that the radius of convergence of the power series $\sum_{n=0}^{\infty} a_n z^n$ is equal to $|q|$.

4.2.2. Evaluate the radius of convergence of:

(i) $\sum_{n=0}^{\infty} \dfrac{(2n)!}{(n!)^2} z^n$;   (ii) $\sum_{n=1}^{\infty} \dfrac{n!}{n^n} z^n$;

(iii) $\sum_{n=0}^{\infty} 2^{-n} z^{2n}$;   (iv) $\sum_{n=0}^{\infty} (n+a^n) z^n$.

4.2.3. If the radii of convergence of $\sum a_n z^n$, $\sum b_n z^n$ are equal R_1, R_2 resp., show that:

(i) the radius of convergence R of $\sum a_n b_n z^n$ satisfies $R \geq R_1 R_2$;

(ii) the radius of convergence R' of

$$\sum \frac{a_n}{b_n} z^n \quad (b_n \neq 0, \ n = 0, 1, 2, \ldots)$$

satisfies $R' \leq R_1/R_2$;

(iii) the radius of convergence R_0 of

$$\sum (a_n b_0 + a_{n-1} b_1 + \ldots + a_0 b_n) z^n$$

satisfies $R_0 \geq \min(R_1, R_2)$.

4.2.4. If the sum of the power series $\sum a_n z^n$ is real in some interval $(-\delta, \delta)$, $\delta > 0$, show that all coefficients a_n are real.

4.2.5. Suppose $R > 0$ is the radius of convergence of the power series $\sum_{n=0}^{\infty} a_n z^n$ and $f(z)$ is its sum. Prove Parseval's identity:

$$\frac{1}{2\pi} \int_0^{2\pi} |f(re^{i\theta})|^2 d\theta = \sum_{n=0}^{\infty} |a_n|^2 r^{2n}, \quad 0 < r < R.$$

Hint: $|f|^2 = f\bar{f}$.

4.2.6. (cont.) If f is bounded: $|f(z)| \leq M$ for all $z \in K(0; R)$, show that
$$\sum_{n=0}^{\infty} |a_n|^2 R^{2n} \leq M^2.$$

4.2.7. (cont.) If $r < R$ and $M(r) = \sup|f(re^{i\theta})|$, $0 \leq \theta \leq 2\pi$, show that
$$|a_n| \leq r^{-n} M(r), \quad n = 0, 1, \ldots$$

4.2.8. If the sum of a power series $\sum a_n z^n$ is defined and bounded in the unit disk, show that $a_n = o(1)$.

4.2.9. If $f(z) = \sum_{n=0}^{\infty} a_n z^n$ in $K(0; r)$ and $|a_1| \geq \sum_{n=2}^{\infty} n|a_n| r^{n-1}$, show that f is univalent in $K(0; r)$ (i.e. $z_1, z_2 \in K(0; r)$, $z_1 \neq z_2$ implies $f(z_1) \neq f(z_2)$).

4.2.10. By using Exercise 4.2.9 find a disk of univalence of $\sum_{n=0}^{\infty} \dfrac{z^n}{n!}$.

4.2.11. If $\sum c_n z^n$ has a positive radius of convergence R, show that the power series $\sum \dfrac{c_n}{n!} z^n$ has an infinite radius of convergence and the sum f of the latter series satisfies
$$|f(z)| \leq M(\theta) \exp(|z|/\theta R), \quad \text{where } 0 < \theta < 1.$$

4.2.12. If $f(z) = \sum_{n=0}^{\infty} a_n z^n$ is univalent in $K(0; 1)$ and $A(r)$ is the area of the image domain of $K(0; r)$, $0 < r < 1$, show that
$$A(r) = \pi \sum_{n=1}^{\infty} n|a_n|^2 r^{2n}.$$

4.3. TAYLOR SERIES

If f is analytic in a domain D and a is an arbitrary point of D, there exists a sequence of complex numbers $\{a_n\}$ such that for any $K(a; r) \subset D$ the power series $\sum_{n=0}^{\infty} a_n(z-a)^n$ is a.u. convergent in $K(a; r)$ its sum being equal $f(z)$. The power series $\sum_{n=0}^{\infty} a_n(z-a)^n$ is called *Taylor series of f center* at a and the Taylor's coefficients a_n can be expressed either by successive derivatives of f at the point a:

$$a_0 = f(a), \quad a_n = \frac{f^{(n)}(a)}{n!} \quad (n = 1, 2, \ldots),$$

4.3. TAYLOR SERIES

or by Cauchy's coefficient formula:

$$a_n = \frac{1}{2\pi i} \int_{C(a;r)} \frac{f(\zeta)}{(\zeta-a)^{n+1}} d\zeta \quad (n = 0, 1, 2, \ldots)$$

where r is such that $\overline{K}(a; r) \subset D$.

The radius of convergence R of Taylor series center a is not less than the distance between $\mathbf{C} \setminus D$ and the point a. On the other hand, the circle of convergence $C(a; R)$ of the power series $\sum_{n=0}^{\infty} a_n(z-a)^n$ contains at least one singular point.

Each point on $C(a; R)$ which is not regular, is called singular. A point $b \in C(a; R)$ is called regular if there exists a power series $\sum_{n=0}^{\infty} b_n(z-b)^n$ convergent in some disk $K(b; \delta)$, $\delta > 0$, such that both series $\sum_{n=0}^{\infty} a_n(z-a)^n$, $\sum_{n=0}^{\infty} b_n(z-b)^n$ have identical sums in $K(a; R) \cap K(b; \delta)$.

4.3.1. Evaluate four initial, non-vanishing coefficients of Taylor series with center at the origin for the following functions and find the corresponding radius of convergence R:

(i) $\dfrac{z}{\mathrm{Log}(1+z)}$;

(ii) $\dfrac{z}{\mathrm{Arctan}\, z}$;

(iii) $\sqrt{\cos z}$ (take the branch corresponding to the value 1 at the origin);

(iv) $\dfrac{1}{\cos z}$;

(v) $\mathrm{Log}(1+e^z)$;

(vi) $\exp e^z$.

4.3.2. By applying Weierstrass theorem on term by term differentiation of a.u. convergent series of analytic functions prove the following theorem:

Suppose $\sum_{n=0}^{\infty} u_n(z)$ is a.u. convergent in $K(0; R)$ and $f(z)$ is its sum; suppose, moreover, $u_n(z) = a_{n0} + a_{n1}z + \ldots + a_{nk}z^k + \ldots$ in $K(0; R)$ for $n = 0, 1, 2, \ldots$

Then all the series $\sum_{n=0}^{\infty} a_{nk}$ are convergent and their sums are equal to corresponding Taylor coefficients of f at the origin.

4.3.3. Evaluate four initial, non-vanishing Taylor coefficients at the origin for the following functions and find the corresponding radius of convergence R:

(i) $\exp \dfrac{z}{1-z}$; (ii) $\sin \dfrac{z}{1-z}$.

4.3.4. Evaluate nth coefficient of Taylor series at the origin and its radius of convergence for following functions:

(i) $\dfrac{1}{2}\left(\operatorname{Log} \dfrac{1}{1-z}\right)^2$;

(ii) $(\operatorname{Arctan} z)^2$;

(iii) $(\operatorname{Arctan} z)\operatorname{Log}(1+z^2)$:

(iv) $\cos^2 z$;

(v) $\operatorname{Log}(1+\sqrt{1+z^2})$.

Hint: Differentiate the given function;

(vi) $\operatorname{Log}(\sqrt{1+z}+\sqrt{1-z})$.

[Take in (v), (vi) these branches of the square root which are equal 1 at the origin.]

4.3.5. Find the radius of convergence for Taylor series with center at the origin for $\sin \pi z^2 / \sin \pi z$.

4.3.6. Find Taylor series with center at the origin for these branches of

(i) $(1+\sqrt{1+z})^{1/2}$; (ii) $(1+\sqrt{1+z})^{-1/2}$

which are equal $2^{1/2}$, $2^{-1/2}$ at the origin.

Hint: Cf. Exercise 1.1.5.

4.3.7. Show that the coefficients c_n of Taylor series of $(1-z-z^2)^{-1}$ with center at the origin satisfy: $c_0 = c_1 = 1$, $c_{n+2} = c_{n+1}+c_n$ $(n \geq 0)$. Find the explicit formula for c_n by representing the given function as a sum of partial fractions. Also find the radius of convergence of $\sum_{n=0}^{\infty} c_n z^n$. The sequence $\{c_n\}$ is so called *Fibonacci sequence*.

4.3.8. Prove that $f(z) = \dfrac{1}{e^z-1} - \dfrac{1}{z} + \dfrac{1}{2}$ has a removable singularity at the origin and is an odd function. Hence

4.3. TAYLOR SERIES

$$\frac{1}{e^z-1} - \frac{1}{z} + \frac{1}{2} = \sum_{k=1}^{\infty} (-1)^{k-1} \frac{B_k}{(2k)!} z^{2k-1}.$$

Evaluate $B_1, B_2, \ldots, B_5$. Find the radius of convergence of the power series on the right and show that $\overline{\lim} \sqrt[n]{|B_n|} = +\infty$. The constants B_k are called *Bernoulli numbers*.

4.3.9. Show that

(i) $\frac{1}{2} z \coth \frac{1}{2} z = 1 + \sum_{k=1}^{\infty} (-1)^{k-1} \frac{B_k}{(2k)!} z^{2k-1}$, $\quad |z| < 2\pi$;

(ii) $z \cot z = 1 - \sum_{k=1}^{\infty} \frac{2^{2k} B_k}{(2k)!} z^{2k}$, $|z| < \pi$.

4.3.10. Express by means of Bernoulli numbers the nth Taylor coefficient at the origin and the radius of convergence of the corresponding Taylor series for the following functions:

(i) $\mathrm{Log} \frac{\sin z}{z}$;

(ii) $\mathrm{Log} \cos z$;

(iii) $\tan z$;

(iv) $(\cos z)^{-2}$;

(v) $\tan^2 z$;

(vi) $\mathrm{Log} \frac{\tan z}{z}$;

(vii) $\frac{z}{\sin z}$.

4.3.11. By using the power series expansions of $\mathrm{Log}(1+z)$, $\sin z$ near the origin evaluate three initial, non-vanishing terms of the power series expansion of $\mathrm{Log} \frac{\sin z}{z}$ and compare the obtained result with Exercise 4.3.10 (i).

4.3.12. Verify that

$$\varphi(z) = \sqrt{\frac{z}{1-z}} \, \mathrm{Arc} \tan \sqrt{\frac{z}{1-z}}$$

is analytic in some neighborhood of $z = 0$. Evaluate the radius of convergence of its Taylor series at the origin.

4.3.13. (cont.) Verify that φ satisfies the differential equation
$$2z(1-z)\varphi'(z) = \varphi(z)+z$$
with the initial condition $\varphi(0) = 0$. Prove the identity:
$$\varphi(z) = z + \frac{2}{3}z^3 + \frac{2\cdot 4}{3\cdot 5}z^5 + \frac{2\cdot 4\cdot 6}{3\cdot 5\cdot 7}z^7 + \ldots, \quad |z| < 1.$$

4.3.14. Prove that
$$(\operatorname{Arc\,sin} z)^2 = z^2 + \frac{1}{2}\cdot\frac{2}{3}z^4 + \frac{1}{3}\cdot\frac{2\cdot 4}{3\cdot 5}z^6 + \ldots, \quad |z| < 1.$$

4.3.15. If f is analytic in $K(0; R)$ and R_n is the remainder of Taylor series center at the origin, i.e.
$$R_n(z) = f(z) - f(0) - zf'(0) - \ldots - \frac{z^n}{n!}f^{(n)}(0),$$
show that
$$R_n(z) = \frac{z^{n+1}}{2\pi i}\int_{C(0;\,r)} \frac{f(\zeta)}{\zeta^{n+1}(\zeta-z)}\,d\zeta, \quad |z| < r < R.$$
Hint: Cf. Exercise 3.1.28.

4.3.16. If f is analytic in $K(0; R)$ and
$$s_n(z) = f(0) + zf'(0) + \ldots + \frac{z^n}{n!}f^{(n)}(0),$$
show that
$$s_n(z) = \frac{1}{2\pi i}\int_{C(0;\,r)} f(\zeta)\frac{\zeta^{n+1}-z^{n+1}}{(\zeta-z)\zeta^{n+1}}\,d\zeta, \quad |z| < r < R.$$

4.3.17. If f is analytic in $K(a; R)$, show that
$$f(z) = f(a) + 2\Big\{\frac{1}{2}(z-a)f'\Big(\frac{z+a}{2}\Big) + \frac{(z-a)^3}{2^3\cdot 3!}f'''\Big(\frac{z+a}{2}\Big) +$$
$$+ \frac{(z-a)^5}{2^5\cdot 5!}f^{(5)}\Big(\frac{z+a}{2}\Big) + \ldots\Big\}, \quad z \in K(a; R).$$

4.3.18. If f is analytic in $K(0; R)$ and $f(re^{i\theta}) = P(\theta) + iQ(\theta)$, $0 < r < R$, show that the coefficients a_n of Taylor series at the origin satisfy
$$a_n = \frac{1}{\pi r^n}\int_0^{2\pi} P(\theta)e^{-in\theta}\,d\theta = \frac{1}{\pi r^n}\int_0^{2\pi} iQ(\theta)e^{-in\theta}\,d\theta.$$

4.3.19. If f is analytic in $K(0; 1)$, $f(0) = 1$ and $\operatorname{re} f(z) > 0$ in $K(0; 1)$, show that Taylor coefficients a_n of f at the origin satisfy: $|a_n| \leqslant 2$, $n = 1, 2, \ldots$

4.3.20. (cont.) Verify that
$$(1-|z|)/(1+|z|) \leqslant |f(z)| \leqslant (1+|z|)/(1-|z|)$$
and show that equality can actually be attained.

4.4. BOUNDARY BEHAVIOR OF POWER SERIES

4.4.1. Discuss the behavior of the following power series on the circle of convergence

(i) $\sum\limits_{n=1}^{\infty} z^n$;

(ii) $\sum\limits_{n=1}^{\infty} n^{-1} z^n$;

(iii) $\sum\limits_{n=1}^{\infty} n^{-2} z^n$;

(iv) $\sum\limits_{n=2}^{\infty} \dfrac{(-1)^n}{\log n} z^{3n-1}$.

4.4.2. Give an example of a divergent series $\sum a_n$ such that the radius of convergence of $\sum a_n z^n$ is equal to 1 and the finite limit $\lim\limits_{r \to 1-} \sum\limits_{n=0}^{\infty} a_n r^n$ exists.

4.4.3. If $\lim z_n = e^{i\theta}$, $|z_n| < 1$ $(n = 1, 2, \ldots)$ and all z_n are situated inside a Stolz angle with vertex $e^{i\theta}$, i.e. an angle made up by two chords of the unit disk with a common vertex $e^{i\theta}$, show that the sequence $\left\{\dfrac{|e^{i\theta} - z_n|}{1 - |z_n|}\right\}$ is bounded. Prove the converse, too.

4.4.4. Prove Abel's Limit Theorem: If $\sum\limits_{n=0}^{\infty} a_n e^{in\theta}$ is convergent and $\{z_n\}$ is any sequence approaching $e^{i\theta}$ inside a Stolz angle with vertex at $e^{i\theta}$, then
$$\lim_{n} \sum_{k=0}^{\infty} a_k z_n^k = \sum_{k=0}^{\infty} a_k e^{ki\theta}.$$

Hint: If $s_n = a_0 + a_1 e^{i\theta} + \ldots + a_n e^{in\theta}$, show that $\sum\limits_{k=0}^{\infty} s_k z^k$ is convergent in $K(0; 1)$ and express $\sum\limits_{k=0}^{\infty} a_k z_n^k$ as a Toeplitz transform of $\{s_n\}$, cf. Exercise 1.1.37.

4.4.5. Prove that
(i) $\sin\theta - \frac{1}{2}\sin 2\theta + \frac{1}{3}\sin 3\theta - \ldots = \frac{1}{2}\theta$, $|\theta| < \pi$;
(ii) $\cos\theta - \frac{1}{2}\cos 2\theta + \frac{1}{3}\cos 3\theta - \ldots = \log(2\cos\frac{1}{2}\theta)$, $\theta \neq (2k+1)\pi$ with k being an integer.

4.4.6. Prove that
(i) $\sin\theta + \frac{1}{3}\sin 3\theta + \frac{1}{5}\sin 5\theta + \ldots = \frac{1}{4}\pi$, $0 < \theta < \pi$;
(ii) $\cos\theta + \frac{1}{3}\cos 3\theta + \frac{1}{5}\cos 5\theta + \ldots = \frac{1}{2}\log(\cot\frac{1}{2}\theta)$ for $\theta \neq k\pi$ with k being an integer.

4.4.7. Prove that
$$1 - \frac{2}{3} + \frac{2\cdot 4}{3\cdot 5} - \frac{2\cdot 4\cdot 6}{3\cdot 5\cdot 7} + \ldots = \frac{1}{\sqrt{2}}\log(1+\sqrt{2}).$$

Hint: Cf. Exercise 4.3.13.

4.4.8. Prove that
$$1 - \frac{1}{2}\cdot\frac{2}{3} + \frac{1}{3}\cdot\frac{2\cdot 4}{3\cdot 5} - \frac{1}{4}\cdot\frac{2\cdot 4\cdot 6}{3\cdot 5\cdot 7} + \ldots = [\log(1+\sqrt{2})]^2.$$

Hint: Cf. Exercise 4.3.14.

4.4.9. Prove that
(i) $1 + \dfrac{1}{2\cdot 4} - \dfrac{1\cdot 3\cdot 5}{2\cdot 4\cdot 6\cdot 8} + \ldots = \left(\dfrac{1+\sqrt{2}}{2}\right)^{1/2}$;
(ii) $\dfrac{1}{2} - \dfrac{1\cdot 3}{2\cdot 4\cdot 6} + \dfrac{1\cdot 3\cdot 5\cdot 7}{2\cdot 4\cdot 6\cdot 8\cdot 10} - \ldots = [2(1+\sqrt{2})]^{-1/2}$.

Hint: Cf. Exercise 4.3.6.

4.4.10. Prove that
$$1 - \frac{1}{3}\cdot\frac{1}{2} + \frac{1}{5}\cdot\frac{1\cdot 3}{2\cdot 4} - \ldots = \log(1+\sqrt{2}).$$

Hint: Consider $\operatorname{Arsinh} z$.

4.5. THE LAURENT SERIES

A functional series of the form $\sum_{n=-\infty}^{\infty} A_n(z-a)^n$ is called the *Laurent series with the center* a. If the Laurent series is convergent for some z_0, then its *regular part*, i.e. the power series $\sum_{n=0}^{\infty} A_n(z-a)^n$ is convergent in $K(a; |z_0-a|)$, whereas

4.5. THE LAURENT SERIES

its *singular* or *principal part*, i.e. the series $\sum_{n=1}^{\infty} A_{-n}(z-a)^{-n}$, is convergent in $\mathbf{C}\setminus \overline{K}(a; |z_0-a|)$ which is an immediate consequence of well-known properties of power series. Hence, if the set of points of convergence contains interior points, there exists a maximal, non-empty annulus of convergence $\{z: r < |z-a| < R\}$ and the sum of Laurent series is analytic in this annulus. Conversely, if f is analytic in $\{z: r < |z-a| < R\}$, where $0 \leqslant r < R \leqslant +\infty$, then it can be expressed as the sum of a Laurent series $\sum_{n=-\infty}^{\infty} A_n(z-a)^n$ which is absolutely and a.u. convergent in this annulus. The development is unique. If $C = C(a; \rho)$ and $r < \rho < R$, then the coefficients A_n of the Laurent development are equal:

$$A_n = \frac{1}{2\pi i} \int_C f(\zeta)(\zeta-a)^{-n-1} d\zeta \qquad (n = 0, \mp 1, \mp 2, \ldots).$$

The Laurent development also holds in case $r = 0$, i.e. when a is an isolated singularity. Since any Laurent series with a non-empty annulus of convergence and $A_{-1} = 0$ has a primitive, it follows that $\text{res}(a; f) = A_{-1}$.

4.5.1. Give an example of a Laurent series with an empty annulus of convergence and a non-empty set of points of convergence.

4.5.2. Show that $\sum_{n=-\infty}^{\infty} a_n z^n$ has a non-empty annulus of convergence, iff

$$\overline{\lim_{n\to+\infty}} \sqrt[n]{|a_{-n}|} < (\overline{\lim_{n\to+\infty}} \sqrt[n]{|a_n|})^{-1}.$$

4.5.3. Find the Laurent development of $(z^2-1)/(z+2)(z+3)$ in:
(i) $\{z: 2 < |z| < 3\}$; (ii) $\{z: |z| > 3\}$.

4.5.4. Show that the Laurent coefficients A_n of

$$(b-a)\left(\text{Log}\frac{z-a}{z-b}\right)^{-1} \quad \text{in} \quad \mathbf{C}\setminus \overline{K}(\tfrac{1}{2}(a+b); \tfrac{1}{2}|b-a|)$$

vanish for $n \geqslant 2$. Find A_1, A_0, A_{-1}.

4.5.5. Find the Laurent development of $(z-a)^{-1}(z-b)^{-1}$, $0 < |a| < |b|$, for (i) $|a| < |z| < |b|$; (ii) $|z| > b$.

4.5.6. Find the Laurent development of $(1+z^2)^{-1}(2+z^2)^{-1}$ for (i) $1 < |z| < \sqrt{2}$; (ii) $|z| > \sqrt{2}$.

4.5.7. Find the Laurent expansion of $\text{Log}[z^2/(z^2-1)]$ for $|z| > 1$.

4.5.8. Express the Laurent coefficients of $\exp(z+z^{-1})$ at the origin in terms of
(i) integrals involving trigonometric functions;
(ii) sums of infinite series by using the identity
$$\exp(z+z^{-1}) = e^z e^{1/z}.$$

4.5.9. The Bessel function $J_n(z)$ is defined as the nth coefficient of Laurent expansion at the origin $(n \geq 0)$:
$$\exp(\tfrac{1}{2}z(\zeta-\zeta^{-1})) = \sum_{n=-\infty}^{\infty} J_n(z)\zeta^n.$$
Prove that
$$J_n(z) = \frac{1}{\pi}\int_0^{\pi} \cos(n\theta - z\sin\theta)\,d\theta = \sum_{k=0}^{\infty} \frac{(-1)^k (\tfrac{1}{2}z)^{2k+n}}{k!(n+k)!}.$$

4.5.10. Suppose f has only the following singularities in the extended plane: a simple pole at -1 with $\operatorname{res}(-1;f) = 1$, a double pole at 2 with $\operatorname{res}(2;f)=2$; moreover, $f(0) = \tfrac{7}{4}$, $f(1) = \tfrac{5}{2}$. Find the Laurent development of f in the annulus $1 < |z| < 2$.

4.5.11. Find the Laurent development of $z^{-1}(1-z)^{-1}$ in annular neighborhoods of: (i) $z=0$; (ii) $z=1$; (iii) $z=\infty$.

4.5.12. Find the Laurent development of $f(z) = \left[\dfrac{z}{(z-1)(z-2)}\right]^{1/2}$ in the annulus $1 < |z| < 2$ assuming $\operatorname{im} f(\tfrac{3}{2}) > 0$.

4.5.13. Find the Laurent development of both branches of
$$[(z-a)(z-b)]^{1/2} \quad \text{in} \quad |z| > \max(|a|,|b|).$$

4.5.14. Show that $\pi^2/\sin^2\pi z$ and $h(z) = \sum_{n=-\infty}^{\infty}(z-n)^{-2}$ are both analytic in $\mathbf{C}\setminus N$, where N is the set of all integers. Also show that the principal parts of both functions in the annular neighborhood of any point of N are the same.

4.5.15. (cont.) Show that $g(z) = \dfrac{\pi^2}{\sin^2\pi z} - h(z)$ is bounded in $\mathbf{C}$. Hence deduce that $g = 0$.

4.5.16. Show that
$$\pi \cot \pi z = \frac{1}{z} + \sum_{n=1}^{\infty} \frac{2z}{z^2 - n^2} = \frac{1}{z} + \sum_{n=-\infty}^{\infty}{}' \left(\frac{1}{z-n} + \frac{1}{n}\right)$$
for $z \neq 0, \mp 1, \mp 2, \ldots$ (here and in the following a prime after a summation sign indicates that the term corresponding to $n=0$ is omitted).

4.5.17. Find the Laurent expansion of $\pi\cot\pi z$ in the annuli:
(i) $0<|z|<1$; (ii) $1<|z|<2$.
Express the Laurent coefficients by means of

$$s_k = \sum_{n=1}^{\infty} n^{-2k} \quad (k=1, 2, \ldots).$$

By using Exercise 4.3.9 (ii) find the relation between s_k and Bernoulli numbers B_k.

4.6. SUMMATION OF SERIES BY MEANS OF CONTOUR INTEGRATION

4.6.1. Suppose f is meromorphic in the finite plane $\mathbf{C}$ and has a finite number of poles $a_1, a_2, \ldots, a_m$ none of which is an integer. Suppose, moreover, that $\lim_{z\to\infty} zf(z) = 0$. Show that the limit $\lim_{N\to+\infty}\sum_{n=-N}^{N} f(n)$ exists and equals to

$$-\sum_{k=1}^{m} \mathrm{res}(a_k;\, \pi f(z)\cot\pi z).$$

Hint: Integrate $\pi f(z)\cot\pi z$ over the boundary ∂Q_N of the square Q_N with corners $(N+\tfrac{1}{2})(\mp 1\mp i)$; also cf. Exercise 2.7.3.

4.6.2. Verify that

$$s = \sum_{n=-\infty}^{\infty} (n^2+n+1)^{-1} = \frac{2\pi}{\sqrt{3}}\tanh\frac{\pi\sqrt{3}}{2}.$$

4.6.3. Show that for all a outside an exceptional set (which should be determined in each case) the following formulas hold:

(i) $\displaystyle\sum_{n=1}^{\infty} (n^2+a^2)^{-1} = \frac{1}{2}\left(\frac{\pi}{a}\coth\pi a - \frac{1}{a^2}\right);$

(ii) $\displaystyle\sum_{n=1}^{\infty} (n^4+a^4)^{-1} = \frac{1}{2}\left(\frac{\pi}{a^3\sqrt{2}}\cdot\frac{\sinh\pi a\sqrt{2}+\sin\pi a\sqrt{2}}{\cosh\pi a\sqrt{2}-\cos\pi a\sqrt{2}} - \frac{1}{a^4}\right);$

(iii) $\displaystyle\sum_{n=1}^{\infty} \frac{n^2}{n^4+a^4} = \frac{\pi}{2a\sqrt{2}}\cdot\frac{\sinh\pi a\sqrt{2}-\sin\pi a\sqrt{2}}{\cosh\pi a\sqrt{2}-\cos\pi a\sqrt{2}};$

(iv) $\displaystyle\sum_{n=1}^{\infty} (n^4-a^4)^{-1} = \frac{1}{2a^4} - \frac{\pi}{4a^3}(\cot\pi a + \coth\pi a).$

4.6.4. If a is not an integer, show that
$$\sum_{n=-\infty}^{\infty}(n-a)^{-2}=\pi^2/\sin^2\pi a.$$

4.6.5. If none of a, b is an integer, show that
$$\sum_{n=-\infty}^{\infty}(n-a)^{-1}(n-b)^{-1}=-\pi^2\frac{\cot\pi a-\cot\pi b}{\pi a-\pi b}.$$

4.6.6. Evaluate $\sum_{n=1}^{\infty}(n^2+1)^{-1}$.

4.6.7. Suppose f is meromorphic in the finite plane $\mathbf{C}$ and has a finite number of poles $a_1, a_2, \ldots, a_m$ none of which is an integer. Suppose, moreover, that $\lim_{z\to\infty} zf(z) = 0$. Show that the limit
$$\lim_{N\to+\infty}\sum_{n=-N}^{N}(-1)^n f(n)$$
exists and is equal to
$$-\sum_{k=1}^{m}\operatorname{res}[a_k;\ \pi f(z)/\sin\pi z].$$

Hint: Integrate $\pi f(z)/\sin\pi z$ over the boundary ∂Q_N of the square Q_N with corners $(N+\tfrac{1}{2})(\mp 1\mp i)$.

4.6.8. Prove that

(i) $\displaystyle\sum_{n=0}^{\infty}\frac{(-1)^n}{n^2+a^2}=\frac{1}{2a^2}+\frac{\pi}{2a\sinh\pi a}$, $a\neq 0, \mp i, \mp 2i, \ldots$;

(ii) $\displaystyle\sum_{n=1}^{\infty}\frac{(-1)^n}{n^4-a^4}=\frac{1}{2a^4}-\frac{\pi}{4a^3}\left(\frac{1}{\sin\pi a}+\frac{1}{\sinh\pi a}\right)$,

$a\neq \mp n, \mp in$ $(n=0,1,2,\ldots)$;

(iii) $\displaystyle\sum_{n=-\infty}^{\infty}\frac{(-1)^n}{n^4+a^4}=\frac{\pi}{a^3\sqrt{2}}\cdot\frac{\sin\dfrac{\pi a}{\sqrt{2}}\cosh\dfrac{\pi a}{\sqrt{2}}+\cos\dfrac{\pi a}{\sqrt{2}}\sinh\dfrac{\pi a}{\sqrt{2}}}{\sin^2\dfrac{\pi a}{\sqrt{2}}+\sinh^2\dfrac{\pi a}{\sqrt{2}}};$

4.6. SUMMATION OF SERIES

(iv) $\displaystyle\sum_{n=-\infty}^{\infty}\frac{(-1)^n n^2}{n^4+a^4}=\frac{\pi}{a\sqrt{2}}\cdot\frac{\cos\dfrac{\pi a}{\sqrt{2}}\sinh\dfrac{\pi a}{\sqrt{2}}-\sin\dfrac{\pi a}{\sqrt{2}}\cosh\dfrac{\pi a}{\sqrt{2}}}{\sin^2\dfrac{\pi a}{\sqrt{2}}+\sinh^2\dfrac{\pi a}{\sqrt{2}}}$;

$a\neq\dfrac{\mp 1\mp i}{2}n$ $(n=0,1,2,\ldots)$ in (iii) and (iv).

4.6.9. By integrating $\dfrac{\pi\sin az}{z^3\sin\pi z}$, where $-\pi<a<\pi$, round the boundary ∂Q_N of the square Q_N with corners $(N+\tfrac{1}{2})(\mp 1\mp i)$, verify the formulas:

(i) $\displaystyle\sum_{n=1}^{\infty}(-1)^{n+1}\frac{\sin an}{n^3}=\frac{a}{12}(\pi^2-a^2)$;

(ii) $\dfrac{1}{1^3}-\dfrac{1}{3^3}+\dfrac{1}{5^3}-\ldots=\dfrac{\pi^3}{32}$.

4.6.10. If x is a complex number different from an integer, $-\pi<a<\pi$ and Q_N is the square with corners $(N+\tfrac{1}{2})(\mp 1\mp i)$, show that

$$I_N=\int_{\partial Q_N}\frac{\cos az}{(x^2-z^2)\sin\pi z}\,dz\to 0 \quad\text{as}\quad N\to+\infty.$$

Verify the formula:

$$\frac{\pi\cos ax}{\sin\pi x}=\frac{1}{x}+2x\sum_{n=1}^{\infty}(-1)^n\frac{\cos\pi a}{x^2-n^2}.$$

4.6.11. If $a\neq 0$ is real and R_N are rectangles with corners $(N+\tfrac{1}{2})\left(\mp 1\mp\dfrac{i}{a}\right)$, show that the integrals

$$I_N=\int_{\partial R_N}\frac{z}{\sin\pi z\sinh\pi az}\,dz\to 0 \quad\text{as}\quad N\to+\infty.$$

By using this show that

(i) $\displaystyle\sum_{n=1}^{\infty}\frac{(-1)^n n}{\sinh\pi an}=-\frac{1}{2\pi a}-\frac{1}{a^2}\sum_{n=1}^{\infty}\frac{(-1)^n n}{\sinh(\pi n/a)}$ for real $a\neq 0$;

(ii) $\dfrac{1}{e^\pi-e^{-\pi}}-\dfrac{2}{e^{2\pi}-e^{-2\pi}}+\dfrac{3}{e^{3\pi}-e^{-3\pi}}-\ldots=\dfrac{1}{8\pi}$.

4.6.12. Verify that the equality of Exercise 4.6.11 (i) also holds for complex a outside the imaginary axis.

4.6.13. If x is complex and not an integer, $-\pi < a < \pi$, and Q_N is the square with corners $(N+\tfrac{1}{2})(\mp 1 \mp i)$, show that the integrals

$$I_N = \int_{\partial Q_N} \frac{z \sin az}{(x^2-z^2)\sin \pi z}\, dz \to 0 \quad \text{as} \quad N \to +\infty$$

and verify the formula

$$\frac{\pi}{2} \cdot \frac{\sin ax}{\sin \pi x} = \sum_{n=1}^{\infty} (-1)^n \frac{n \sin an}{x^2 - n^2}.$$

4.6.14. Show that

(i) $\displaystyle \frac{\pi}{\cos \pi z} = \sum_{n=0}^{\infty} \frac{(-1)^n (2n+1)}{(n+\tfrac{1}{2})^2 - z^2}$, $z \neq \mp\tfrac{1}{2}, \mp\tfrac{3}{2}, \mp\tfrac{5}{2}, \ldots$;

(ii) $\displaystyle \frac{\pi}{\cosh \pi z} = \sum_{n=0}^{\infty} \frac{(-1)^n (2n+1)}{z^2 + (n+\tfrac{1}{2})^2}$, $z \neq \mp\tfrac{1}{2}i, \mp\tfrac{3}{2}i, \mp\tfrac{5}{2}i, \ldots$

4.6.15. By integrating $f(z) = \pi z^{-7} \cot \pi z \coth \pi z$ round a suitable contour show that

$$\frac{\coth \pi}{1^7} + \frac{\coth 2\pi}{2^7} + \frac{\coth 3\pi}{3^7} + \ldots = \frac{19\pi^7}{56700}.$$

4.7. INTEGRALS CONTAINING A COMPLEX PARAMETER THE GAMMA FUNCTION

Suppose $W(z, \zeta)$ is a complex-valued function of two complex variables z, ζ defined and continuous on $G \times \{\Gamma\}$, where G is a domain and $\{\Gamma\}$ is the set of points of a regular curve Γ. If $W'_z(z, \zeta)$ exists and is continuous on $G \times \{\Gamma\}$, then $H(z) = \int_\Gamma W(z, \zeta)\,d\zeta$ is analytic in G; moreover, $H'(z) = \int_\Gamma W'_z(z, \zeta)\,d\zeta$. In particular Γ can be a segment $[a, b]$ of the real axis. If the limit $\lim_{b \to \infty} \int_{[a,b]} W(z, t)\,dt$ exists, it will be denoted $\int_a^{+\infty} W(z, t)\,dt$.

4.7.1. The integral $\int_a^{+\infty} W(z, t)\,dt$ is said to be *almost uniformly* (a.u.) convergent in a domain G, if for any compact subset $F \subset G$ and any $\varepsilon > 0$ there exists A such that for any b, B with $A < b < B$ we have

4.7. THE GAMMA FUNCTION

$$\left|\int_b^B W(z,t)\,dt\right| < \varepsilon \quad \text{for all } z \text{ in } F.$$

If both $W(z,t)$, $W'_z(z,t)$ are continuous on $G \times [a, +\infty)$ and $H(z) = \int_a^{+\infty} W(z,t)\,dt$ is a.u. convergent in G, show that H is analytic in G.
Hint: Use Weierstrass theorem on a.u. convergent series of analytic functions.

4.7.2. Prove that $\int_0^{+\infty} e^{-zt}\,dt$ is a.u. convergent in the right half-plane $\{z: \operatorname{re} z > 0\}$ and $\int_0^{+\infty} e^{-zt}\,dt - \dfrac{1}{z} = 0$ for all z with $\operatorname{re} z > 0$. By separating real and imaginary parts in both terms find the values of two real integrals.

4.7.3. Prove that $H(z) = \int_1^{+\infty} e^{-t} t^{z-1}\,dt$ is analytic in the open plane and $H'(0) = \int_1^{+\infty} t^{-1} e^{-1}\,dt$.

4.7.4. If φ is a complex-valued function of a real variable $x \in (-\infty, +\infty)$ and $\int_{-\infty}^{+\infty} |\varphi(x)|\,dx < +\infty$, show that

$$f(z) = \int_{-\infty}^{+\infty} (x-z)^{-1} \varphi(x)\,dx$$

is analytic in the upper half-plane.

4.7.5. The gamma function as defined by the equality

$$\Gamma(z) = \int_0^{+\infty} t^{z-1} e^{-t}\,dt, \quad \operatorname{re} z > 0,$$

is analytic in the right half-plane and satisfies $z\Gamma(z) = \Gamma(z+1)$. Show that this functional equation defines a function meromorphic in $\mathbf{C}$ whose only singularities are simple poles at $z = 0, -1, -2, \ldots$ with $\operatorname{res}(-n; \Gamma(z)) = (-1)^n/n!$.

4.7.6. Prove that a function G meromorphic in the finite plane $\mathbf{C}$ satisfies the functional equation $G(z+1) = zG(z)$, iff $G(z) = \Gamma(z)P(z)$, where P is meromorphic in $\mathbf{C}$ and has the period 1.

4.7.7. Prove that

$$\int_0^{+\infty} \exp(-x^\alpha)\,dx = \alpha^{-1}\Gamma(\alpha^{-1}), \quad \alpha > 0.$$

4.7.8. By integrating $e^{-z} z^{s-1}$ round the boundary of

$$D(\delta, R) = \{z: \delta < |z| < R, 0 < \arg z < \pi/2\},$$

show that

$$\int_0^{+\infty} y^{s-1}e^{-iy}dy = \Gamma(s)\exp(-\tfrac{1}{2}\pi i s), \quad 0 < \mathrm{re}\, s < 1.$$

4.7.9. If $0 < \alpha < 1$, show that

(i) $\displaystyle\int_0^{+\infty} x^{-\alpha}\cos x\, dx = \Gamma(1-\alpha)\sin\tfrac{1}{2}\alpha\pi;$

(ii) $\displaystyle\int_0^{+\infty} x^{-\alpha}\sin x\, dx = \Gamma(1-\alpha)\cos\tfrac{1}{2}\alpha\pi.$

Also verify the left-hand side continuity of the integral (ii) at $\alpha = 1$.

4.7.10. Verify that

$$\int_0^{\pi/2} \sin^{p-1}\theta \cos^{q-1}\theta\, d\theta = \frac{1}{2}\cdot\frac{\Gamma(\tfrac{1}{2}p)\Gamma(\tfrac{1}{2}q)}{\Gamma(\tfrac{1}{2}(p+q))}.$$

4.7.11. Show that

(i) $\displaystyle\int_0^{\pi/2}(\tan\theta)^\alpha d\theta = \tfrac{1}{2}\pi(\cos\tfrac{1}{2}\pi\alpha)^{-1}, \quad -1 < \alpha < 1;$

(ii) $\displaystyle\int_0^{\pi/2}\sqrt{\sin\theta}\,d\theta = (2\pi)^{3/2}[\Gamma(\tfrac{1}{4})]^{-2}.$

Hint: Use the formula $\Gamma(z)\Gamma(1-z) = \pi/\sin\pi z$.

4.7.12. Express the following elliptic integrals by means of gamma function:

(i) $\displaystyle\int_0^{\pi/2}(1-\tfrac{1}{2}\sin^2\theta)^{-1/2}d\theta;$

(ii) $\displaystyle\int_0^{\pi/0}(1-\tfrac{1}{2}\sin^2\theta)^{1/2}d\theta.$

Hint: Put $\sin\theta = \sqrt{1-\sqrt{u}}$.

4.7.13. Prove that

$$\Gamma\left(\frac{1}{n}\right)\Gamma\left(\frac{2}{n}\right)\ldots\Gamma\left(\frac{n-1}{n}\right) = n^{-1/2}(2\pi)^{(n-1)/2}.$$

Hint: Show that $n = \displaystyle\prod_{k=1}^{n-1}\left(1-\exp\frac{2k\pi i}{n}\right)$ and hence deduce the formula

$$\prod_{k=1}^{n-1}\sin\frac{k\pi}{n} = n\cdot 2^{1-n}.$$

4.7.14. Prove that
$$\int_0^1 \log \Gamma(t)\,dt = \tfrac{1}{2}\log 2\pi.$$

4.7.15. Evaluate the integral
$$I(a) = \int_a^{a+1} \log \Gamma(t)\,dt, \quad a > 0.$$

4.8. NORMAL FAMILIES

A family $\mathscr{F}$ of functions f analytic in a domain D is said to be *normal*, if every sequence $\{f_n\}$ of functions $f_n \in \mathscr{F}$ contains a subsequence $\{f_{n_k}\}$ which either converges a.u. in D, or diverges to ∞ a.u. in D. The limit function $f = \lim_{k\to\infty} f_{n_k}$ is analytic, unless it reduces to the constant ∞. If any sequence $\{f_n\}$ of functions $f_n \in \mathscr{F}$ contains an a.u. convergent subsequence, then $\mathscr{F}$ is said to be *compact*.

A necessary and sufficient condition for compactness of the family $\mathscr{F}$ is the existence of a common, finite upper bound of the absolute values of all $f \in \mathscr{F}$ on each compact subset of D (Stieltjes–Osgood theorem, also called Montel's compactness condition). Clearly each compact family is normal. The real-valued function $\rho(z,f) = 2|f'(z)|(1+|f(z)|^2)^{-1}$ is called the *spherical derivative* of f and has an obvious geometrical meaning (cf. Ex. 4.8.1). Now, a family $\mathscr{F}$ of functions f analytic in a domain D is normal, if there exists in every compact subset of D a common finite upper bound for the spherical derivative of all functions of the family. This criterion is due to F. Marty.

Another sufficient condition for normality is due to Montel and its proof is based on the properties of the modular function: if all functions f of a family $\mathscr{F}$ are analytic in a domain D and every $f \in \mathscr{F}$ does not take in D two fixed, finite values α, β ($\alpha \neq \beta$), then f is normal.

The concept of normal family is very important in the existence questions for solutions of extremal problems.

4.8.1. Explain the geometrical meaning of the spherical derivative.

4.8.2. Verify that the family of all similarity transformations $az+b$ is not a normal family in the finite plane.

4.8.3. If $\mathscr{F}$ is normal in a domain D and there exists a point $z_0 \in D$ and a real constant $M_0 < +\infty$ such that $|f(z_0)| \leq M_0$ for all $f \in \mathscr{F}$, show that $\mathscr{F}$ is a compact family.

4.8.4. If $\mathscr{F}$ is a compact family of functions analytic in a domain D, show that also $\mathscr{F}_1 = \{g: g = f', f \in \mathscr{F}\}$ is compact. By considering the sequence $\{n(z^2 - n^2)\}$ verify that the derivatives of functions of a normal family not necessarily form a normal family.

4.8.5. Suppose $F(w)$ is an entire function (i.e. a function analytic in the finite plane $\mathbf{C}$) and $\mathscr{F}$ is a compact family of functions analytic in a domain D. Verify that the functions $F \circ f$, $f \in \mathscr{F}$ also form a compact family in D.

4.8.6. Suppose $\mathscr{F}$ is the family of all functions $f(z) = az$, where a is a complex constant and $F(w) = e^w \sin w$. Verify that $\mathscr{F}$ is normal in $\mathbf{C} \setminus \bar{K}(0; 1)$, whereas the functions $F \circ f$ do not form a normal family.

4.8.7. Prove Hurwitz's theorem: If $\{f_n\}$ is an a.u. convergent sequence in a domain D and $f_n(z) \neq 0$ for all $z \in D$ and all f_n, then the limiting function f is either identically 0 in D, or does not vanish in D.

Hint: Verify that $\rho(z, f) \equiv \rho(z, 1/f)$. Also prove that the terms of an a.u. sequence form a normal family.

4.8.8. Suppose $\{f_n\}$ is an a.u. convergent sequence of univalent functions in a domain D and $g = \lim f_n$. Show that either g is univalent in D, or is a constant.

Hint: Consider $f(z) - f(a)$ in $D \setminus a$.

4.8.9. Show that the family of all functions f analytic in the unit disk and such that $f(0) = 0$, $f'(0) = 1$, is not a normal family.

4.8.10. Let T_0 be the family of all functions f analytic in the unit disk and such that $f(0) = 0$, $f'(0) = 1$, $f(z) \neq 0$ for $z \neq 0$. Show that T_0 is compact.

Hint: Consider $\log(f(z)/z)$.

4.8.11. (cont.) Show that there exists a constant $\alpha > 0$ such that any $f \in T_0$ takes in the unit disk any value $w_0 \in K(0; \alpha)$.

Hint: If f_n does not take the value α_n and $\alpha_n \to 0$, consider the sequence $g_n(z) = \log(1 - f_n(z)/\alpha_n)$.

4.8.12. Show that the family S_0 of all functions analytic and univalent in the domain D omitting one fixed value α is normal.

4.8.13. Suppose $G(M)$ is the family of functions analytic in a domain D and such that $\iint_D |f(z)|^2 \, dx \, dy \leq M$. Show that $G(M)$ is compact in D.

Hint: If $\bar{K}(z_0; R) \subset D$, verify that

$$|f(z_0)|^2 \leq \frac{1}{2\pi} \int_0^{2\pi} |f(z_0 + re^{i\theta})|^2 \, d\theta, \quad 0 < r \leq R,$$

and deduce that $\pi R^2 |f(z_0)|^2 \leq M$.

CHAPTER 5

Meromorphic and Entire Functions

5.1. MITTAG–LEFFLER'S THEOREM

Let $\{a_n\}$ be an arbitrary sequence of complex numbers such that $a_0 = 0 < |a_1| < |a_2| < \ldots$ and $\lim a_n = \infty$ and let $\{G_n(w)\}$ be an arbitrary sequence of polynomials with vanishing constant terms. There exists a meromorphic function F which is analytic in the finite plane except for the poles $a_0, a_1, a_2, \ldots$ and has $G_n(1/(z-a_n))$ as singular parts at a_n.

Let $\sum_{n=1}^{\infty} u_n$ be an arbitrary, convergent series with positive terms and let $\{H_n(z)\}$ be a sequence of polynomials such that

$$\left| G_n\left(\frac{1}{z-a_n}\right) - H_n(z) \right| \leq u_n \quad (n = 1, 2, \ldots)$$

for $z \in \bar{K}(0; r_n)$, where $r_n < |a_n|$ and $\lim r_n = +\infty$. In particular H_n can be a suitable partial sum of Taylor's expansion of the singular part which is analytic in the open disk $K(0; |a_n|)$. Then the function

$$F(z) = H(z) + G_0\left(\frac{1}{z}\right) + \sum_{n=1}^{\infty}\left[G_n\left(\frac{1}{z-a_n}\right) - H_n(z) \right],$$

where H is analytic in the finite plane (i.e. H is an entire function), has all the required properties. This representation due to Mittag–Leffler enables us to find a meromorphic function with given poles and given singular parts at these poles up to an entire function.

5.1.1. Find the most general meromorphic function F whose only singularities are:

(i) poles $1, 2, 3, \ldots$ of first order with $\text{res}(n; F) = n$;
(ii) poles a^n of first order ($|a| > 1$, $n = 1, 2, \ldots$) with $\text{res}(a^n; F) = a^n$;
(iii) poles $\sqrt{n}$ of first order ($n = 0, 1, 2, \ldots$) with residues 1;

(iv) poles $1, 2, 3, \ldots$ of second order with singular parts $n(z-n)^{-2}$;
(v) poles $1, 2, 3, \ldots$ of second order with singular parts $n^2(z-n)^{-2} + (z-n)^{-1}$;
(vi) poles $\mp 1, \mp 2, \mp 3, \ldots$ with $\text{res}(n; F) = |n|$;
(vii) poles $0, \mp 1, \mp 2, \ldots$ with residues 1;
(viii) poles $0, -1, -2, \ldots$ with residues 1.

5.1.2. Find the most general meromorphic function F whose only singularities are simple poles $-1, -2, -3, \ldots$ with $\text{res}(-n; F) = (-1)^n$.

5.1.3. Find the most general meromorphic function F whose only singularities are simple poles $w = m+ni$ ($m, n = 0, 1, 2, \ldots$) with residues 1.

5.1.4. Show that the function $\int_0^1 e^{-t} t^{z-1} dt$ is analytic for $\text{re}\, z > 0$ and has the expansion

$$\int_0^1 e^{-t} t^{z-1} dt = \sum_{n=0}^{\infty} \frac{(-1)^n}{n!} \cdot \frac{1}{z+n}, \quad \text{re}\, z > 0.$$

Hint: Use the power series expansion of e^{-t}.

5.1.5. Find the partial fractions expansion of $\Gamma(z)$.

5.2. PARTIAL FRACTIONS EXPANSIONS OF MEROMORPHIC FUNCTIONS

5.2.1. Suppose f is a meromorphic function whose only singularities are simple poles $a_1, a_2, a_3, \ldots$ with $0 < |a_1| \leqslant |a_2| \leqslant |a_3| \leqslant \ldots$, $\lim a_n = \infty$ and $\text{res}(a_n; f) = A_n$. Suppose there exists a sequence $\{C_n\}$ of contours such that

1° C_n omits all the poles a_j;
2° each C_n lies inside C_{n+1};
3° $\min_{z \in C_n} |z| = R_n \to +\infty$ as $n \to +\infty$;
4° the length L_n of C_n is $O(R_n)$;
5° $\max_{z \in C_n} |f(z)| = o(R_n)$.

By considering the integral $\int_{C_n} \zeta^{-1}(\zeta-z)^{-1} f(\zeta)\, d\zeta$ show that

$$f(z) = f(0) + \sum_{n=1}^{\infty} A_n \left(\frac{1}{z-a_n} + \frac{1}{a_n} \right),$$

where the terms of the series are formed of all summands corresponding to the poles situated between C_n and C_{n+1}.

5.2. EXPANSIONS OF MEROMORPHIC FUNCTIONS

5.2.2. Find the expansion of f into partial fractions in case f satisfies the conditions 1°–5° and has a pole $z = 0$ with a singular part $G(1/z)$.

5.2.3. Derive the following expansions:

(i) $\dfrac{\pi}{\sin \pi z} = \dfrac{1}{z} + 2z \sum_{n=1}^{\infty} \dfrac{(-1)^{n+1}}{n^2 - z^2}$, $\quad z \neq 0, \mp 1, \mp 2, \ldots$;

(ii) $\dfrac{\pi}{\cos \pi z} = 2 \sum_{n=0}^{\infty} \dfrac{(-1)^n (n+\tfrac{1}{2})}{(n+\tfrac{1}{2})^2 - z^2}$, $\quad z \neq \mp \tfrac{1}{2}, \mp \tfrac{3}{2}, \mp \tfrac{5}{2}, \ldots$;

(iii) $\pi \tan \pi z = 2z \sum_{n=0}^{\infty} [(n+\tfrac{1}{2})^2 - z^2]^{-1}$, $\quad z \neq \mp \tfrac{1}{2}, \mp \tfrac{3}{2}, \mp \tfrac{5}{2}, \ldots$

5.2.4. If $\alpha \neq 0$, $\beta/\alpha \neq \mp 1, \mp 2, \ldots$, show that

$$\frac{\pi}{\alpha} \cot \frac{\pi \beta}{\alpha} = \sum_{n=0}^{\infty} \left(\frac{1}{n\alpha + \beta} - \frac{1}{n\alpha + (\alpha - \beta)} \right).$$

Also show that

$$\frac{1}{1 \cdot 2} + \frac{1}{4 \cdot 5} + \frac{1}{7 \cdot 8} + \ldots = \frac{\pi}{3\sqrt{3}}.$$

5.2.5. Find the expansions of hyperbolic functions analogous to the expansions in Exercise 5.2.3.

5.2.6. Find the expansion of $(\sin z \sinh z)^{-1}$ into partial fractions.

5.2.7. Show that

$$(\cosh z - \cos z)^{-1} = \frac{1}{z^2} + \pi z^2 \sum_{n=0}^{\infty} \frac{(-1)^n n}{\sinh \pi n} (\tfrac{1}{4} z^4 + n^4 \pi^4)^{-1}$$

for $z \neq n\pi(\mp 1 \mp i)$, $n = 0, 1, 2, \ldots$

5.2.8. If λ_n are positive roots of the equation $\tan z = z$ (cf. Ex. 3.9.16), show that

$$\frac{z \sin z}{\sin z - z \cos z} = \frac{3}{z} + \sum_{n=1}^{\infty} \frac{2z}{z^2 - \lambda_n^2}.$$

5.2.9. By considering the integral

$$\int_{\partial Q_n} \frac{\sin \zeta}{\zeta(\zeta - z) \cos^2 \zeta} \, d\zeta,$$

where ∂Q_n is the square with vertices $n\pi(\mp 1, \mp i)$, show that

$$\frac{\sin z}{\cos^2 z} = \sum_{n=-\infty}^{\infty} (-1)^n [z-(n+\tfrac{1}{2})\pi]^{-2}.$$

5.2.10. If $0 < a < 1$, show that

$$\frac{e^{az}}{e^z-1} = \frac{1}{z} + \sum_{n=1}^{\infty} \frac{2z\cos 2\pi na - 4\pi n\sin 2\pi na}{z^2+4\pi^2 n^2}.$$

5.3. JENSEN'S FORMULA. NEVANLINNA'S CHARACTERISTIC

5.3.1. If f is meromorphic in $K(0; R)$ and has m zeros $a_1, a_2, \ldots, a_m$ ($0 < |a_1| \leqslant |a_2| \leqslant \ldots \leqslant |a_m|$) and n poles $b_1, b_2, \ldots, b_n$ ($0 < |b_1| \leqslant |b_2| \leqslant \ldots \leqslant |b_n|$) in the closed disk $\bar{K}(0; r)$, $0 < r < R$, show that the following Jensen's formula holds:

$$\frac{1}{2\pi}\int_0^{2\pi} \log|f(re^{i\theta})|\,d\theta = \log|f(0)| + \log\frac{r^m}{|a_1 a_2 \ldots a_m|} - \log\frac{r^n}{|b_1 b_2 \ldots b_n|}.$$

Hint: The factor $\dfrac{r^2-\bar{a}z}{r(z-a)}$ removes a zero $z = a$ of the function f and does not change $|f(re^{i\theta})|$.

5.3.2. Show that Jensen's formula is valid also in case there are some zeros, or poles of f situated on $C(0; r)$.

5.3.3. Write Jensen's formula for f having a zero, or pole of order λ at the origin.

5.3.4. Evaluate the integral

$$\Phi(r) = \frac{1}{2\pi}\int_0^{2\pi} \log|f(re^{i\theta})|\,d\theta$$

as a function of r for:
 (i) $f(z) = z^3 + 2z + 3$;
 (ii) $f(z) = \sin z$;
 (iii) $f(z) = \cot z$.

5.3.5. Let f be meromorphic in $K(0; R)$, $0 < R \leqslant +\infty$. We define:

1° $\qquad \log^+ x = \begin{cases} \log x & \text{for } x \geqslant 1, \\ 0 & \text{for } 0 \leqslant x \leqslant 1; \end{cases}$

2° $n(r,f)$ = the number of poles of f in $K(0; r)$ taken with due multiplicity $(0 < r \leq R)$;

3° $$m(r,f) = \frac{1}{2\pi} \int_0^{2\pi} \log^+|f(re^{i\theta})|\,d\theta;$$

4° $$T(r,f) = m(r,f) + \int_0^r \frac{n(t,f)}{t}\,dt.$$

Show that Jensen's formula can be written in the form:
$$T(r,f) - T(r,1/f) = \log|f(0)|$$
under the assumption that f is analytic and $\neq 0$ at $z = 0$. $T(r,f)$ is called *Nevanlinna's characteristic* of f. It plays a fundamental role in the theory of meromorphic functions.

5.3.6. Let f be meromorphic in $K(0; R)$. Find a necessary and sufficient condition involving the distribution of zeros and poles that the integral
$$\int_0^{2\pi} \log|f(re^{i\theta})|\,d\theta$$
has a constant value.

5.3.7. Determine $n(r,f)$, $n(r, 1/f)$ for $f(z) = z^{-1}\tan z$.

5.3.8. Determine $T(r,f)$ for $f(z) = e^z$.

5.3.9. If $P(z) = az^n + \ldots + a_0$ is a polynomial of degree n, $f(z) = \exp P(z)$, show that
$$T(r,f) \sim \frac{|a|}{\pi} r^n \quad \text{as} \quad r \to +\infty.$$

5.3.10. If $f(z) = \exp\{\pi i/(1-z)\}$, $z \in K(0; 1)$, show that
$$T(r,f) = \frac{1}{2\pi} \log \frac{1+r}{1-r}.$$

5.3.11. If f is analytic in $K(0; R)$, show that
$$\Phi(r) = \frac{1}{2\pi} \int_0^{2\pi} \log|f(re^{i\theta})|\,d\theta$$
is a convex and increasing function of $\log r$.

5.3.12. Suppose f is analytic and bounded in the unit disk, $f(0) \neq 0$ and $\mu(r) = n(r, 1/f)$ is the number of zeros of f in $K(0; r)$. Show that
$$\lim_{r \to 1-} \mu(r) \log r = 0.$$

5.3.13. Let f be an analytic function which is bounded in the unit disk and has zeros $a_1, a_2, a_3, \ldots$, where $0 < |a_1| \leq |a_2| \leq |a_3| \leq \ldots$ and each zero is written as many times as its multiplicity. Show that

(i) $\lim |a_1 a_2 \ldots a_n|$ exists and is $\neq 0$;

(ii) the series $\sum_{n=1}^{\infty} (1 - |a_n|)$ is convergent.

5.3.14. Suppose $\{a_n\}$ is a sequence of complex numbers such that $|a_n| < 1$ ($n = 1, 2, \ldots$), $0 < |a_1| \leq |a_2| \leq \ldots$ and $\sum_{n=1}^{\infty} (1 - |a_n|) = +\infty$. If f, g are analytic and bounded in the unit disk and $f(a_n) = g(a_n)$, $n = 1, 2, \ldots$, show that $f = g$.

5.4. INFINITE PRODUCTS

An *infinite product* $p_1 p_2 \ldots p_n \ldots = \prod_{n=1}^{\infty} p_n$ is said to be convergent if at most a finite number of the factors p_n are zero and if the sequence of products of the non-zero factors converges to a finite non-zero limit.

Thus we have for a convergent infinite product $\lim p_n = 1$. For this reason we usually write an infinite product in the form $\prod_{n=1}^{\infty} (1 + a_n)$. The infinite product $\prod_{n=1}^{\infty} (1 + a_n)$, $1 + a_n \neq 0$, is convergent and divergent together with the series $\sum_{n=1}^{\infty} \text{Log}(1 + a_n)$. A sufficient condition for convergence of the product $\prod_{n=1}^{\infty} (1 + a_n)$ is the convergence of the series $\sum_{n=1}^{\infty} |a_n|$.

If $\{u_n(z)\}$ is a sequence of functions which are analytic in a domain G and $|u_n(z)| \leq A_n$ for every $z \in G$ with $\sum_{n=1}^{\infty} A_n < +\infty$, then the product $\prod_{n=1}^{\infty} [1 + u_n(z)]$ is convergent in G and represents a function which is analytic in G.

5.4.1. If $\sum_{n=1}^{\infty} |u_n|^2 < +\infty$, show that the necessary and sufficient condition for convergence of the product $\prod_{n=1}^{\infty} (1 + u_n)$ is the convergence of the series $\sum_{n=1}^{\infty} u_n$.

5.4. INFINITE PRODUCTS

5.4.2. If $u_n = (-1)^n n^{-1/2}$, show that the series $\sum_{n=1}^{\infty} u_n$ is convergent, whereas the product $\prod_{n=1}^{\infty}(1+u_n)$ is divergent.

5.4.3. Show that

(i) $\prod_{n=1}^{\infty}\left(1-\dfrac{1}{n^2}\right) = \dfrac{1}{2}$;

(ii) $\prod_{n=2}^{\infty} \dfrac{n^3-1}{n^3+1} = \dfrac{2}{3}$;

(iii) $\prod_{n=1}^{\infty}\left[1+\dfrac{(-1)^{n+1}}{n}\right] = 1$.

5.4.4. If $|z| < 1$, show that the product $\prod_{n=0}^{\infty}(1+z^{2^n})$ is convergent to $(1-z)^{-1}$.

5.4.5. Determine the domains of convergence of the infinite products

(i) $\prod_{n=1}^{\infty}\left[1+\left(\dfrac{z}{n}\right)^2\right]$;

(ii) $\prod_{n=1}^{\infty}\left[1+\left(1+\dfrac{1}{n}\right)^{n^2} z^n\right]$;

(iii) $\prod_{n=1}^{\infty} \cos \dfrac{z}{n}$;

(iv) $\prod_{n=1}^{\infty}\left(1+\dfrac{1}{z+n}\right) e^{-1/n}$.

5.4.6. Suppose $\{a_n\}$ is a sequence of complex numbers such that: $|a_n| < 1$ and $a_n \neq 0$ $(n = 1, 2, \ldots)$, $a_m \neq a_n$ for $m \neq n$, $\sum_{n=1}^{\infty}(1-|a_n|^2) < +\infty$. Show that the product $\prod_{n=1}^{\infty} \dfrac{z-a_n}{z-\bar{a}_n^{-1}}$ is convergent in the unit disk and represents a function F analytic in this disk, equal to zero only at $z = a_n$ $(n = 1, 2, \ldots)$ and bounded by 1 in absolute value.

5.4.7. Show that the product $\prod_{n=1}^{\infty}\left(1+\dfrac{z}{n}\right) e^{-z/n}$ is convergent in the whole plane and that its limit is an entire function.

Hint: Prove that $|(1-z)e^z - 1| \leq |z|^2$ for $|z| \leq 1$.

5.4.8. Show that the sequence of functions
$$h_n(z) = \frac{z(z+1)(z+2)\cdots(z+n)}{n!\exp(z\log n)}$$
is convergent in the whole plane and that its limit $h(z)$ represents an entire function equal to zero only for $z = 0, -1, -2, \ldots$

5.4.9. (cont.) If $\Gamma(z) = [h(z)]^{-1}$, prove that

(i) $z\Gamma(z) = \Gamma(z+1)$;

(ii) $\Gamma(n) = (n-1)!$ for positive integers n.

5.5. FACTORIZATION OF AN ENTIRE FUNCTION

Every polynomial with zeros $0, a_1, \ldots, a_n$ ($a_k \neq 0$ for $k = 1, \ldots, n$) can be represented as a product:
$$Az^k\left(1 - \frac{z}{a_1}\right)\cdots\left(1 - \frac{z}{a_n}\right).$$

Weierstrass proved the existence of an analogous decomposition for an arbitrary entire function and he showed that it is possible to construct an entire function whose zeros form an arbitrary sequence with no limiting-point other than infinity. Let $\{a_n\}$ be an arbitrary sequence with $\lim a_n = \infty$. Moreover, let
$$E(z, m) = (1-z)\exp\left(z + \frac{1}{2}z^2 + \ldots + \frac{1}{m}z^m\right) \quad \text{for} \quad m \geq 1,$$
$$E(z, 0) = 1 - z;$$
the functions $E(z, m)$ are called the Weierstrass primary factors. There exists always a sequence of positive integers $\{m_n\}$ such that
$$\sum_{n=1}^{\infty}\left|\frac{z}{a_n}\right|^{m_n+1}$$
is convergent in the whole plane and every entire function f with zeros at the points $0, a_1, \ldots a_n, \ldots$ and otherwise different from zero can be written in the form of a Weierstrass product:
$$f(z) = z^m \exp g(z) \prod_{n=1}^{\infty} E\left(\frac{z}{a_n}, m_n\right),$$
where g is an entire function. In particular, if there exists an integer q such that $\sum_{n=1}^{\infty} |a_n|^{-q} < +\infty$, whereas $\sum_{n=1}^{\infty} |a_n|^{-q+1} = +\infty$, then q is called the genus of the sequence $\{a_n\}$. Then we may choose $m_n = q-1$ for all n. Not all the terms of

5.6. FACTORIZATION OF ELEMENTARY FUNCTIONS

the sequence $\{a_n\}$ must be different; if a number z_0 appears k times in it, then the corresponding primary factor in the Weierstrass product appears k times.

5.5.1. Verify if the following sequences have a finite genus q and possibly, determine it:

(i) $-1, 1, -2, 2, \ldots;$ (ii) $\{\log(n+1)\}$;

(iii) $\{[n/4] i^n\}$, where $[x]$ denotes the greatest integer $n \leqslant x$;

(iv) $\{s_n + i t_n\}$, where $(s_n; t_n)$ is the sequence of all pairs of integers so ordered that $\max(|s_n|, |t_n|)$ increases with n;

(v) $\{(-1)^n \sqrt{n}\}$;

(vi) $1; 2, 2; 3, 3, 3; 4, 4, 4, 4; 5, \ldots$

5.5.2. Determine the Weierstrass product for sequences (i), (ii), (vi) in Exercise 5.5.1.

5.5.3. Derive Weierstrass theorem from Mittag–Leffler theorem.

5.5.4. Show that every meromorphic function f is a quotient of two entire functions.

5.5.5. Let $\{a_n\}$ be a sequence of complex numbers such that $a_0 = 0$, $a_m \neq a_n$ for $m \neq n$ and $\lim a_n = \infty$, and let $\{\eta_n\}$ be an arbitrary sequence of complex numbers. Let ω be the Weierstrass product for the sequence $\{a_n\}$ of the form:

$$\omega(z) = z \prod_{n=1}^{\infty} E\left(\frac{z}{a_n}, m_n\right).$$

Show that there exists a sequence $\{q_n\}$ of complex numbers such that the series

(A) $\qquad \dfrac{\eta_0 \omega(z)}{z \omega'(0)} + \displaystyle\sum_{n=1}^{\infty} \dfrac{\eta_n}{\omega'(a_n)} \cdot \dfrac{\omega(z)}{z-a_n} \left(\dfrac{z}{a_n}\right)^{q_n}$

is a.u. convergent in the whole plane and its sum Φ is an entire function such that $\eta_k = \Phi(a_k)$, $k = 0, 1, 2, \ldots$ This is the so-called *Pringsheim's interpolation formula*.

5.5.6. Find an entire function F such that $F(n) = (n-1)!$ for all positive integers n.

5.6. FACTORIZATION OF ELEMENTARY FUNCTIONS

5.6.1. Derive the formula

$$\sin \pi z = \pi z \prod_{n=1}^{\infty} \left(1 - \frac{z^2}{n^2}\right).$$

5.6.2. Derive the formulas:

(i) $\sinh \pi z = \pi z \prod_{n=1}^{\infty} \left(1 + \frac{z^2}{n^2}\right);$

(ii) $\cosh z - \cos z = z^2 \prod_{n=1}^{\infty} \left(1 + \frac{z^4}{4n^4 \pi^4}\right);$

(iii) $e^z - 1 = z e^{z/2} \prod_{n=1}^{\infty} \left(1 + \frac{z^2}{4\pi^2 n^2}\right);$

(iv) $e^{az} - e^{bz} = (a-b) z \exp\left[\tfrac{1}{2}(a+b) z\right] \prod_{n=1}^{\infty} \left[1 + \frac{(a-b)^2 z^2}{4 n^2 \pi^2}\right].$

5.6.3. Derive the formula

$$\cos \pi z = \prod_{n=0}^{\infty} \left[1 - \left(\frac{z}{n+\tfrac{1}{2}}\right)^2\right].$$

5.6.4. Evaluate infinite products:

(i) $\prod_{n=2}^{\infty} \left(1 + \frac{z^4}{n^4}\right);$

(ii) $\prod_{n=2}^{\infty} \left(1 - \frac{z^4}{n^4}\right).$

Show that

$$\prod_{n=2}^{\infty} \left(1 - \frac{1}{n^4}\right) = \frac{1}{8\pi} (e^\pi - e^{-\pi}).$$

5.6.5. Evaluate infinite products:

(i) $\prod_{n=1}^{\infty} \left(1 + \frac{z^2}{n^2} + \frac{z^4}{n^4}\right);$

(ii) $\prod_{n=1}^{\infty} \left(1 + \frac{z^6}{n^6}\right);$

(iii) $\prod_{n=1}^{\infty} \left(1 - \frac{z^6}{n^6}\right).$

Hint: Exercise 5.6.1.

5.6.6. Show that

(i) $\prod_{n=1}^{\infty} \left(1 + \frac{1}{n^2} + \frac{1}{n^4}\right) = \frac{1 + \cosh \pi \sqrt{3}}{2\pi^2};$

5.6. FACTORIZATION OF ELEMENTARY FUNCTIONS

(ii) $\prod_{n=1}^{\infty}\left(1+\dfrac{1}{n^6}\right) = \dfrac{1}{2}\cdot\dfrac{\sinh\pi}{\pi^3}(\cosh\pi - \cos\pi\sqrt{3})$.

5.6.7. If λ_n is the root of the equation $\tan z = z$ contained in the interval $(n\pi, (n+\tfrac{1}{2})\pi)$, $n = 1, 2, \ldots$, show that

$$\sin z - z\cos z = \frac{1}{3}z^2 \prod_{n=1}^{\infty}\left(1 - \frac{z^2}{\lambda_n^2}\right).$$

5.6.8. Verify the formula

$$\frac{\sin\pi z}{\pi z(1-z)} = \prod_{n=1}^{\infty}\left(1 + \frac{z-z^2}{n+n^2}\right).$$

5.6.9. Verify the formula

$$\frac{\sin 3z}{\sin z} = -\prod_{n=-\infty}^{\infty}\left(1 - \frac{4z^2}{(n\pi+z)^2}\right) \quad (z \neq n\pi,\ n = 0, \mp 1, \mp 2, \ldots).$$

5.6.10. Suppose F is an entire function whose zeros $a_1, a_2, \ldots, a_n, \ldots$ are all simple and satisfy $0 < |a_1| \leqslant |a_2| \leqslant \ldots$, $\lim a_n = \infty$. Suppose, moreover, there exists a sequence of contours $\{C_n\}$ satisfying the conditions 1°–4° of Exercise 5.2.1 and such that

$$\max_{\zeta \in C_n}\left|\frac{F'(\zeta)}{F(\zeta)}\right| = o(R_n).$$

Prove that

$$F(z) = F(0)\exp\left(\frac{zF'(0)}{F(0)}\right)\prod_{n=1}^{\infty}\left(1 - \frac{z}{a_n}\right)e^{z/a_n},$$

where all the factors corresponding to zeros situated between C_n and C_{n+1} are considered as one term.

5.6.11. Show that

$$\frac{\sin\pi(z+a)}{\sin\pi a} = \left(1+\frac{z}{a}\right)\prod_{n=-\infty}^{\infty}{}'\left(1+\frac{z}{a+n}\right)e^{-z/n} \quad (a \neq 0, \mp 1, \mp 2, \ldots).$$

The prime after a product sign indicates that the factor corresponding to $n = 0$ should be omitted.

5.6.12. Show that

$$\cos\frac{\pi z}{4} - \sin\frac{\pi z}{4} = (1-z)\left(1+\frac{z}{3}\right)\left(1-\frac{z}{5}\right)\ldots$$

5.6.13. Use the formula
$$[\Gamma(z)]^{-1} = z e^{\gamma z} \prod_{n=1}^{\infty} \left(1 + \frac{z}{n}\right) e^{-z/n}$$
to show that
$$\Gamma(z)\Gamma(1-z) = \frac{\pi}{\sin \pi z}.$$

5.6.14. Express the following products in terms of the function Γ:

(i) $\prod_{n=1}^{\infty} \left(1 - \frac{1}{2n}\right) e^{1/2n}$;

(ii) $\prod_{n=1}^{\infty} \left(1 + \frac{z}{2n-1}\right)\left(1 - \frac{z}{2n}\right)$;

(iii) $(1+\frac{1}{2})(1-\frac{1}{4})(1+\frac{1}{6}) \cdots$

5.7. ORDER OF AN ENTIRE FUNCTION

An entire function f is said to be *of finite order* if there exists a positive number A such that $f(z) = O(\exp r^A)$ as $|z| = r \to +\infty$. Then the number

$$\rho = \varlimsup_{r \to +\infty} \frac{\log \log M(r)}{\log r}, \quad \text{where} \quad M(r) = M(r, f) = \sup_{|z| \leqslant r} |f(z)|$$

which is finite, is called the *order of the entire function* f. If $f(z) = c_0 + c_1 z + c_2 z^2 + \ldots$, then

(5.7A) $$\rho = \varlimsup \frac{n \log n}{\log(1/|c_n|)}.$$

5.7.1. If k is a positive integer, show that $\exp z^k$ is an entire function of order k.

5.7.2. Show that $\exp(e^z)$ is not a function of finite order.

5.7.3. Show that

(i) $\cos \sqrt{z}$ is an entire function of order $\frac{1}{2}$;

(ii) an arbitrary polynomial is an entire function of order 0.

5.7.4. If f is an entire function which is not a polynomial, show that
$$\lim_{r \to +\infty} \frac{\log M(r, f)}{\log r} = +\infty.$$

5.7.5. If $f(z) = z^k g(z)$ is of order $\rho > 0$, show that g is also of order ρ.

5.7. ORDER OF AN ENTIRE FUNCTION

5.7.6. If f is an entire function of order ρ, show that f' is also an entire function of order ρ, and conversely.

5.7.7. Suppose f is an entire function of order ρ and $m(r)$ denotes the number of zeros of f situated inside $K(0; r)$. Prove that $m(r) = O(r^{\rho+\varepsilon})$, where $\varepsilon > 0$ can be arbitrary.
Hint: Cf. Exercise 5.3.1.

5.7.8. (cont.) If f is an entire function of finite order ρ and $a_1, a_2, \ldots, a_n, \ldots$ are its zeros, $0 < |a_1| \leqslant |a_2| \leqslant \ldots$ show that for every $\alpha > \rho$ the series $\sum_{n=1}^{\infty} |a_n|^{-\alpha}$ is convergent.

5.7.9. Show that an entire function f of finite order ρ has a representation of the form

$$f(z) = z^k \exp(g(z)) \prod_{n=1}^{\infty} E\left(\frac{z}{a_n}, m\right)$$

where g is an entire function and $m \leqslant \rho$ does not depend on n.

5.7.10. Determine the order of the following entire functions:

(i) $\displaystyle\sum_{n=1}^{\infty} \left(\frac{z}{n}\right)^n$;

(ii) $\displaystyle\sum_{n=1}^{\infty} \left(\frac{\log n}{n}\right)^{n/a} z^n \quad (a > 0)$;

(iii) $\displaystyle\sum_{n=0}^{\infty} e^{-n^2} z^n$;

(iv) $\displaystyle\sum_{n=0}^{\infty} \frac{\cosh \sqrt{n}}{n!} z^n$.

CHAPTER 6

The Maximum Principle

6.1. THE MAXIMUM PRINCIPLE FOR ANALYTIC FUNCTIONS

6.1.1. If f is analytic and not a constant in a domain D, show that $|f|$ cannot have in D a local maximum.
Hint: Cf. Exercise 4.2.5.

6.1.2. If f is analytic and not a constant in a domain D, show that $|f|$ cannot have a local minimum at $z_0 \in D$, unless $f(z_0) = 0$.

6.1.3. Suppose f is analytic and not a constant in a domain D and $|f|$ has a continuous extension on $\bar{D}$. If $f(z) \neq 0$ in D and $m = \inf|f(z)|$, $M = \sup|f(z)|$, $z \in \operatorname{fr} D$, show that $m < |f(z)| < M$ in D.

6.1.4. Suppose f is analytic in a domain D and $|f|$ has a continuous extension on $\bar{D}$ without being a constant.
If $|f|$ has a constant value on the boundary of D, show that $f(z_0) = 0$ at some point $z_0 \in D$.

6.1.5. If P is a polynomial of degree n, show that $\{z: |P(z)| = A\}$ cannot have more than n components for any fixed A.

6.1.6. If f is a nonconstant, analytic function in $K(0; R)$, show that $M(r, f) \sup_{|z|=r} |f(z)|$ is a strictly increasing function of $r \in (0, R)$.

6.1.7. Prove Hadamard's three circles theorem:
Suppose f is analytic in $B = \{z: r_1 < |z| < r_2\}$, continuous in $\bar{B}$ and $M_k = \sup_{|z|=r_k} |f(z)|$, $k = 1, 2$.
If $M(r) = \sup_{|z|=r} |f(z)|$, then $\log M(r)$ is a convex function of $\log r$, i.e.

$$\log M(r) \leqslant \frac{\log r_2 - \log r}{\log r_2 - \log r_1} \log M_1 + \frac{\log r - \log r_1}{\log r_2 - \log r_1} \log M_2.$$

Hint: Consider $[f(z)]^p z^{-q}$, where p, q are suitably chosen integers.

6.1.8. If f is analytic in $K(\infty; 1)$, continuous in $\bar{K}(\infty; 1)$ and has a finite limit at ∞, show that $|f|$ attains a maximum at a point of $C(0; 1)$. Also prove that $M(r,f) = \sup_{|z|=r} |f(z)|$ strictly decreases, unless f is a constant.

6.1.9. If P is a polynomial of degree n and $|P(z)| \leqslant M$ in $K(0; 1)$, show that $|P(z)| \leqslant M|z|^n$ in $K(\infty; 1)$.

6.1.10. If P is a polynomial of degree n and $|P(z)| \leqslant M$ for $z \in [-1, 1]$, show that for all z situated inside an ellipse with semiaxes a, b and foci $-1, 1$ we have: $|P(z)| \leqslant M(a+b)^n$.
Hint: Consider the mapping $z = \frac{1}{2}(w+w^{-1})$.

6.1.11. Suppose f is analytic and bounded by 1 in absolute value in $K(0; 1)$ and tends uniformly to 0 in the angle $\alpha \leqslant \arg z \leqslant \beta$ as $|z| \to 1$. Show that $f = 0$.
Hint: Consider the product $\varphi(z) = f(z)f(\omega z)f(\omega^2 z) \ldots f(\omega^{n-1}z)$, where $\omega = \exp(2\pi i/n)$ and n is a suitably chosen integer.

6.1.12. Suppose f is analytic in $S = \{z: |\operatorname{re} z| < a\}$ and there exist two real constants C, λ such that $|f(z)| \leqslant \exp \lambda |y|$ for $z = x+iy$, $|x| < a$, and $\overline{\lim} |f(z)| \leqslant C$ for $z \to a+iy_0$, resp. $z \to -a+iy_0$ with arbitrary y_0. Show that for all $z \in S$ we have: $|f(z)| \leqslant C$.
Hint: Consider $\varphi(z) = f(z)\exp(\varepsilon z^2)$, $\varepsilon > 0$.

6.2. SCHWARZ'S LEMMA

6.2.1. If f is analytic and less than 1 in absolute value in $K(0; 1)$ and if $f(0) = 0$, show that either $|f(z)| < |z|$ in $K(0; 1) \setminus 0$, or else $f(z) = e^{i\alpha}z$ with some real α.
Hint: Consider $z^{-1}f(z)$.

6.2.2. If f is analytic in $K(0; 1)$ and $|f(z)| < 1$ for all $z \in K(0; 1)$, show that
$$\left|\frac{f(z)-f(0)}{z}\right| \leqslant |1 - \overline{f(0)}f(z)|.$$

6.2.3. Under the assumptions of Exercise 6.2.2 show that either $|f'(0)| < 1$, or $f(z) = e^{i\alpha}z$ (α is real).

6.2.4. If f maps $K(0; 1)$ 1:1 conformally onto a domain D containing the unit disk and $f(0) = 0$, show that $|f'(0)| \geqslant 1$ with the sign of equality for $f(z) = e^{i\alpha}z$ only.

6.2.5. If $f(z) = \sum_{n=0}^{\infty} a_n z^n$ is analytic in $K(0; 1)$ and $|f(z)| \leqslant M$ in $K(0; 1)$, show that $M|a_1| \leqslant M^2 - |a_0|^2$.

6.2.6. (cont.) If m, n are integers, $0 \leqslant 2m < n$, show that $M|a_n| \leqslant M^2 - |a_m|^2$.

6.2.7. If ω is analytic in $K(0; 1)$ and $|\omega(z)| < 1$ in $K(0; 1)$, $\omega(0) = \alpha > 0$, show that the set of all values taken by ω in $\bar{K}(0; r)$ is contained in $\bar{K}(z_0; \rho)$, where

$$z_0 = \alpha \frac{1-r^2}{1-\alpha^2 r^2} \quad \text{and} \quad \rho = \frac{r(1-\alpha^2)}{1-\alpha^2 r^2}.$$

Hint: Consider $\Omega(z) = \dfrac{\omega(z) - \alpha}{1 - \alpha \omega(z)}$.

6.2.8. (cont.) Show that any point of $\bar{K}(z_0; \rho)$ is a value of a certain function ω taken at a point of $\bar{K}(0; r)$.

6.2.9. (cont.) Find the set Ω_r of all possible values taken in $K(0; r)$ by functions ω analytic in $K(0; 1)$ and such that $\omega(0) \geqslant 0$, $|\omega(z)| \leqslant 1$.

6.2.10. If Q is a polynomial of degree n with complex coefficients and $|xQ(x)| \leqslant M$ for $x \in [-1, 1]$, show that

$$|Q(z)| \leqslant M(1 + \sqrt{2})^{n+1}, \quad z \in K(0; 1).$$

Hint: Cf. Exercise 6.1.10.

6.2.11. If Q is a polynomial of degree n and

$$|(z-\eta)Q(z)| \leqslant M, \quad z \in C(0; 1) \quad (\eta \text{ is a constant, } |\eta| = 1),$$

show that

$$|Q(z)| \leqslant \tfrac{1}{4}(n+2)^2 M, \quad z \in K(0; 1).$$

6.3. SUBORDINATION

Suppose f, F are analytic in $K(0; R)$. The function f is said to be *subordinate* to F in $K(0; r)$, $r \leqslant R$, if there exists a function ω analytic in $K(0; r)$ such that $\omega(0) = 0$, $|\omega(z)| < r$ in $K(0; r)$ and $f = F \circ \omega$ in $K(0; r)$. If f is subordinate to F in $K(0; r)$, we write $f \prec_r F$. In the particular case $r = 1$ we shall write $f \prec F$. If $f \prec_r F$, then obviously any value of f taken in $K(0; r)$ is also a value of F taken in the same disk.

6.3.1. If $r_1 < r_2 < R$ and $f \prec_{r_2} F$, show that $f \prec_{r_1} F$.

6.3.2. If $f \prec_r F$, show that $|f'(0)| \leqslant |F'(0)|$.

6.3.3. If $f \prec_R F$, show that $M(r, f) \leqslant M(r, F)$, $r \in (0, R)$.

6.3.4. Suppose F is analytic and univalent in $K(0; 1)$ and $D = F(K(0; 1))$, $w_0 = F(0)$. If f is analytic in $K(0; 1)$, $f(K(0; 1)) \subset D$ and $f(0) = w_0$, show that $f \prec F$.

6.3. SUBORDINATION

6.3.5. If f is analytic in $K(0; 1)$, $f(0) = 0$ and $|\operatorname{re} f(z)| < 1$ in $K(0; 1)$, show that

(i) $|f'(0)| \leq \dfrac{4}{\pi}$;

(ii) $|f(z)| \leq \dfrac{2}{\pi} \log \dfrac{1+|z|}{1-|z|}$.

6.3.6. If f is meromorphic in $K(0; 1)$, $f(0) = 0$ and all the values of f lie outside $\bar{K}(a; r)$, show that
$$|f'(0)| \leq r \cot^2 \alpha, \qquad \alpha = \arcsin(r/|a|).$$

6.3.7. Let $\mathscr{P}$ be the class of all functions p analytic in $K(0; 1)$ and such that $p(0) = 1$, $\operatorname{re} p(z) > 0$ in $K(0; 1)$. Show that for a given, fixed z ($0 < |z| = r < 1$) the values $p(z)$ cover the whole disk with diameter $[(1-r)(1+r)^{-1}, (1+r)(1-r)^{-1}]$ as p ranges over $\mathscr{P}$.

6.3.8. Given $z \in K(0; 1)$ find precise estimates of $|p(z)|$ and $\arg p(z)$ valid for all $p \in \mathscr{P}$.

6.3.9. Given $z \in K(0; 1)$ find precise estimates of $\operatorname{re} p(z)$ and $\operatorname{im} p(z)$ valid for all $p \in \mathscr{P}$.

6.3.10. If f is analytic, does not vanish and $|f(z)| < M$ in $K(0; 1)$, show that
$$|f(z)| \leq |f(0)|^{\frac{1-|z|}{1+|z|}} M^{\frac{2|z|}{1-|z|}}, \qquad z \in K(0; 1).$$
Hint: Consider $\log[M/f(z)]$.

6.3.11. Suppose $\{f_n\}$ is a sequence of functions analytic and non-vanishing in $K(0; 1)$ and such that $|f_n(z)| < M$ for all n and all $z \in K(0; 1)$. If $\sum_{n=1}^{\infty} |f_n(0)| < +\infty$, show that $\sum [f_n(z)]^2$ is uniformly and absolutely convergent in $\bar{K}(0; \tfrac{1}{3})$.

6.3.12. Suppose f is analytic in $K(0; 1)$ and $f(0) = 0$. If f does not take the value α ($0 < \alpha < 1$) and satisfies $|f(z)| < 1$ in $K(0; 1)$, show that
$$|f'(0)| \leq \left(2\alpha \log \dfrac{1}{\alpha}\right) \Big/ (1-\alpha^2).$$

6.3.13. Let F be an analytic function which maps the unit disk 1:1 onto a convex domain D so that $F(0) = 0$, $F'(0) = 1$. If f is an odd function subordinate to F in $K(0; 1)$, show that
$$|f'(z)| \leq \dfrac{1}{1-|z|^2}.$$

Hint: Consider $\varphi(z) = \dfrac{1}{2}\left[f\left(\dfrac{z+\alpha}{1+\bar{\alpha}z}\right) + f\left(\dfrac{z-\alpha}{1-\bar{\alpha}z}\right)\right]$, $0 < |\alpha| < 1$.

6.3.14. If $f(z) = a_1 z + a_2 z^2 + \ldots$, $F(z) = A_1 z + A_2 z^2 + \ldots$ are analytic in $K(0; 1)$ and $f \prec F$, show that

$$|a_2| \leqslant \max(|A_1|, |A_2|).$$

Hint: Cf. Exercise 6.2.5.

6.4. THE MAXIMUM PRINCIPLE FOR HARMONIC FUNCTIONS

6.4.1. If u is harmonic in a domain D and has a local extremum at $z_0 \in D$, show that $u = \text{const}$ in D.

6.4.2. If u is harmonic in the finite plane and has a finite limit as $z \to \infty$, show that $u = \text{const}$.

6.4.3. Suppose u is a non-constant, harmonic function in a domain D which has a continuous extension on $\bar{D}$. If $m = \inf u(z)$, $M = \sup u(z)$, show that $m < u(z) < M$ in D.

6.4.4. Suppose u is a non-constant, harmonic function in a domain D and there exists a constant M such that for any $\varepsilon > 0$ and any ζ on the boundary of D there exists a neighborhood $K(\zeta; \rho(\varepsilon))$ of ζ such that $u(z) < M + \varepsilon$ for any $z \in D \cap K(\zeta; \rho(\varepsilon))$. Show that $u(z) < M$ for all $z \in D$.

6.4.5. Suppose $f(z) = z^{-1} + a_0 + a_1 z + \ldots$ is analytic and univalent in $K(0; 1) \setminus 0$ and D_f is the set of all values taken by f. Suppose $F(z) = Az^{-1} + A_0 + A_1 z + \ldots$ is meromorphic in $K(0; 1)$ and $D_F \subset D_f$. Show that either $|A| > 1$, or else $F(z) = f(ze^{i\alpha})$.

Hint: Consider $h(z) = \log|z| - \log|f^{-1}(F(z))|$.

6.4.6. If u is harmonic in $\{z: R_1 < |z| < R_2\}$ and $A(r) = \sup\limits_{\theta} u(re^{i\theta})$, show that $A(r)$ is a convex function of $\log r$.

Hint: Find a function harmonic in $\{z: r_1 < |z| < r_2\}$ equal to $A(r_k)$ on $C(0; r_k)$, $k = 1, 2$.

6.4.7. Let f be analytic and univalent in $K(0; R)$ with $f(0) = 0$. If the image curve of $C(0; r)$, $0 < r < R$, is starshaped w.r.t. 0 (i.e. $\arg f(re^{i\theta})$ is an increasing function of θ), show that also the image curves of all $C(0; \rho)$, $\rho < r$, are starshaped w.r.t. 0.

Hint: Cf. Exercise 1.1.20 (i).

CHAPTER 7

Analytic Continuation. Elliptic Functions

7.1. ANALYTIC CONTINUATION

A function f analytic in a domain D defines an *analytic element*, or a *function element* (f, D). If (f_1, D_1), (f_2, D_2) are such that $D_1 \cap D_2$ is non-empty and $f_1(z) = f_2(z)$ for all $z \in D_1 \cap D_2$, then the function elements (f_1, D_1), (f_2, D_2) are said to be *direct analytic continuations* of each other and $(f, D_1 \cup D_2)$ with $f(z) = f_j(z)$ for $z \in D_j$ ($j = 1, 2$) sets up a new analytic element. It may, however, happen that $D_1 \cap D_2$ has several components and $f_1 = f_2$ in one component, whereas $f_1 \neq f_2$ in some other component. Then we say that (f_j, D_j), $j = 1, 2$, are *single-valued branches* of a certain global (generally multi-valued) analytic function.

Two function elements (f, D), (F, Δ) are said to be *analytic continuations* of each other, if there exists a finite chain $\{(f_k, D_k)\}$, $k = 0, 1, \ldots, n$, of function elements such that $(f_0, D_0) = (f, D)$, $(f_n, D_n) = (F, \Delta)$ and (f_k, D_k) is a direct analytic continuation of (f_{k-1}, D_{k-1}), $k = 1, \ldots, n$. All possible analytic continuations of a certain function element set up a *complete analytic function*.

7.1.1. Show that the sum of the power series

$$\log 2 - \frac{1-z}{2} - \frac{1}{2}\left(\frac{1-z}{2}\right)^2 - \frac{1}{3}\left(\frac{1-z}{2}\right)^3 - \cdots$$

is a direct analytic continuation of the sum of the power series $z - \frac{1}{2}z^2 + \frac{1}{3}z^3 - \cdots$

7.1.2. Verify that the function elements corresponding to the sums of the following power series and their disks of convergence:

$$1+z+z^2+\cdots, \qquad \frac{1}{1-i}\left[1+\frac{z-i}{1-i}+\left(\frac{z-i}{1-i}\right)^2+\cdots\right]$$

are direct analytic continuations of each other.

7.1.3. Verify that the sums of power series:
$$f_1(z) = z + \tfrac{1}{2}z^2 + \tfrac{1}{3}z^3 + \ldots,$$
$$f_2(z) = \pi i - (z-2) + \tfrac{1}{2}(z-2)^2 - \tfrac{1}{3}(z-2)^3 + \ldots$$
have disjoint disks of convergence D_1, D_2 but the corresponding function elements (f_1, D_1), (f_2, D_2) are analytic continuations of each other.

7.1.4. Give an example of two function elements (f_1, D_1), (f_2, D_2) such that $D_1 \cap D_2$ has two components and $f_1 = f_2$ in one component, whereas $f_1 \neq f_2$ in the other component.

7.1.5. Show that there exists an analytic continuation of the analytic element determined by the power series $f(z) = 1 + \tfrac{1}{2}z + \tfrac{1}{3}z^2 + \ldots$ and its disk of convergence which is analytic in $K(0; 1) \setminus 0$ and has 0 as a simple pole.

7.1.6. If (f, D) is an analytic continuation of the function element $(\sum_{n=0}^{\infty} z^{2^n}, K(0; 1))$, show that $D \subset K(0; 1)$.

7.1.7. If the radius of convergence of $\sum_{n=0}^{\infty} a_n z^n$ equals to 1 and all a_n are non-negative, show that $(\sum_{n=0}^{\infty} a_n z^n, K(0; 1))$ has no direct analytic continuation (f, D) with $1 \in D$.

7.1.8. If $\sum c_n$ is a convergent series with positive terms and $\{w_n\}$ is the sequence of all rational numbers, show that
$$f(z) = \sum_{n=1}^{\infty} c_n (z - w_n)^{-1}$$
is analytic in the upper half-plane H_+ and also in the lower half-plane H_-. Also prove that the function elements (f, H_+), (f, H_-) are not analytic continuations of each other.

7.1.9. Give an example of a function f analytic in $K(0; 1)$ and continuous in $\overline{K(0; 1)}$ such that for any direct analytic continuation (f_1, D) of $(f, K(0; 1))$ we have $D \subset K(0; 1)$.

7.2. THE REFLECTION PRINCIPLE

Suppose D is a domain situated in the upper half-plane whose boundary contains an open segment γ of the real axis and D^* is the domain obtained by reflection of D w.r.t. the real axis. If f is analytic in D and has a continuous

7.2. THE REFLECTION PRINCIPLE

extension on $D \cup \gamma$ such that f assumes real values of γ, then the function: $F(z) = f(z)$ for $z \in D \cup \gamma$, $F(z) = \overline{f(\bar{z})}$ for $z \in D^*$ is a direct analytic continuation of f on $D \cup \gamma \cup D^*$. If γ and $f(\gamma)$ are circular arcs, then after suitable linear transformations we can reduce this case to the case just considered, it may, however, happen that the analytic continuation becomes a meromorphic function. If D is a Jordan domain, f is univalent in D and $f(D)$ is again a Jordan domain, then the assumption of continuity of f on $D \cup \gamma$ can be dropped since f has necessarily a homeomorphic extension on $\overline{D}$.

7.2.1. If f is an entire function and takes real values on the real axis and imaginary values on the imaginary axis, show that f is odd.

7.2.2. If f is meromorphic in $K(0; 1+h)$, $h > 0$, and $|f(z)| = 1$ on $C(0; 1)$, show that f is a rational function.

7.2.3. Suppose f is meromorphic in a domain D symmetric w.r.t. the real axis and real on it. If a is a pole of f and $A = \mathrm{res}(a; f)$, show that $\mathrm{res}(\bar{a}; f) = \overline{A}$.

7.2.4. Suppose f is meromorphic in a domain D symmetric w.r.t. an arc C of $C(a; R)$ and real on C. If $b \ne a$ is a pole of order n with singular part $\sum_{k=1}^{n} c_k(z-b)^{-k}$, show that

$$\mathrm{res}(b^*; f) = \sum_{k=1}^{n} (-1)^k \frac{k\bar{c}_k(b^*-a)^{k+1}}{R^{2k}},$$

b^* being the reflection of b w.r.t. C.

7.2.5. If f is analytic in a domain D symmetric w.r.t. the real axis, show that $f = f_1 + if_2$, where f_1, f_2 are analytic in D and real on the real axis.

7.2.6. If f is analytic in $K(0; 1)$ except for a simple pole z_1, continuous in $\overline{K}(0; 1)$ and real on $C(0; 1)$, show that $f(z) = (Az^2 + Bz + \overline{A})[(z-z_1)(1-\bar{z}_1 z)]^{-1}$ with real B.

7.2.7. Find the 1:1 conformal mapping of the upper half of $K(0; 1)$ onto the lower half-plane such that the points $-2, \infty, 2$ are image points of $-1, 0, 1$ resp.

7.2.8. Find the 1:1 conformal mapping of the upper half of $K(0; 1)$ onto the upper half-plane $\{w: \mathrm{im}\, w > 0\}$ carrying the diameter $(-1, 1)$ into the ray $(-\frac{1}{4}, +\infty)$ and its center $z = 0$ into $w = 0$.

7.2.9. If f is analytic in $K(0; 1)$, continuous in $\bar{K}(0; 1)$ and maps $K(0; 1)$ onto a n-sheeted unit disk (i.e. the equation $f(z) = a$ has exactly n roots in $K(0; 1)$ for any $a \in K(0; 1)$), show that f is rational. Find its general form.

7.2.10. If f maps $1:1$ conformally the rectangle R onto a rectangle R' so that the corners of R and R' correspond, show that f is a similarity and the ratio of sides is the same for both rectangles.

7.2.11. If f maps $1:1$ conformally the annulus $\{z: 1 < |z| < r\}$ onto $\{w: 1 < |w| < r'\}$ so that $|f(z)| \to 1$ as $|z| \to 1$, show that $f(z) = e^{i\alpha}z$ with real α and $r' = r$.

7.2.12. Suppose f is analytic and univalent in the unit disk and maps it onto a domain D symmetric w.r.t. the straight line L through the origin. Show that $L \cap D$ is the image of a certain diameter of $K(0; 1)$, if $f(0) = 0$.

7.2.13. If f maps $1:1$ conformally the upper half-plane $H_+ = \{z: \operatorname{im} z > 0\}$ onto a circular polygon D so that the points $z_1, \ldots, z_n$ of the real axis correspond to the corners of D, show that f can be continued analytically along any path omitting the points $z_1, \ldots, z_n$.

Also prove that any two branches (f_1, H_+), (f_2, H_+) of the complete analytic function so obtained satisfy: $f_1 = (af_2 + b)/(cf_2 + d)$, where a, b, c, d are complex constants.

7.2.14. (cont.) Suppose $\{f, z\} = \left(\dfrac{f''}{f'}\right)' - \dfrac{1}{2}\left(\dfrac{f''}{f'}\right)^2$ is the Schwarzian derivative of f. Verify that the Schwarzian derivative of the multi-valued function f of Exercise 7.2.13 is single-valued.

Hint: Show that $\{(af+b)/(cf+d), z\} = \{f, z\}$.

7.3. THE MONODROMY THEOREM

Suppose D is a domain in the finite plane and a complete analytic function f has the property that a certain function element (f_0, D_0) of f with $D_0 \subset D$ can be continued along any path γ contained in D. Then, according to the monodromy theorem, the continuation of the same element (f_0, D_0) along any two paths γ_1, γ_2 with common end points (one of the common end points being situated in D_0) leads to the same terminal element, whenever γ_1, γ_2 are homotopic with respect to D. Since any two paths γ_1, γ_2 with common end points situated in a simply connected domain D are homotopic w.r.t. D (i.e. can be continuously deformed into each other within D), we obtain for simply connected domains the following theorem: If a function element (f_0, D_0) of a complete

7.3. THE MONODROMY THEOREM

analytic function f can be continued along any path γ contained in a simply connected domain D, there exists in D a single-valued, analytic function coinciding with f_0 in D_0.

Perhaps the most interesting applications of the monodromy theorem are connected with the so-called elliptic modular function $\lambda = \lambda(\tau)$. If T_0 is the domain $\{\tau: 0 < \mathrm{re}\,\tau < 1; |\tau - \frac{1}{2}| > \frac{1}{2}; \mathrm{im}\,\tau > 0\}$, there exists a unique $1:1$ conformal mapping $\lambda = \lambda(\tau)$ of T_0 onto the upper half-plane $\mathrm{im}\,\lambda > 0$ such that the points $\tau = \infty, 0, 1$ correspond to the points $\lambda = 0, 1, \infty$. By using the reflection principle one can continue $\lambda(\tau)$ onto the whole upper half-plane $\mathrm{im}\,\tau > 0$ whose image will be the infinitely many sheeted λ-plane punctured at $0, 1$. Since $\lambda'(\tau) \neq 0$, there exists locally at any point $\lambda \neq 0, 1$ an inverse function element $\tau = \tau(\lambda)$ which can be continued along any arc in the λ-plane not passing through $0, 1$.

7.3.1. If a function element $(f, D), D \subset K(0; 1) \setminus 0$, can be continued along any path situated in the punctured unit disk $K(0; 1) \setminus 0$, show that each function element obtained by continuations within this domain has the form $F(\log z)$ where $F(w)$ is analytic in the left half-plane $\{w: \mathrm{re}\,w < 0\}$.

7.3.2. If an entire function g taking all finite values has a non-vanishing derivative, show that it is a similarity transformation, i.e. $g(z) = az+b$ $(a \neq 0)$.

7.3.3. (cont.) Suppose f is an entire function such that $f(z)f'(z) \neq 0$ for all z. Prove that $f(z) = \exp(az+b)$ $(a \neq 0)$.

7.3.4. Let f be analytic in a domain D and let $K = K(z_0; r) \subset D$. Prove that the function element $\left(\int_{[z_0, z]} f(\zeta)\,d\zeta, K \right)$ can be continued along any regular arc γ starting at z_0 and situated in D. If γ joins z_0 to Z in D, show that the value $F(Z)$ of the terminal branch equals $\int_\gamma f(\zeta)\,d\zeta$.

7.3.5. (cont.) Prove Cauchy's theorem in homotopic form: If f is analytic in D and the arcs γ_1, γ_2 with common end points $z_0, Z \in D$, are homotopic w.r.t. D, then $\int_{\gamma_1} f(z)\,dz = \int_{\gamma_2} f(z)\,dz$.

7.3.6. Let f be a function analytic and non-vanishing in a simply connected domain D. Verify the existence of an analytic logarithm of f in D by using the monodromy theorem.

7.3.7. Show that any entire function g omitting the values $0, 1$ reduces to a constant.

7.3.8. Prove that any nonconstant entire function assumes all values in the finite plane except at most one.

7.3.9. Show that $Q(w) = \lambda\left(\dfrac{\log w}{\pi i}\right)$ is analytic in the unit disk $\{w: |w| < 1\}$, does not depend on the choice of $\arg w$ and has the form: $Q(w) = \sum_{k=1}^{\infty} A_k w^k$ with $A_1 > 0$ and all A_k real. Also prove that $\lambda(\tau)e^{-i\pi\tau} \to A_1$ as $\operatorname{im}\tau \to +\infty$.

7.3.10. (cont.) Verify that Q has in some disk $K(0; \delta)$ a univalent inverse function which can be continued along any path in the open plane omitting the points 0, 1. Also verify that $A_1 > 1$.
[*Remark*: one can prove that $A_1 = 16$, cf. [1], or [24]].

7.3.11. Suppose f is analytic in the unit disk, $f(0) = 0$, and $f(z) \neq 0, 1$ for $0 < |z| < 1$. Show that f is subordinate to Q in the unit disk.

7.3.12. Let $f(z) = a_1 z + a_2 z^2 + \dots$ be analytic in the unit disk and let $f(z)$ be different from 0, 1 in $0 < |z| < 1$. Show that $|a_1| \leqslant 16$.

7.4. THE SCHWARZ–CHRISTOFFEL FORMULAE

Let D be a simply connected domain whose boundary is a closed polygonal line L without self-intersections. There exists a function f analytic in the upper half-plane $\{z: \operatorname{im} z > 0\}$, continuous on the real axis (in spherical metric) and mapping the upper half-plane onto D. It has the following form:

$$(7.4\text{A}) \qquad f(z) = C \int_0^z \prod_{k=1}^{n} (\zeta - x_k)^{\alpha_k - 1} d\zeta + C_1,$$

where C, C_1 are complex constants, $x_1 < x_2 < \dots < x_n$ are points on the real axis corresponding to consecutive vertices $w_1, w_2, \dots, w_n$ and $\alpha_k \pi$ are the interior angles of D at w_k, $k = 1, 2, \dots, n$. The integral is a line integral along any regular curve with end points $0, z$ situated (apart from the end point 0) in the upper half-plane. On the other hand, the function F mapping $\{z: \operatorname{re} z > 0\}$ on the outside of D and such that $F(b) = \infty$, has the form:

$$(7.4\text{B}) \qquad F(z) = C \int_0^z (\zeta - b)^{-2} (\zeta - \bar{b})^{-2} \prod_{k=1}^{n} (\zeta - x_k)^{\beta_k - 1} d\zeta + C_1$$

where $x_1 < x_2 < \dots < x_n$ are the points corresponding to the vertices $w_1, w_2, \dots, w_n$ and $\beta_k \pi$ are exterior angles of D.

We have obviously: $\sum_{k=1}^{n} \alpha_k = n-2$, $\sum_{k=1}^{n} \beta_k = n+2$. The formula (7.4A) also holds when D is an unbounded domain whose boundary is a polygonal line L

7.4. THE SCHWARZ–CHRISTOFFEL FORMULAE

without self-intersections and two sides of L are half-lines.

Sometimes a formal application of Schwarz–Christoffel formulas in a degenerate case (e.g. $\alpha_k = 2$, or there exist more than two boundary rays) enables us to guess the mapping function. The obtained mapping needs verification which is occasionally possible.

7.4.1. If one of the vertices of D is the image point of $x_n = \infty$, show that the mapping function has the form

$$f(z) = C \int_0^z \prod_{k=1}^{n-1} (\zeta-x_k)^{\alpha_k-1} d\zeta + C_1.$$

7.4.2. Find the function mapping the upper half-plane onto the inside of an equilateral triangle. Using the formula

$$\int_0^1 u^{p-1}(1-u)^{q-1} du = \Gamma(p)\Gamma(q)/\Gamma(p+q) \quad (p>0, q>0)$$

express the length a of the sides in terms of Γ function.

7.4.3. Map the upper half-plane onto a rhombus with angles $\pi\alpha$, $\pi(1-\alpha)$ so that its vertices correspond to $z = 0, \mp 1, \infty$. Find the side-length.

7.4.4. (cont.) Verify that the upper semicircle and the positive imaginary axis correspond to the diagonals of the rhombus. Also find the preimage of its center.

7.4.5. Map 1:1 conformally the upper half-plane $\{z: \mathrm{im}\, z > 0\}$ onto the part of the upper half-plane lying outside the rectangle: $-k \leqslant \mathrm{re}\, w \leqslant 0$, $0 \leqslant \mathrm{im}\, w \leqslant h$, so that the points $\mp a, \mp \delta$ correspond to the vertices. Discuss the limiting case: $\delta \to 0$.

7.4.6. Map 1:1 conformally the upper half-plane $\mathrm{im}\, z > 0$ onto the domain containing the first quadrant whose boundary consists of two rays: $\{w: \mathrm{re}\, w \leqslant 0, \mathrm{im}\, w = 1\}$, $\{w: \mathrm{re}\, w \geqslant 0, \mathrm{im}\, w = -1\}$ and a segment $[-i, i]$. The points $z = -1, 1$ should correspond to $w = i, -i$.

7.4.7. Map 1:1 conformally the upper half-plane onto a part of it with the boundary consisting of two rays: $\{w: \mathrm{re}\, w \leqslant 0, \mathrm{im}\, w = \pi\}$, $[-\pi\cot\delta, +\infty)$ and a segment $[-\pi\cot\delta, \pi i]$, where $0 < \delta < \tfrac{1}{2}\pi$. The points $z = -1, 0$ should correspond to $w = \pi i, -\pi\cot\delta$. Discuss the limiting case $\delta \to 0$.

7.4.8. (cont.) Find the mapping of the right half-plane $\mathrm{re}\, \zeta > 0$ onto the w-plane slit along the rays: $\mathrm{im}\, w = \mp\pi$, $\mathrm{re}\, w \leqslant 0$, which carries the points $\zeta = \mp i$ into $w = \mp\pi i$.

7.4.9. Find the mapping of the upper half-plane $\operatorname{im} z > 0$ onto the strip domain $0 < \operatorname{im} w < \pi$ slit along the ray $\operatorname{im} w = v_1$, $\operatorname{re} w \leqslant 0$ $(0 < v_1 < \pi)$.

7.4.10. Find the mapping of the upper half-plane $\operatorname{im} z > 0$ onto the first quadrant slit along the ray: $\operatorname{im} w = \pi$, $\operatorname{re} w \geqslant h$, which carries the points $z = 0$, a^2 into $w = 0$, $h + \pi i$ $(a > 1, h > 0)$.

7.4.11. Suppose F is a function analytic and non-vanishing in $K(0; \delta)$ which has a constant argument on $(-\delta, \delta)$. Prove that the function

$$W(z) = \int_{z_0}^{z} \zeta^{-1} F(\zeta) d\zeta$$

where $z_0 \in K(0; \delta)$, $\operatorname{im} z_0 > 0$, is analytic in $K(0; \delta) \setminus (-i\delta, 0]$ and maps the radii $(-\delta, 0)$, $(0, \delta)$ onto two parallel rays the distance between them being equal to $\pi |F(0)|$.

7.4.12. Find the image domain of the upper half-plane under the mapping

$$w = H(z) = \int_{z_0}^{z} \frac{d\zeta}{\zeta(\zeta - b)\sqrt{\zeta - a}} \quad (0 < a < b, \operatorname{im} z_0 > 0).$$

Find the distance between two parallel rays in each pair of boundary rays and evaluate b so as to make both distances equal.

7.4.13. Show that the mapping

$$w = \int_{0}^{z} \zeta^{-1}(\zeta - a)^{-1/3} d\zeta$$

carries the upper half-plane $\operatorname{im} z > 0$ into a domain bounded by a polygonal line consisting of two parallel rays and a segment such that interior angles are equal $\tfrac{2}{3}\pi$, $\tfrac{1}{3}\pi$. Evaluate the distance between the rays.

7.4.14. Find the mapping of the upper half-plane $\operatorname{im} z > 0$ onto the domain $D = \mathbf{C} \setminus (\overline{K}(i; 1) \cup \overline{Q}_1)$ where Q_1 is the first quadrant.
Hint: Map first D under $W = w^{-1}$.

7.4.15. Find the mapping of the upper half-plane $\operatorname{im} z > 0$ onto
$$D_\mu = \{w: 0 < \operatorname{im} w < 1\} \cup \{w: 0 \leqslant \arg(w - i) < \mu\pi\} \quad \text{where} \quad 0 < \mu < 1.$$
Assume that $z = -1, 0$ correspond to $w = i, \infty$ resp.

7.4.16. (cont.) Find the mapping of the right half-plane $\operatorname{re} Z > 0$ onto $D_\mu \cup (-\infty, +\infty) \cup D_\mu^*$, where D_μ^* is a reflection of D_μ w.r.t. the real axis.

7.4.17. Show that the function mapping the inside of the unit disk onto the

7.4. THE SCHWARZ–CHRISTOFFEL FORMULAE

inside of a simple, closed polygonal line has the same form as in (7.4A), x_k being replaced by the points on the unit circle.

7.4.18. Show that the mapping

$$w = \int_0^t \frac{d\tau}{(1-\tau^n)^{2/n}}$$

carries the unit disk into the inside of a regular n-angle. Evaluate its perimeter.

7.4.19. Find the mapping of the unit disk $|t| < 1$ onto the n-pointed star (i.e. $2n$-angle whose all sides and alternate angles are equal).

7.4.20. Map 1:1 conformally the unit disk $|t| < 1$
 (i) onto a pentagram (a five-pointed star obtained by extending the sides of a regular pentagon);
 (ii) onto a Solomon's seal (a six-pointed star formed of two congruent equilateral triangles placed one upon another).

7.4.21. Show that the mapping

$$w = A \int_0^z \prod_{k=1}^n \left[\frac{1-\bar{z}_k \zeta}{(1-\bar{\zeta}_k \zeta)^{1+\beta_k}} \right] d\zeta + B$$

where $z_k = \exp i\varphi_k$, $\zeta_k = \exp i\psi_k$, $0 \leqslant \varphi_1 < \psi_1 < \varphi_2 < \psi_2 < \ldots < \psi_n < 2\pi$, $\beta_k > 0$, $\sum_{k=1}^n \beta_k = 2$, carries the unit disk $|z| < 1$ into the w-plane slit along the rays $l_1, l_2, \ldots, l_n$. Verify that z_k correspond to the end-points of l_k and β_k are angles between l_{k+1} and l_k ($l_{n+1} = l_1$).

7.4.22. Show that the mapping

$$w = \int_0^z \frac{dt}{(1-t^4)\sqrt{1+t^4}}$$

carries the unit disk $|z| < 1$ into the domain

$$\{w \colon |\operatorname{re} w| < \tfrac{1}{2}a\} \cup \{w \colon |\operatorname{im} w| < \tfrac{1}{2}a\}.$$

Evaluate a.

7.4.23. Prove that the function f mapping 1:1 conformally the unit disk $|z| < 1$ onto a convex domain bounded by a polygonal line and such that $f(0) = 0$ has the form

$$f(z) = A \int_0^z \prod_{k=1}^n (1-t \exp i\theta_k)^{-\alpha_k} dt + B$$

where $0 \leqslant \theta_1 < \theta_2 < \ldots < \theta_n < 2\pi$, $\alpha_k > 0$, $\sum_{k=1}^{n} \alpha_k \leqslant 2$.

Verify that
$$|f'(z)| \leqslant |f'(0)|(1-|z|)^{-2}, \quad |f(z)| \leqslant |f'(0)||z|(1-|z|)^{-1}.$$

7.4.24. A function f analytic in a disk K is said to be close-to-convex, if there exists in K a univalent, analytic function Φ mapping K onto a convex domain and such that $\mathrm{re} f'(z)/\Phi'(z) > 0$ in K.

(i) Show that each close-to-convex function is univalent in K.

Hint: Consider the mapping $f \circ \Phi^{-1}$ and cf. Exercise 3.1.14.

(ii) If $\sum_{k=1}^{n} (\psi_k - \varphi_k) = \pi$ in Exercise 7.4.21, show that the mapping function is close-to-convex.

[*Remark*: Exercise 7.4.24 (i), as well as its counterpart for a half-plane is a convenient tool in proving the univalence of mappings obtained by a formal application of the Schwarz–Christoffel formulas.]

7.4.25. If $|\zeta_1| = |\zeta_2| = \ldots = |\zeta_n| = 1$ and $\alpha_k > -1$ $(k = 1, 2, \ldots, n)$, $\sum_{k=1}^{n} \alpha_k = 2$, $\sum_{k=1}^{n} \alpha_k \zeta_k = 0$, show that

$$W(z) = \int_{1}^{z} \zeta^{-2} \prod_{k=1}^{n} (1 - \zeta \overline{\zeta}_k)^{\alpha_k} d\zeta$$

is continuous on $C(0; 1)$ and maps it onto a closed polygonal line L with interior angles $(1-\alpha_k)\pi$. If L has no self-intersections, show that W is meromorphic and univalent in $K(0; 1)$ and maps it onto the outside of L.

7.4.26. Find the 1:1 conformal mapping of the outside of the unit disk onto the outside of a triangle with exterior angles $\alpha_k \pi$ ($k = 1, 2, 3$) such that the points η_k on $C(0; 1)$ correspond to the vertices and the points at infinity to each other. Find the relation between α_k and η_k.

7.4.27. If f maps 1:1 conformally the outside of the unit disk onto the outside of an equilateral triangle T so that $f(\infty) = \infty$, show that the preimages of vertices of T also form an equilateral triangle.

7.4.28. Find the 1:1 conformal mapping f of the outside of the unit disk onto the outside of a regular n-angle such that $f(\infty) = \infty$ and $f(\eta^k) = \eta^k$ ($\eta = \exp(2\pi i/n)$, $k = 0, 1, \ldots, n-1$).

7.5. JACOBIAN ELLIPTIC FUNCTIONS sn, cn, dn

If $0 < k < 1$, then the Schwarz–Christoffel integral

$$u = u(z) = \int_0^z [(1-t^2)(1-k^2t^2)]^{-1/2} dt$$

maps 1:1 conformally the upper half-plane $\operatorname{im} z > 0$ onto the inside of the rectangle with vertices $\mp K$, $\mp K + iK'$. We have: $u(0) = 0$, $u(\mp 1) = \mp K$, $u(\mp k^{-1}) = \mp K + iK'$, $u(\infty) = iK'$. The inverse function $z = \operatorname{sn}(u, k)$ can be continued analytically by reflections all over the open plane and becomes a meromorphic function with two periods $4K, 2iK'$. We have:

$$K = K(k) = \int_0^1 [(1-t^2)(1-k^2t^2)]^{-1/2} dt = \int_0^{\pi/2} [1-k^2\sin^2\varphi]^{-1/2} d\varphi = F(k, \pi/2)$$

F being the complete Legendre elliptic integral.

The functions cn, dn are defined by the equations:

$$\operatorname{sn}^2 u + \operatorname{cn}^2 u = 1, \quad k^2 \operatorname{sn}^2 u + \operatorname{dn}^2 u = 1.$$

All the roots of the equations $\operatorname{sn}^2 u = 1$, $\operatorname{sn}^2 u = k^{-2}$ are double, hence cn, dn have no branch points and are meromorphic in the finite plane.

7.5.1. If $u = \int_0^z (6 - 5t^2 + t^4)^{-1/2} dt$, show that

$$z = \sqrt{2} \operatorname{sn}(u\sqrt{3}, \sqrt{\tfrac{2}{3}}).$$

7.5.2. If $u = \int_0^z (1 - t^4)^{-1/2} dt$, show that

$$z = \operatorname{cn}[\sqrt{2}\,(u-K),\, 2^{-1/2}].$$

7.5.3. Let k' be the complementary modulus w.r.t. k, i.e. $k' = \sqrt{1-k^2}$. Show that $K(k') = K'(k)$ and $K(k) = K'(k')$.

7.5.4. Derive the expansions:
(i) $\operatorname{sn}(u, k) = u - (1+k^2)u^3/3! + (1+14k^2+k^4)u^5/5! - \ldots$;
(ii) $\operatorname{cn}(u, k) = 1 - u^2/2! + (1+4k^2)u^4/4! + \ldots$;
(iii) $\operatorname{dn}(u, k) = 1 - k^2u^2/2! + k^2(4+k^2)u^4/4! + \ldots$;
and find the radius of convergence in each case.

7.5.5. Express $K(k)$ as the sum of a power series. Verify that $K'(k)/K(k)$ is a strictly decreasing function of $k \in (0, 1)$ which decreases from $+\infty$ to 0.

7.5.6. Show that $\operatorname{sn}(u+iK') = (k\operatorname{sn} u)^{-1}$. Find the initial terms of the Laurent expansion of $\operatorname{sn} u$ in the neighborhood of $u = iK'$.

7.5.7. Determine the periods, zeros and poles of $\operatorname{sn}(iv, k')$ as depending on k.

7.5.8. Verify the identity $\operatorname{sn}^{-2}(u, k) + \operatorname{sn}^{-2}(iu, k') \equiv 1$.

7.5.9. Find an elliptic function $f(u)$ with periods $2, 2i$ which has simple zeros at points whose coordinates are either both odd, or both even and has simple poles at points whose one coordinate is even and the other one odd. Discuss the uniqueness.

7.5.10. If z moves round the rectangle with vertices $iK', O, K, K+iK'$, show that $w = \operatorname{cn} z/(1+\operatorname{sn} z)$ moves round a quadrant of the unit disk.

7.6. THE FUNCTIONS σ, ζ, $\wp$ OF WEIERSTRASS

The function $\sigma(z) = \sigma(z; \omega_1, \omega_2)$ of Weierstrass is an entire function defined as the infinite product

$$\sigma(z) = \sigma(z; \omega_1, \omega_2) = z \prod_k{}' \left(1 - \frac{z}{\Omega_k}\right) \exp\left[\frac{z}{\Omega_k} + \frac{1}{2}\left(\frac{z}{\Omega_k}\right)^2\right],$$

where $\Omega_k = m_k\omega_1 + n_k\omega_2$, $(m_k; n_k)$ is the sequence of all pairs of integers and the prime after the product sign indicates that the pair $(0; 0)$ should be omitted; ω_1, ω_2 are two complex numbers with $\operatorname{im}(\omega_1/\omega_2) \neq 0$. The logarithmic derivative $\sigma'(z)/\sigma(z)$ is the meromorphic function $\zeta(z)$ of Weierstrass. The function $\wp(z) = -\zeta'(z)$ is a meromorphic, doubly-periodic (or elliptic) function with periods ω_1, ω_2 whose only singularities are double poles $m\omega_1+n\omega_2$. We have:

$$\wp(z) = \wp(z; \omega_1, \omega_2) = z^{-2} + \sum_k{}' [(z-\Omega_k)^{-2} - \Omega_k^{-2}].$$

7.6.1. Show that $\sigma(z) = z + c_5 z^5 + c_7 z^7 + \ldots$

7.6.2. If ω_1 is real and ω_2 purely imaginary, show that σ is real on the real axis and purely imaginary on the imaginary axis.

7.6.3. If $\operatorname{im} \omega_1 = \operatorname{re} \omega_2 = 0$, show that $\sigma(z) = \overline{\sigma(\bar{z})}$.

7.6.4. Verify that

$$\sigma(u+\omega_k)/\sigma(u) = -\exp[(2u+\omega_k)\zeta(\tfrac{1}{2}\omega_k)] \quad (k = 1, 2).$$

7.6. FUNCTIONS $\sigma, \zeta, \wp$

7.6.5. Show that

(i) $\dfrac{\sigma(2u)}{\sigma^4(u)} = -\wp'(u)$;

(ii) $2\zeta(2u) - 4\zeta(u) = \dfrac{\wp''(u)}{\wp'(u)}$.

7.6.6. If $\zeta(u)$ has poles at $2m\omega + 2n\omega'$ and $\tau = \omega'/\omega$, $h = e^{i\pi\tau}$, $\eta = \zeta(\omega)$, $\operatorname{im} \tau > 0$, show that

$$2\eta\omega = \frac{\pi^2}{6}\left(1 - 24\sum_{k=1}^{\infty}\frac{h^{2k}}{(1-h^{2k})^2}\right).$$

7.6.7. If $1 < |z| < q$, show that

$$\wp(\log z; 2\log q, 2\pi i)$$
$$= \sum_{n=-\infty}^{\infty}(q^{-2n}z + q^{2n}z^{-1} - 2)^{-1} - \frac{1}{12} - 2\sum_{n=1}^{\infty}(q^{2n} + q^{-2n} - 2)^{-1}.$$

7.6.8. Suppose $\wp$ has periods $2a$, $2bi$, where a, b are real and positive. Show that

(i) $\wp$ is real on both axes;

(ii) $\wp$ is also real on straight lines $\operatorname{re} z = an$, $\operatorname{im} z = bn$ $(n = 0, \mp 1, \mp 2, \ldots)$;

(iii) $\wp$ maps $1:1$ conformally any rectangle R: $na < \operatorname{re} z < (n+1)a$, $mb < \operatorname{im} z < (m+1)b$, $m, n = 0, \mp 1, \mp 2, \ldots$ onto the lower, or the upper half-plane.

7.6.9. If $\wp(z) = \wp(z; 2\omega, 2\omega')$, $\operatorname{im}\omega = \operatorname{re}\omega' = 0$ and $\wp(\omega) = e_1$, $\wp(\omega + \omega') = e_3$, $\wp(\omega') = e_2$, show that $e_2 < e_3 < e_1$ and $e_2 < 0 < e_1$. Prove that the converse also holds.

7.6.10. (cont.) Show that

$$[\wp(u+\omega) - e_1][\wp(u) - e_1] = (e_1 - e_2)(e_1 - e_3).$$

7.6.11. If $e_2 < e_3 < e_1$, $k = \sqrt{\dfrac{e_3 - e_2}{e_1 - e_2}}$ and $\wp$ has periods $\dfrac{2K(k)}{\sqrt{e_1 - e_2}}, \dfrac{2iK'(k)}{\sqrt{e_1 - e_2}}$, verify the identity:

$$\wp\left(u(e_1 - e_2)^{-1/2}\right) = e_2 + (e_1 - e_2)\operatorname{sn}^{-2}(u, k).$$

7.6.12. Derive Legendre formula:

$$\eta_1\omega_2 - \eta_2\omega_1 = \mp \pi i$$

where $\eta_k = \zeta(\tfrac{1}{2}\omega_k)$, $k = 1, 2$.

Hint: Integrate $\zeta(z)$ round the parallelogram with center 0 and sides ω_1, ω_2.

7.6.13. Suppose $\sigma(z)$ is the Weierstrass σ-function with zeros $m+n\tau$ (im $\tau > 0$; $m, n = 0, \mp 1, \mp 2, \ldots$) and
$$\vartheta(z) = \exp\{-\eta_1 z^2\}\sigma(z), \quad \text{where} \quad \eta_1 = \zeta(\tfrac{1}{2}) = \sigma'(\tfrac{1}{2})/\sigma(\tfrac{1}{2}).$$
Show that
(i) ϑ is an entire function with the same zeros as σ;
(ii) $\vartheta(z+1) = -\vartheta(z)$, hence ϑ has the period 2;
(iii) $\vartheta(z+\tau) = -\vartheta(z)\exp[-\pi i(2z+\tau)]$.

7.7. CONFORMAL MAPPINGS ASSOCIATED WITH ELLIPTIC FUNCTIONS

7.7.1. Suppose a, b are real and positive, $\{z: -a < \text{re } z < 0, 0 < \text{im } z < b\} = R$ and $w = f(z)$ is the univalent mapping of R onto $\{w: \text{im } w > 0\} \setminus \bar{K}(0; 1)$ such that the points $z = 0$, ib, $ib-a$, $-a$ and $w = \infty$, $-h$, -1, 1 ($h > 1$ depends on a, b) correspond to each other. Use the reflection principle to show that f has a double pole at the origin and is doubly-periodic with periods $4a$, $2ib$.

If $\wp(z) = \wp(z; 4a, 2ib)$ and $e_1 = \wp(2a)$, $e_2 = \wp(ib)$, $e_3 = \wp(2a+ib)$, show that
$$f(z) = [(e_1-e_2)(e_1-e_3)]^{-1/2}[\wp(z)-e_1], \quad h = \sqrt{\frac{e_1-e_2}{e_1-e_3}}.$$

7.7.2. Find the function mapping 1:1 conformally the annulus $1 < |\zeta| < Q$ onto $K(\infty; 1) \setminus (-\infty, -h]$ where $h > 1$. Also find the relation between Q and h.

7.7.3. Show that the mapping $w = \sqrt{k}\,\text{sn}(u, k)$ carries the open segment $(-K+\tfrac{1}{2}iK', K+\tfrac{1}{2}iK')$ into the upper semicircle of $C(0; 1)$.

7.7.4. (cont.) Show that the image domain of the rectangle with corners $\mp K \mp \tfrac{1}{2}iK'$ is the unit disk $|w| < 1$ slit along the radial segments $(-1, -\sqrt{k}]$, $[\sqrt{k}, 1)$.

7.7.5. If $0 < b < a$ and k is such that $\dfrac{K'(k)}{2K(k)} = \dfrac{1}{\pi}\log\dfrac{a+b}{a-b}$ show that the function
$$w = w(z) = \sqrt{k}\,\text{sn}\left[\frac{2K(k)}{\pi}\arcsin\frac{z}{\sqrt{a^2-b^2}}, k\right]$$
is analytic in H, where H is the ellipse with boundary $\dfrac{x^2}{a^2} + \dfrac{y^2}{b^2} = 1$.

Also prove that $w = w(z)$ maps H 1:1 conformally onto the unit disk.

7.7. CONFORMAL MAPPINGS

7.7.6. Given k, $0 < k < 1$, show that the function

$$w = w(z) = \sqrt{k}\,\operatorname{sn}\left(\frac{2Ki}{\pi}\log z, k\right)$$

is analytic in the annulus $A = \{z\colon 1 < |z| < R\}$, $R = \exp(\pi K'/4K)$, and maps it 1:1 onto $K(0;1)\setminus[-\sqrt{k},\sqrt{k}]$.

7.7.7. If $A = \{z\colon 1 < |z| < Q\}$ and k is such that $\dfrac{\pi K'(k)}{2K(k)} = \log Q$, show that

$$W = \frac{1}{k}\left[\operatorname{sn}\left(\frac{iK}{\pi}\log z, k\right)\right]^{-2}$$

maps A 1:1 conformally onto $K(\infty;1)\setminus[k^{-1},+\infty)$.

7.7.8. Find the mapping of the right half-plane $\operatorname{re} w > 0$ slit along $[1, k^{-1}]$ onto $\{z\colon 1 < |z| < Q\}$. Find the relation between k and Q $(0 < k < 1)$.

7.7.9. Find the conformal mapping of the extended plane slit along two circular arcs with end points a_1, a_2 and b_1, b_2 situated on the same circle onto a concentric circular annulus.

7.7.10. Suppose $w = f(z)$ is a 1:1 conformal mapping of the triangle with vertices 0, $2K$, $(1+i)K$ $(K = K(1/\sqrt{2}))$ onto the lower semi-disk of $K(0;1)$ such that the points $z = 0, K, 2K$ and $w = 1, 0, -1$ correspond to each other. Show that $w = \operatorname{cn}(z, 1/\sqrt{2})$.

7.7.11. Given the rectangle $R\colon 0 < x < a$, $0 < y < b$ and a point $\xi+i\eta \in R$. Express in terms of $\sigma(z; 2a, 2ib)$ the function mapping 1:1 conformally R onto $K(0;1)$ so that the origin is the image point of $\xi+i\eta$.

Hint: Use the reflection principle and evaluate zeros and poles of the mapping function.

CHAPTER 8

The Dirichlet Problem

8.1. THE RIEMANN MAPPING THEOREM

If D is a simply connected domain whose boundary contains at least two points, there exists a function f analytic in D which maps D 1:1 conformally onto the unit disk.

This theorem, due to Riemann, has the following completion, first conjectured by Osgood and then proved by Osgood, Taylor and Carathéodory: If D is a Jordan domain (i.e. a domain bounded by a closed Jordan curve) then the mapping function f admits a homeomorphic extension to $\bar{D}$.

8.1.1. Show that the finite plane $\mathbf{C}$ cannot be mapped 1:1 conformally onto the unit disk.

8.1.2. Give an example of a locally univalent mapping f of the unit disk such that $f(0) = 0$, $f'(0) = 1$ and $f[K(0; 1)] = \mathbf{C}$.
Hint: Verify that $q(w) = we^w$ takes every finite value in $G = \mathbf{C} \setminus (-\infty, -1]$.

8.1.3. Suppose D is a Jordan domain and g is a real-valued function harmonic in $D \setminus z_0$, continuous vanishing on fr D and tending to $+\infty$ as $z \to z_0$. Find the univalent function mapping D onto $K(0; 1)$ in terms of g.

8.1.4. Suppose D is a Jordan domain and f_1, f_2 are two mapping functions of Riemann's theorem such that $f_1(z_k) = f_2(z_k)$ at three boundary points z_k of D. Show that $f_1 = f_2$.

8.1.5. Suppose D_k are Jordan domains symmetric w.r.t. circular arcs γ_k ($k = 1, 2$). Suppose, moreover, f is a univalent mapping of D_1 onto D_2 which carries two boundary points z_1, z_2 symmetric w.r.t. γ_1 into w_1, w_2 symmetric w.r.t. γ_2 and one end point of γ_1 into an end point of γ_2. Show that $f(\gamma_1) = \gamma_2$.

8.1.6. Find the images of arcs of $C(1; 1)$, $C(0; 1)$, $\operatorname{re}\tau = \frac{1}{2}$ contained in T_0 under the mapping $\lambda = \lambda(\tau)$, where λ is the modular function. Also evaluate $\lambda(\frac{1}{2}+i\frac{1}{2}\sqrt{3})$, $\lambda(i)$, $\lambda(1+i)$.

8.1.7. Construct by means of Riemann's mapping theorem a univalent function having $C(0; 1)$ as its natural boundary.

8.1.8. Suppose D is a simply connected domain whose boundary contains at least two points and $a \in D$. Show that D can be mapped onto $K(0; r)$ under a univalent function φ so that $\varphi(a) = 0$, $|\varphi'(a)| = 1$. Verify that $r = r(a; D)$ is uniquely determined by D and a.

The number $r(a; D)$ is called the inner radius of D at a.

8.1.9. If f is univalent in $K(0; 1)$, $D = f[K(0; 1)]$ and $a = f(z_0)$, show that $r(a; D) = (1-|z_0|^2)|f'(z_0)|$.

8.1.10. Find the inner radius of:

(i) the disk $K(0; R)$ at a, $|a| < R$;

(ii) the upper half-plane at ih $(h > 0)$;

(iii) the plane slit along a ray at a point on its prolongation whose distance from the end of the ray is equal to d.

8.1.11. If φ is a univalent mapping of a simply connected domain D_0 onto D and $a = \varphi(a_0)$, show that $r(a; D) = |\varphi'(a_0)|r(a_0; D_0)$. Otherwise, the metric $|dw|/r(w; D)$ remains unchanged under $1:1$ conformal mappings. It is called the hyperbolic metric.

8.1.12. Consider all univalent functions in a simply connected domain D_0 which satisfy: $\varphi(a_0) = 0$, $\varphi'(a_0) = 1$, where a_0 is a fixed point of D_0. Show that the area of the image domain has a minimum, if $\varphi(D_0) = K(0; r)$, $r = r(a_0; D_0)$.
Hint: Exercise 4.2.12.

8.1.13. Suppose that G is a domain in the extended plane $\hat{C}$ and $\hat{C} \setminus G$ is a finite union $\bigcup_{k=1}^{n} \Gamma_k$ of n disjoint continua. Prove that G can be mapped $1:1$ conformally onto a bounded domain whose boundary consists of n closed, analytic Jordan curves.

8.2. POISSON'S FORMULA

Let U be a real, bounded and piece-wise continuous function of θ, $0 \leqslant \theta \leqslant 2\pi$, $U(0) = U(2\pi)$. The function

(8.2A) $$u(z) = u(r, \varphi) = \frac{1}{2\pi} \int_0^{2\pi} \frac{(R^2-r^2)U(\theta)}{R^2-2Rr\cos(\theta-\varphi)+r^2} \, d\theta$$

is harmonic in the disk $K(0; R)$ w.r.t. $x = r\cos\varphi$, $y = r\sin\varphi$ and satisfies $\lim_{r \to R-} u(r, \theta) = U(\theta)$ at any continuity point of U. In case U is continuous, the Poisson formula (8.2A) gives the solution of the Dirichlet problem for a disk.

8.2.1. Show that the Poisson kernel
$$J = \frac{R^2 - r^2}{R^2 - 2Rr\cos(\theta - \varphi) + r^2}$$
can be written as $\operatorname{re}[(z+\zeta)/(z-\zeta)]$ with $z = Re^{i\theta}$, $\zeta = re^{i\varphi}$.

8.2.2. Given a continuous, real-valued and periodic function U ($U(0) = U(2\pi)$), find a function f analytic in $K(0; R)$ and continuous in $\overline{K}(0; R)$ such that $\operatorname{re} f(Re^{i\theta}) = U(\theta)$.

8.2.3. If u is harmonic in $K(0; R)$, continuous in $\overline{K}(0; R)$ and nonnegative on $C(0; R)$, show that
$$u(0)\frac{R-|z|}{R+|z|} \leqslant u(z) \leqslant u(0)\frac{R+|z|}{R-|z|}.$$

8.2.4. Find the function u harmonic in $K(0; 1)$, continuous in $\overline{K}(0; 1)$ and such that $u(e^{i\theta}) = \cos^2\theta$.

8.2.5. Given two complementary arcs α, β of $C(0; 1)$ and a function U vanishing on β and equal to 1 on α. Find a corresponding function u according to (8.2A). Show that $2\pi u(z)$ is equal to the length of an arc γ of $C(0; 1)$ cut off by the straight lines through z and the end points of α.

8.2.6. Show that the limit function u of an a.u. convergent sequence $\{u_n\}$ of harmonic functions is harmonic.

8.2.7. Derive the formula (8.2A) under the assumption that U has in $[0, 2\pi]$ a Fourier series representation:
$$U(\theta) = \tfrac{1}{2}a_0 + \sum_{n=1}^{\infty}(a_n\cos n\theta + b_n\sin n\theta).$$
Hint: $a_n\left(\dfrac{r}{R}\right)^n\cos n\varphi$, $b_n\left(\dfrac{r}{R}\right)^n\sin n\varphi$ are harmonic functions of $z = re^{i\varphi}$ in $K(0; R)$.

8.2.8. Derive in a similar way a formula analogous to (8.2A) for a function u harmonic in $\overline{\mathbf{C}}\setminus \overline{K}(0; R)$ with given boundary values U on $C(0; R)$.

8.3. THE DIRICHLET PROBLEM

8.2.9. Give a Fourier series representation of a bounded, harmonic function $u(re^{i\varphi})$ with boundary values: $u(Re^{i\theta}) = 1$ for $|\theta| < \alpha$, $u(Re^{i\theta}) = 0$ for $\alpha < \theta < 2\pi - \alpha$ $(0 < \alpha < \pi, r < R)$.

8.3. THE DIRICHLET PROBLEM

A domain G is said to be *regular with respect to the Dirichlet problem*, if any real-valued function U defined and continuous on the boundary fr$G$ of G has a continuous extension u to $\bar{G}$ which is harmonic in G. The function u is called the solution of Dirichlet's problem for G with given boundary values U. If we can associate with any point $\zeta \in \text{fr } G$ a continuum H_ζ (i.e. a closed, connected set containing at least two points) such that $\zeta \in H_\zeta \subset \mathbb{C}\setminus G$, then G is regular w.r.t. Dirichlet's problem, cf. [19], p. 93.

8.3.1. Show that the domain $K(0; 1)\setminus 0$ is not regular w.r.t. the Dirichlet problem.

8.3.2. Find the solution of Dirichlet's problem for the annulus $R_1 < |z| < R_2$ with boundary values

(i) constant on each boundary component: $U(R_1 e^{i\theta}) = A$, $U(R_2 e^{i\theta}) = B$;

(ii) equal $U_k(\theta)$ on $C(0; R_k)$, $k = 1, 2$, where $U_k(\theta)$ are represented by uniformly convergent Fourier series.

Hint: Put

$$u(r, \varphi) = \frac{a_0 + b_0 \log r}{2} + \sum_{k=1}^{\infty} [(a_k r^k + b_k r^{-k})\cos k\varphi + (c_k r^k + d_k r^{-k})\sin k\varphi].$$

8.3.3. Show that any Jordan domain is regular with respect to the Dirichlet problem. If $z = \Phi(w)$ maps G 1:1 conformally onto the unit disk $|z| < 1$, express the solution by means of Φ.

8.3.4. Find the solution h of the Dirichlet problem for the domain $G = K(-1; \sqrt{2}) \cap K(1; \sqrt{2})$ with boundary values $H(\omega)$, $\omega \in \text{fr } G$.

Hint: Exercise 2.9.6.

8.3.5. Prove that the visual angle $\omega(z)$ of the segment $[a, b]$ of the real axis in a point $z = x + iy$ of the upper half-plane is a harmonic function. Find $\lim_{z \to \zeta} \omega(z)$ for ζ situated on the real axis.

8.3.6. Given a system of $n+1$ points on the real axis $x_0 < x_1 < x_2 < \ldots < x_n$ and n real numbers $u_1, u_2, \ldots, u_n$. Find the function u bounded and harmonic in the upper half-plane whose boundary values on the real axis form a step function equal to 0 on $(-\infty, x_0)$, $(x_n, +\infty)$ and equal to u_k on (x_{k-1}, x_k), $k = 1, 2, \ldots, n$.

8.3.7. Find the solution of Dirichlet's problem for the upper half-plane and continuous boundary values U, where U vanishes outside a finite interval $[a, b]$.

8.3.8. Suppose U is a real-valued function of $t \in (-\infty, +\infty)$ such that

$$\int_{-\infty}^{+\infty} \frac{|U(t)|}{1+|t|} \, dt < +\infty.$$

Show that

$$u(z) = \frac{y}{\pi} \int_{-\infty}^{+\infty} \frac{U(t)}{(x-t)^2+y^2} \, dt$$

is harmonic in the upper half-plane and $\lim_{z \to t_0} u(z) = U(t_0)$ for any continuity point t_0 of U.

8.3.9. Suppose H is the interior of the ellipse $3x^2+4y^2 = 12$ and L is the segment with end points at its foci. Find the function u harmonic in $H \setminus L$, continuous in $\overline{H}$, equal to 1 on $\mathrm{fr}\, H$ and equal to 0 on L.

8.3.10. Find the function u harmonic and bounded in $\mathbf{C} \setminus \{(-\infty, -1] \cup [1, +\infty)\}$ which is equal to $-h$ on $(-\infty, -1]$ and equal to h on $[1, +\infty)$.

8.3.11. Find the function u harmonic outside the ellipse

$$\frac{x^2}{a^2} + \frac{y^2}{b^2} = 1 \quad (a > b > 0)$$

such that $u(z) \to A$ as z approaches the ellipse and $u(z) - \log|z| = O(1)$ as $z \to \infty$. Discuss the uniqueness. Find the value A for which $u(z) - \log|z| = o(1)$ as $z \to \infty$.

Hint: Consider first an analogous problem for the unit disk.

8.3.12. Suppose u is harmonic and bounded in the upper half-plane slit along $[i, +i\infty)$ and has the boundary values 0 on the real axis and boundary values 1 on the ray $[i, +i\infty)$. Find the derivative $\dfrac{\partial u}{\partial n}$ on the real axis.

8.4. HARMONIC MEASURE

Let G be a domain such that the spherical image of $\Gamma = \mathrm{fr}\, G$ is a finite system of simple, closed curves. If γ is a finite system of open arcs on Γ, then the function $\omega(z; \gamma, G)$ harmonic and bounded in G which tends to 1 as $z \to \zeta \in \gamma$ and tends to 0 as $z \to \zeta \in \Gamma \setminus \bar{\gamma}$ is said to be *harmonic measure of γ at z*.

8.4. HARMONIC MEASURE

8.4.1. Find a function harmonic in $K(0; 1)$ and tending to 0 as $z \to \zeta$, $\zeta \in C(0; 1) \setminus 1$ which is not identically 0. Verify that the boundedness condition in the definition of harmonic measure is essential for its uniqueness.

8.4.2. Determine the lines of constant harmonic measure $\omega(z; \gamma, G)$, if:
 (i) $\gamma = (a, b)$, $G = \{z: \text{im } z > 0\}$;
 (ii) $\gamma = \{z: |z| = 1, \theta_1 < \arg z < \theta_2\}$, $G = K(0; 1)$.

8.4.3. Find the harmonic measure of each of the boundary rays l_1, l_2 of the angular domain $G = \{z: 0 < \arg z < \alpha\}$, $\alpha < 2\pi$.

8.4.4. Find the harmonic measure of the diameter l and of the upper semi-circle γ with respect to the upper semi-disk G of $K(0; 1)$. Determine the lines of constant harmonic measure.

8.4.5. Find the asymptotic value of the harmonic measure of the boundary semi-circle of the upper semi-disk of $K(0; r)$ as $r \to +\infty$ and z is fixed.

8.4.6. Find the harmonic measure of the circular arc γ_0 on the boundary of $G_0 = K(\infty; 1) \cap (+; +)$.

8.4.7. Find the harmonic measure ω of the arc: $|z| = 1$, $0 < \arg z < 2\pi\lambda$, w.r.t. the annulus $\{z: 1 < |z| < R\}$, $0 < \lambda < 1$.
Hint: Exercise 8.3.2 (ii).

8.4.8. The boundary of G consists of two circular arcs γ_1, γ_2 with common end points a, b (a, b real, $a < b$) both γ_k situated in the upper half-plane. Determine $\omega(z; \gamma_k, G)$ ($k = 1, 2$).

8.4.9. Prove the monotoneity of harmonic measure:
 (i) $\gamma_1 \subset \gamma_2$ implies $\omega(z; \gamma_1, G) \leqslant \omega(z; \gamma_2, G)$;
 (ii) if the boundaries Γ_1, Γ_2 of G_1, G_2 have a non-empty intersection and $G_1 \subset G_2$, then for any $\gamma \subset \Gamma_1 \cap \Gamma_2$ and any $z \in G_1$ we have: $\omega(z; \gamma, G_1) \leqslant \omega(z; \gamma, G_2)$.

8.4.10. Prove the following "two constants theorem": Let f be analytic and bounded by M in absolute value in a domain G whose boundary Γ consists of a finite system of Jordan curves.
 Suppose that for a system of open arcs $\alpha \subset \Gamma$ we have: $\overline{\lim}_{z \to \zeta} |f(z)| \leqslant m < M$ for any $\zeta \in \alpha$. If $\beta = \Gamma \setminus \bar{\alpha}$, then
$$\log |f(z)| \leqslant \omega(z; \alpha, G) \log m + \omega(z; \beta, G) \log M.$$

8.4.11. Use Exercises 8.4.5 and 8.4.10 to prove the following Phragmén–Lindelöf theorem:

Suppose f is analytic in the upper half-plane and $\overline{\lim_{z \to x}} |f(z)| \leq 1$ for any real and finite x. If $M(r) = \sup_{0 < \theta < \pi} |f(re^{i\theta})|$, then either $\lim_{r \to +\infty} \dfrac{\log M(r)}{r} > 0$, or else $|f(z)| \leq 1$ in the whole upper half-plane.

8.5. GREEN'S FUNCTION

Given a domain G in the extended plane and a point $z_0 \in G$. A function g harmonic in $G \setminus z_0$, continuous in $\overline{G} \setminus z_0$, vanishing on fr$G$ and such that $g(z) + \log|z - z_0| = O(1)$, as $z \to z_0 \neq \infty$ (resp., $g(z) - \log|z| = O(1)$ as $z \to z_0 = \infty$) is called (classical) *Green's function* $g(z, z_0; G)$ of G with pole z_0.

If a domain G has a Green's function, the function is unique. If the boundary of G is a finite system of analytic Jordan curves, then Green's function of G exists and has at each point ζ of fr$G$ a normal derivative $\partial g / \partial n_\zeta$ and the solution u of the Dirichlet problem for G with boundary values $U(\zeta)$ has the following form:

$$(8.5\text{A}) \qquad u(z_0) = \frac{1}{2\pi} \int_{\partial G} U(\zeta) \frac{\partial g(\zeta, z_0; G)}{\partial n_\zeta} \, ds.$$

8.5.1. Show that any bounded domain regular w.r.t. the Dirichlet problem has Green's function.

Hint: Consider Dirichlet's problem with boundary values $\log|\zeta - z_0|$.

8.5.2. Show that any simply connected domain G whose boundary contains at least two points, possesses Green's function. Express it by means of the univalent function mapping G onto $K(0; 1)$.

8.5.3. If the domain G_z has the classical Green's function and $z = z(w)$ is a 1:1 conformal mapping of G_w onto G_z show that also G_w has the classical Green's function and $g(w, w_0; G_w) = g(z(w), z(w_0); G_z)$.

8.5.4. Suppose G is a simply connected domain whose boundary is an analytic Jordan curve Γ and f is a univalent mapping of G onto $K(0; 1)$.

If $z_0 \in G$ and $\Phi(z) = [1 - \overline{f(z_0)} f(z)] / [f(z) - f(z_0)]$, show that the interior normal derivative of $g(z, z_0; G)$ is equal to $|\Phi'(\zeta)|$, $\zeta \in \text{fr} G$.

8.5.5. Determine $g(z, z_0; K(0; 1))$ and derive Poisson's formula from (8.5A).

8.5.6. Determine Green's function for the upper half-plane and derive the formula of Exercise 8.3.8 from (8.5A).

8.5. GREEN'S FUNCTION

8.5.7. Find the Green's function $g(z, \infty; G)$ for
   (i) $G = K(\infty; 1)$;
   (ii) $G = \mathbf{C} \setminus [-1, 1]$;
   (iii) $G = \left\{ z = x+iy: \dfrac{x^2}{a^2} + \dfrac{y^2}{b^2} > 1 \right\}$.

8.5.8. Suppose $0 < h < c < 1$ and take the constant τ determining the function ϑ of Exercise 7.6.13 equal to $\log h/\pi i$. Verify that the function

$$F(w) = \vartheta\left(\frac{1}{2\pi i} \log \frac{w}{c}\right) : \vartheta\left(\frac{1}{2\pi i} \log cw\right)$$

has the following properties:
   (i) F is single-valued and analytic in $A = \{w: h < |w| < 1\}$;
   (ii) F has exactly one simple zero in A;
   (iii) $|F|$ has constant values on each of two components of fr$A$.

8.5.9. (cont.) Find the Green's function $g(w, c; A)$.

8.5.10. (cont.) Show that the interior normal derivative

$$\frac{\partial g}{\partial n} = \left| \frac{\Phi'(w)}{\Phi(w)} \right|, \quad \text{where} \quad \Phi(w) = w^{\frac{\log c}{\log h}}/F(w).$$

Express $\dfrac{\partial g}{\partial n}$ in terms of Weierstrass ζ-function.

8 5.11. Suppose G is a simply connected domain in the finite plane whose complement contains at least 2 different points and $\rho(w, w_0; G)$ is the hyperbolic distance of two points $w, w_0 \in G$ (which is defined as hyperbolic distance of their image points after a univalent mapping onto the unit disk).
  Find the relation between $\rho(w, w_0; G)$ and the Green's function of G.

8.5.12. Let G be a domain whose boundary Γ is a finite system of analytic Jordan curves. Show that $g(z, z_1; G)$ is a harmonic function of z_1 in $G \setminus z$. Also verify the symmetry of the Green's function: $g(z, z_1; G) \equiv g(z_1, z; G)$.
  Hint: Cf. formula (8.5A) and Exercise 8.5.1.

8.5.13. If $G_0 \subset G$, show that $g(z, z_1; G_0) \leqslant g(z, z_1; G)$ for any $z, z_1 \in G_0$.

8.5.14. Let G be a domain whose boundary $G = \bigcup\limits_{k=1}^{n} \Gamma_k$, where Γ_k are analytic Jordan curves with positive orientation w.r.t. G. If $h(z, z_0)$ is the (multi-valued) conjugate of $g(z, z_0; G)$, show that the increment $\Delta h(\zeta, z_0)$ as ζ describes Γ_k

is equal to $-2\pi\omega_k(z_0)$, where $\omega_k(z_0) = \omega(z_0; \Gamma_k, G)$ is the harmonic measure of Γ_k w.r.t. G.

Hint: Express ω_k by means of (8.5A).

8.5.15. Find the increment of $\log F(w)$, where F is the function of Exercise 8.5.8 and w moves over ∂A.

8.5.16. Suppose a, b, c are complex numbers different from each other and such that $c \notin [a, b]$. Determine the domain yielding the maximal value of $g(a, b; D)$, among all convex domains D such that $a, b \in D$, $c \in \mathbf{C} \setminus D$.

8.5.17. Let G be a domain such that $H = \mathbf{C} \setminus G$ is a finite system of bounded continua. Given two real constants A, B, show that there exists a unique function φ harmonic in G, continuous in $\mathbf{C}$, which takes the value B on H and has the form $A\log|z| + O(1)$ as $z \to \infty$.

8.6. BERGMAN KERNEL FUNCTION

In this chapter G denotes a domain in the finite plane whose complementary set is a finite system of disjoint continua and $L_2(G)$ is the family of all functions f analytic in G and such that the integral $\iint_G |f(x+iy)|^2 dx\,dy$ (taken in the sense of Lebesgue) is finite.

8.6.1. Verify that $L_2(G)$ is a complex, linear space with the usual definitions of addition and multiplication by numbers.

8.6.2. If $f, g \in L_2(G)$, show that the integral $\iint_G f\bar{g}\,dx\,dy = (f, g)$ is finite and has usual properties of an inner product.

8.6.3. Show that for any $\zeta \in G$ there exists $f \in L_2(G)$ such that $f(\zeta) = 1$.

Hint: If Γ is the unbounded component of $\mathbf{C} \setminus G$ consider the univalent function φ mapping $\mathbf{C} \setminus \Gamma$ onto a disk.

8.6.4. If $A(M)$ is the family of all $f \in L_2(G)$ such that

$$(f, f) = \|f\|^2 \leqslant M^2 \quad \text{and} \quad \bar{K}(a; \rho) \subset G,$$

show that there exists a uniform bound of $|f(z)|$ for $f \in A(M)$, $z \in \bar{K}(a; \rho)$.

Hint: Cf. Exercise 4.8.13.

8.6.5. Evaluate a common bound for $|f(z)|$ in terms of M, a, ρ ($|a|+\rho < 1$), when $G = K(0; 1)$.

8.6. BERGMAN KERNEL FUNCTION

8.6.6. Show that for any $\zeta \in G$ there exists $f_0 \in L_2(G)$ which has the minimal norm $(f,f)^{1/2}$ among all $f \in L_2(G)$ taking the value 1 at the point ζ.

8.6.7. (cont.) Prove that for any $g \in L_2(G)$ with $g(\zeta) = 0$ we have $(g, f_0) = 0$.
Hint: Consider $f^* = f_0 + \varepsilon e^{i\theta} g$ ($\varepsilon > 0$, $0 \leq \theta \leq 2\pi$).

8.6.8. An analytic function k of $z \in G$ depending on a complex parameter $\zeta \in G$ and defined by the formula

$$k(z, \zeta) = \|f_0\|^{-2} f_0(z, \zeta),$$

where f_0 is the solution of the extremal problem considered in Exercise 8.6.6 is called *Bergman kernel function*.

Express $k(z, \zeta)$ in case G is simply connected, by means of the univalent function w mapping G onto $K(0; 1)$ and such that $w(\zeta) = 0$. Also verify the uniqueness.
Hint: Exercise 8.1.12.

8.6.9. Determine $k(z, \zeta)$ for: (i) $G = K(0; R)$; (ii) $G = \{z: \operatorname{im} z > 0\}$.

8.6.10. Given $k(z, \zeta)$ for a simply connected domain G, find the univalent function w mapping G onto $K(0; 1)$ so that $w(\zeta) = 0$, $w'(\zeta) > 0$.

8.6.11. Verify the reproducing property of k:

$$f(\zeta) = \iint_G f(z)\overline{k(z, \zeta)}\,dx\,dy, \quad \zeta \in G, f \in L_2(G).$$

Hint: Consider $f(z)/f(\zeta) - f_0(z, \zeta)$; (cf. Ex. 8.6.7).

8.6.12. Show that

$$k(\zeta, \zeta) = \iint_G |k(\zeta, z))|^2 dx\,dy \quad \text{and} \quad |k(\zeta, \eta)|^2 \leq k(\zeta, \zeta) k(\eta, \eta).$$

8.6.13. Verify that the reproducing property of Exercise 8.6.11 implies the uniqueness of k.

8.6.14. Suppose $\{\varphi_n\}$ is a complete, orthonormal set of functions in $L_2(G)$ (i.e. $(\varphi_j, \varphi_k) = \delta_{jk}$ with $\delta_{jk} = 0$ for $j \neq k$, $\delta_{kk} = 1$ and the linear combinations $\sum_{k=0}^{n} c_k \varphi_k$ form a dense set in $L_2(G)$).

If $f, g \in L_2(G)$ and a_k, b_k are Fourier coefficients of f and g, resp., show that $\sum_{n=0}^{\infty} a_n \bar{b}_n$ is convergent and the Parseval identity $(f, g) = \sum_{n=0}^{\infty} a_n \bar{b}_n$ holds.

8.6.15. (cont.) Prove that $k(z, \zeta) = \sum_{n=0}^{\infty} \varphi_n(z)\overline{\varphi_n(\zeta)}$.

Hint: Exercise 8.6.11.

8.6.16. Show that the functions

$$\varphi_n(z) = \sqrt{\frac{n}{\pi}(1-h^{2n})^{-1}}\, z^{n-1} \quad (n = \mp 1, \mp 2, \ldots),$$

$$\varphi_0(z) = \left(2\pi \log \frac{1}{h}\right)^{-1/2} z^{-1}$$

form a complete, orthonormal system for the annulus $A = \{z: h < |z| < 1\}$.

8.6.17. Find Bergman kernel function for the annulus $A = \{z: h < |z| < 1\}$.

8.6.18. If $A = \{z: h < |z| < 1\}$ and f has the Laurent expansion $f(z) = \sum_{n=-\infty}^{\infty} b_n z^n$ in A, find the conditions for coefficients b_n in order that f belong to the class $L_2(A)$.

8.6.19. Examine the behavior of $k(z, \zeta)$ under conformal mapping. Show that $\sqrt{k(\zeta, \zeta)}\, |d\zeta|$ is a conformal invariant.

CHAPTER 9

Two-Dimensional Vector Fields

Potential theory in two dimensions is usually concerned with fields of force in which the vectors of the field are always parallel to a distinguished plane. Moreover, the vectors associated with the points of any straight line perpendicular to the distinguished plane are equal; hence the whole field is characterized by the field in the distinguished plane. If the field does not depend on time, it is called stationary.

9.1. STATIONARY TWO-DIMENSIONAL FLOW OF INCOMPRESSIBLE FLUID

Stationary two-dimensional flow of incompressible fluid is characterized by two harmonic functions: the *velocity potential* φ and the *flow function* ψ. If $w = u+iv$ is the vector expressing the velocity of a particle of the fluid past a given point $z = x+iy$, then $u = \varphi_x$, $v = \varphi_y$. On the other hand, the difference $\psi(x_1, y_1) - \psi(x_0, y_0)$ yields the volume of fluid passing in one second through a face of height 1 whose basis is any arc joining x_0+iy_0 to x_1+iy_1. In absence of singularities and in case the velocity field is a simply connected domain, $f = \varphi+i\psi$ is an analytic function which is called *complex potential of flow*. The lines $\varphi = \text{const}$ are called *equipotential lines* and their orthogonal trajectories $\psi(x, y) = \text{const}$ are called the *lines of flow*. In a stationary velocity field the lines of flow are the paths of the particles of the fluid.

9.1.1. Express the velocity w and its absolute value $|w|$ in terms of the complex potential of flow.

9.1.2. Prove the following uniqueness theorem for the velocity potential: if the domain D is swept out by regular arcs γ and on each γ the functions u, U have both a constant value (depending on γ), then $U = Au+B$, where A, B are real constants.

Hint: Consider $u_x - iu_y$.

9.1.3. The lines of flow are circles $x^2+y^2-2ax = 0$. Find the complex potential of flow. Evaluate the ratio of velocity of two particles of fluid passing the points $2a$, $a(1+i)$.

9.1.4. The lemniscates $r^2 = a^2\cos 2\theta$ (r, θ are polar coordinates) are the lines of flow. Evaluate the ratio of velocity of particles passing the points $z = a$, $z = a(\sqrt{\frac{3}{8}} + i\sqrt{\frac{1}{8}})$ (a is real).

9.1.5. Find the loci of constant velocity for the flow pattern as considered in Exercise 9.1.3.

9.1.6. Discuss the equipotential lines and the lines of flow and evaluate the velocity, if the complex potential of flow is equal: (i) az; (ii) aiz; (iii) z^{-1}; (iv) $\log\dfrac{z-b}{z-c}$; (v) z^2; (vi) $a\log z$; (vii) $ai\log z$ ($a > 0$).

9.1.7. If the volume of fluid flowing outside across a cylindrical face of height 1 and basis $C(a; r)$ is equal Q for all r sufficiently small, the point a is called a *source of intensity* Q. If the flux Q is negative, the point a is called a *sink of intensity* $|Q|$.

The line integral $\int_\gamma w_s ds$, where w_s is the tangent component of velocity and γ is a contour, is called the *circulation along* γ. If the circulation along all contours containing the point a inside and sufficiently close to a is equal $\Gamma \neq 0$, then a is called a vortex.

If γ is a contour containing inside one singular point and f is the complex potential of flow, show that

$$\Gamma + iQ = \int_\gamma f'(z)\,dz.$$

9.1.8. Given a flow with complex potential $\log(z^2+z^{-2})$. Evaluate sources and sinks and the corresponding intensities.

9.1.9. Discuss the flow pattern with the complex potential $2i\log(z^2-a^2)$. Evaluate the circulation along the circles $|z \mp a| = a$.

9.1.10. The complex potential of flow is equal $\log\sinh \pi z$. Evaluate the flux across the circle $C(0; \frac{3}{2})$ and the circulation along it.

9.1.11. Evaluate the complex potential of flow of fluid from the lower half-plane into the upper half-plane through an orifice in the real axis between 1 and -1 supposed the velocity at ∞ is equal 0 and the flux across the orifice is equal Q. Find the pressure at various points of the orifice.

9.1.12. Evaluate the complex potential of flow past a circular cylinder with cross-section $C(0; 1)$ immersed in a parallel stream, assuming that the velocity

9.1. FLOW OF INCOMPRESSIBLE FLUID

w_∞ at infinity is parallel to the real axis. Evaluate the velocity at $z = i$ and $z = 2$, as well as the difference of pressure at these points.

9.1.13. Evaluate the complex potential of flow past an elliptical cylinder $\frac{x^2}{a^2} + \frac{y^2}{b^2} = 1$ $(a > b)$ immersed in a parallel stream, if the velocity at infinity is equal $w_\infty e^{i\alpha}$ (α real, $w_\infty > 0$).

9.1.14. Consider analogous problems for cylinders with cross-sections:
(i) $\bar{K}(-i; \sqrt{2}) \cap \bar{K}(i; \sqrt{2})$;
(ii) $[-ih, ih]$.

9.1.15. Evaluate the complex potential of flow with a source of intensity Q at $z = a$, assuming $w_\infty = 0$.

9.1.16. Evaluate the complex potential of flow with a source of intensity Q at -1 and a sink of the same intensity at $z = 0$, assuming $w_\infty = 0$.

9.1.17. Evaluate the complex potential of flow in the first quadrant with a source of intensity Q at $1+i$ and a sink of the same intensity at $z = 0$ assuming $w_\infty = 0$.

9.1.18. Discuss the limiting case as $h \to 0$ of flow with a source of intensity p/h at $z = h$ and a sink of the same intensity at $z = -h$ (a dipole of the momentum p). Find the loci of constant velocity.

9.1.19. Evaluate the complex potential of flow with a vortex at $z = a$ involving the circulation Γ. Assume $w_\infty = 0$.

9.1.20. Evaluate the complex potential of flow in the upper half-plane under the assumption there exists a source at $z = ai$ of intensity Q and the velocity at ∞ is parallel to the real axis. Find the velocity at $z = 0$.

9.1.21. Consider analogous problems as in Exercise 9.1.20 assuming that:
(i) $z = ai$ is a dipole of the momentum p and $w_\infty > 0$;
(ii) $z = ai$ is a source-vortex of intensity Q involving the circulation Γ.

9.1.22. Determine the complex potential of flow past a cylinder with the cross-section $\bar{K}(0; 1)$ with a vortex $z = ai$ $(a > 1)$ involving a circulation Γ. Assume $w_\infty = 0$.

9.1.23. Evaluate the complex potential of flow past a cylinder with the cross-section $\bar{K}(0; 1)$ assuming that $z = a$ $(a > 1)$ is a source of intensity Q and $w_\infty = 0$.

9.1.24. Consider an analogous problem, the source being replaced by a vortex with circulation Γ.

9.1.25. Show that the complex potential function

$$f(z) = a\left(z + \frac{R^2}{z}\right) + \frac{\Gamma}{2\pi i} \log z$$

corresponds to an asymmetric flow past a circular cylinder with the cross-section $\overline{K}(0; R)$. Determine the real constant a so that $z = iR$ is the only point where the stream lines enter (and exit) the cylinder cross-section. Find the velocity at ∞ in this case and evaluate the force acting on the cylinder according to the Joukovski lift formula.

9.2. TWO-DIMENSIONAL ELECTROSTATIC FIELD

Two-dimensional electrostatic field is produced by a system of charged, long, parallel wires and cylindrical conductors. The potential is constant throughout each conductor and there are no charges in the interiors of conductors. The electrostatic field can be characterized by an analytic function $f = \varphi + i\psi$ or the *complex potential of the electrostatic field*. The real, single-valued harmonic function ψ is called the (real) *electrostatic potential*. The field vector w is equal $-\operatorname{grad} \psi = -\overline{if'(z)}$. The complex z-plane playing the role of a system of coordinates may be any plane perpendicular to all conductors involved. One single wire with a charge of q units per 1 cm of length which intersects the coordinate plane at $z = a$ gives rise to the electrostatic field with complex potential $-2iq\log(z-a)$ (cf. Ex. 9.2.2). Given the complex potential $f = \varphi + i\psi$, the lines $\psi = \text{const}$ correspond to equipotential lines, whereas the lines $\varphi = \text{const}$ are the lines of force.

9.2.1. An infinite, straight-line shaped wire l is charged with a positive charge of q electrostatic units per 1 cm of length. Show that the Coulomb force w produced by the wire and acting on a unit charge at a distance r cm from the wire is perpendicular to the wire and $|w| = 2qr^{-1}$.

9.2.2. (cont.) Show that w can be derived from the complex potential $f(z) = -2qi\log z$, where the z-plane is perpendicular to l and cuts it at the origin.

9.2.3. Evaluate the complex potential of the electrostatic field due to two long parallel wires each bearing a positive charge q per unit of length, the distance of wires being $2h$. Discuss the equipotential lines.

9.2.4. Evaluate the complex potential of an electrostatic dipole, i.e. the limiting case of electrostatic field due to two long parallel wires each bearing charges of opposite sign $\mp q$ per unit of length, the distance of wires being $2h$, $2qh \to M$ as $h \to 0$. Find the field strength and the loci of a constant field strength.

9.2.5. If σ is the density of charge on the surface of a cylindrical conductor and ψ is the real electrostatic potential, show that $\sigma = -\dfrac{1}{4\pi} \cdot \dfrac{\partial \psi}{\partial n}$.

Hint: The flux of the electrostatic vector field across a closed surface S is equal to 4π times the total charge inside S (Gauss theorem).

9.2.6. Given the potential ψ outside a charged cylindrical conductor with the cross-section Γ, evaluate the total charge on the conductor.

9.2.7. A n-tuply connected domain G whose boundary consists of two disjoint systems Γ_0, Γ_1, of closed analytic curves determines an electrostatic condenser. The systems Γ_0, Γ_1 correspond to the cross-sections with the plane of reference of two systems of conductors (outer and inner coatings), the potential of conductors of either system being kept on the same level. Usually the inner coating is charged and the outer one is grounded so that its potential is equal to zero. Show that the total charges q_0, q_1 on both coatings are equal but have opposite signs.

Hint: Consider the integral $\displaystyle\int_{\Gamma_0+\Gamma_1} \dfrac{\partial \psi}{\partial n}\, ds$ and apply the Green's formula:

$$\int_{\partial G} \dfrac{\partial \psi}{\partial n} h\, ds = -\iint_G (h_x \psi_x + h_y \psi_y)\, dx\, dy - \iint_G h \Delta \psi\, dx\, dy \quad \text{with} \quad h = 1.$$

9.2.8. (cont.) If ψ_0, ψ_1 are potentials of Γ_0, Γ_1 resp., and q is the charge on Γ_1, the ratio $c = q/(\psi_1 - \psi_0)$ does not depend on ψ_0, ψ_1 and is called the capacity of a condenser. If $\omega(z) = \omega(z; \Gamma_1, G)$ is the harmonic measure of Γ_1 w.r.t. the domain G, show that

$$4\pi c = \iint_G (\omega_x^2 + \omega_y^2)\, dx\, dy.$$

Hint: Consider $\displaystyle\int_{\partial G} \omega \dfrac{\partial \omega}{\partial n}\, ds$.

9.2.9. (cont.) Show that the capacity of a two-dimensional electrostatic condenser is invariant under conformal mapping.

9.2.10. Find the complex potential of the electrostatic field between two coaxial circular conducting cylinders of radii r and R. Show that the capacity of such a condenser is equal to $\tfrac{1}{2}(\log(R/r))^{-1}$ cm per unit of length of generatrix.

9.2.11. Show that the capacity c of a pair of long conducting parallel wires of radius a separated by a distance b is equal

$$c = \frac{1}{2}\left(\log \frac{b+\sqrt{b^2-4a^2}}{b-\sqrt{b^2-4a^2}}\right)^{-1}.$$

9.2.12. Find the complex potential of the electrostatic field between two conducting circular cylinders with cross-sections $\overline{K}(0;1)$, $\overline{K}(1;4)$ assuming that the potential on the first cylinder is equal 1, whereas the second one is grounded. Evaluate the capacity of the cylinder.

9.2.13. The potential on the cylinder with cross-section $\overline{K}(5;4)$ is equal 1, whereas the cylinder corresponding to $\overline{K}(-5;4)$ is grounded. Find the extremal densities of charge on both cylinders.

9.2.14. A horizontal wire of negligible radius bearing a charge λ per unit length is suspended at a distance h above the surface of a conducting plane. Find the complex potential and the charge per unit area in the plane.

Hint: The field will not change, if we replace the plane by a symmetric wire bearing a charge $-\lambda$ per unit length.

9.2.15. Show that the electrostatic potential outside a system of long, conducting cylinders with cross-sections Γ_k ($k = 1, 2, \ldots, n$) bearing jointly a positive charge λ per unit length and kept on the same potential B, has the form $\psi(z) = 2\lambda \log|z|^{-1} + o(1)$ as $z \to \infty$, if B is suitably chosen. Prove that the ratio $B/2\lambda = \gamma$ does not depend on λ. Express γ by means of the Green's function $g(z)$ of the exterior of $\bigcup_{k=1}^{n} \Gamma_k$. The constant γ is called Robin's constant.

9.2.16. Prove that Robin's constant for the exterior of the ellipse $\frac{x^2}{a^2} + \frac{y^2}{b^2} = 1$ is equal to $-\log \frac{1}{2}(a+b)$.

CHAPTER 10

Univalent Functions

10.1. FUNCTIONS OF POSITIVE REAL PART

10.1.1. If f is a complex-valued function of a real variable $t \in [a, b]$, $f = u+iv$, and g is a real-valued function of bounded variation, the Stieltjes integral $\int_a^b f dg$ is defined as $\int_a^b u\,dg + i \int_a^b v\,dg$. By using the well-known estimate for real-valued f:
$$\left| \int_a^b f\,dg \right| \leq V_a^b(g) \max |f|,$$
where $V_a^b(g)$ denotes the total variation of g on the interval $[a, b]$ show that an analogous estimate also holds for complex-valued f.

Hint: Take real α such that $\left| \int_a^b f\,dg \right| = e^{i\alpha} \int_a^b f\,dg$ and transform the last expression.

10.1.2. If the sequence of complex-valued functions $\{f_n\}$ of a real variable $t \in [a, b]$ is uniformly convergent in $[a, b]$, and the sequence $\{g_n\}$ of real-valued functions with uniformly bounded total variation on $[a, b]$ converges to g, show that the limit $\lim \int_a^b f_n\,dg_n$ exists and is equal to $\int_a^b f\,dg$.

10.1.3. Using the Schwarz formula (Ex. 8.2.2) prove the following theorem due to Herglotz:

Let f be analytic in $K(0; R)$ with $f(0) = 1$ and $\mathrm{re}f(z) > 0$ for all $z \in K(0; R)$. There exists an increasing function μ of $t \in [0, 2\pi]$, $\mu(0) = 0$, $\mu(2\pi) = 1$ such that
$$f(z) = \int_0^{2\pi} \frac{Re^{it}+z}{Re^{it}-z}\,d\mu(t).$$

Hint: Consider the sequence $\mu_n(t) = (2\pi)^{-1} \int_0^t u(R_n e^{i\theta})\,d\theta$, where $u(z) = \mathrm{re}f(z)$, $R_n = \left(1 - \dfrac{1}{n}\right) R$, and write Schwarz's formula in a Stieltjes integral form.

10.1.4. Let $\mathscr{P}$ be the class of all functions analytic in $K(0; 1)$ and such that $f(0) = 1$, $\operatorname{re} f(z) > 0$ in $K(0; 1)$. Show that the function $(1+z)/(1-z)$ belongs to $\mathscr{P}$ and cannot be represented by the Schwarz formula in the unit disk. Write a corresponding Herglotz representation.

10.1.5 If H is a fixed, continuous, complex-valued function of $t \in [a, b]$ and μ is a variable, increasing function of $t \in [a, b]$ subject to the conditions $\mu(0) = 0$, $\mu(1) = 1$, show that the set of all possible values of the Stieltjes integral $\int_a^b H(t) d\mu(t)$ is identical with the convex hull of the curve $\Gamma\colon w = H(t)$, $a \leqslant t \leqslant b$.

10.1.6. If c_n is the nth Taylor coefficient at $z = 0$ of $f \in \mathscr{P}$, express c_n in terms of the function μ of Exercise 10.1.3. Show that $c_n \in \overline{K}(0; 2)$. Also show that for any $w_0 \in \overline{K}(0; 2)$ and any positive integer n there exists $h \in \mathscr{P}$ such that $h^{(n)}(0)/n! = w_0$.

10.1.7. Using Exercises 10.1.3 and 10.1.5, find the region of variability of the point $f(z)$ for fixed $z \in K(0; 1)$ and f ranging over $\mathscr{P}$.

10.1.8. If $f \in \mathscr{P}$ and $\arg f(z_0) = \alpha$, show that

$$\left|\frac{f'(z_0)}{f(z_0)}\right| \leqslant \frac{2\cos\alpha}{1-|z_0|^2}.$$

Hint: Consider $\varphi(z) = \left[f\left(\frac{z+z_0}{1+\bar{z}_0 z}\right) - f(z_0)\right]\left[f\left(\frac{z+z_0}{1+\bar{z}_0 z}\right) + \overline{f(z_0)}\right]^{-1}$.

10.1.9. Let f be analytic in $K(0; R)$ and $\neq 0$ in $K(0; R) \setminus 0$. Suppose ψ is a positive, differentiable function of $r \in (0, R)$ such that

$$\left|\frac{f'(z)}{f(z)}\right| \leqslant \frac{d}{dr}\log\psi(r) = \frac{\psi'(r)}{\psi(r)}, \quad |z| = r.$$

Show that for any real, fixed θ the function $|f(re^{i\theta})|/\psi(r)$ decreases in $(0, R)$ and the finite limit $\lim_{r \to R-} |f(re^{i\theta})|/\psi(r)$ exists.

10.1.10. If $f \in \mathscr{P}$, show that for any fixed θ the finite limit $\lim_{r \to 1-}(1-r)f(re^{i\theta})$ exists.

10.1.11. If the function μ of Exercise 10.1.3 is continuous at $t = \theta$ ($\theta \neq 0, 2\pi$), show that

$$\lim_{r \to 1-}(1-r)f(re^{i\theta}) = 0.$$

10.1.12. If μ has a jump h at $t = \theta$ ($\theta \neq 0, 2\pi$), i.e. $\mu(\theta+) - \mu(\theta-) = h$, show that
$$\lim_{r \to 1-} (1-r)f(re^{i\theta}) = 2h.$$

10.1.13. If $\theta = 0, 2\pi$, show that
$$\lim_{r \to 1-} (1-r)f(r) = 2[1 - \mu(2\pi-) + \mu(0+)].$$

10.1.14. If $f \in \mathscr{P}$, show that for any fixed, real θ the finite limit $\lim_{r \to 1-} (1-r)f(re^{i\theta}) = \alpha(\theta)$ exists and is real, nonnegative. Also show that the set $\{\theta : \alpha(\theta) > 0\}$ is at most countable; if θ_k are its elements, show that $\sum_{k=1}^{\infty} \alpha(\theta_k) \leq 2$.

10.2. STARSHAPED AND CONVEX FUNCTIONS

10.2.1. If f is analytic and does not vanish in the annulus $\{z : r - \delta < |z| < r + \delta\}$, show that
$$\frac{\partial}{\partial \theta} \arg f(re^{i\theta}) = \operatorname{re} \frac{zf'(z)}{f(z)}.$$

10.2.2. If f is analytic in $K(0; R)$, does not vanish in $K(0; R) \setminus 0$, $f(0) = 0$, $f'(0) \neq 0$, and $\operatorname{re}\{zf'(z)/f(z)\} > 0$ for any $z \in K(0; R)$, show that f is univalent in $K(0; R)$.
Hint: The argument principle.

10.2.3. Show that
(i) $f_1(z) = z(1-z)^{-3}$ is univalent in $K(0; \frac{1}{2})$;
(ii) $f(z) = z(1-z)^{-\alpha}$ is univalent in $K(0; 1)$ for $0 \leq \alpha \leq 2$.

10.2.4. Let S^* be the family of all normalized starlike univalent functions, i.e. $f \in S^*$ means that f is analytic in $K(0; 1)$, $f(0) = 0$, $f'(0) = 1$ and $zf'(z)/f(z)$ has positive real part in $K(0; 1)$. Show that the domain $D_f = f[K(0; 1)]$ is starlike (or starshaped) with respect to the origin, i.e. $w \in D_f$ implies $[0, w] \subset D_f$. Also prove the converse.

10.2.5. Using Herglotz's formula derive structural formula for $f \in S^*$.

10.2.6. Find precise estimates of $|f|$ and $|f'|$ on $C(0; r)$ for $f \in S^*$.
Hint: Exercise 10.2.5, or Exercise 10.1.8.

10.2.7. Find the region of variability G_z of $[z/f(z)]^{1/2}$ for a fixed z ($0 < |z| < 1$) and f ranging over S^*.
Hint: Exercise 10.2.5 and Exercise 10.1.5.

10.2.8. If $f \in S^*$, show that $\operatorname{re}[f(z)/z]^{1/2} > \frac{1}{2}$.

10.2.9. If f maps the unit disk 1:1 conformally onto a convex domain B, show that any smaller disk $K(0; r)$, $0 < r < 1$, is also mapped onto a convex domain B_r.

Hint: Consider the function $\psi = f^{-1} \circ \varphi$, where

$$\varphi(z) = tf\left(\frac{z_1}{z_2}z\right) + (1-t)f(z), \quad z \in K(0; 1) \text{ and } 0 < t < 1, \ |z_1| \leqslant |z_2|;$$

assume $f(0) = 0$.

10.2.10. Let S^c be the family of all normalized convex univalent functions, i.e. $f \in S^c$ means that f is univalent in $K(0; 1)$, $f(0) = 0$, $f'(0) = 1$ and $f[K(0; 1)]$ is a convex domain. Show that $f \in S^c$, iff $zf' \in S^*$.

10.2.11. If $f \in S^c$, show that $\operatorname{re}\sqrt{f'(z)} > \frac{1}{2}$.

10.2.12. If $z_1, z_2 \in K(0; 1)$, show that

$$\operatorname{re}\left[(z_2-z_1)^{-1}\operatorname{Log}\frac{1-z_1}{1-z_2}\right] > \frac{1}{2}.$$

Hint: Integrate $F(z) = (1+z)/(1-z)$ over the segment $[z_1, z_2]$.

10.2.13. If $f \in S^c$, show that $\operatorname{re}[f(z)/z] > \frac{1}{2}$.

Hint: Using Exercises 10.2.11, 10.1.3 find a formula expressing f' as a double Stieltjes integral. Integrate under the sign of integral and use Exercise 10.2.12.

10.2.14. If $f \in S^c$, show that $|1 - z/f(z)| \leqslant |z|$, $z \in K(0; 1)$.

10.2.15. If $f \in S^c$, show that $\operatorname{re}[zf'(z)/f(z)] > \frac{1}{2}$.

Hint: Consider $g(z) = [f'(\zeta)(1-|\zeta|^2)]^{-1}\left[f(\zeta) - f\left(\frac{\zeta-z}{1-\bar\zeta z}\right)\right]$.

10.2.16. If $f \in S^c$, show that:
(i) $(1+|z|)^{-1} \leqslant |zf'(z)/f(z)| \leqslant (1-|z|)^{-1}$;
(ii) $|\arg[zf'(z)/f(z)]| \leqslant \arcsin|z|$.

10.2.17. Find sharp estimates of $|f(z)|$ and $\arg[f(z)/z]$ for $f \in S^c$.

10.3. UNIVALENT FUNCTIONS

10.3.1. Suppose that Γ is a closed Jordan curve with parametric representation $R = R(\theta)$, $\Phi = \Phi(\theta)$, $0 \leqslant \theta \leqslant 2\pi$, where R, Φ are polar coordinates; $R(\theta)$, $\Phi(\theta)$ are supposed to be continuous; moreover, $\Phi(\theta)$ is piecewise monotonic

10.3. UNIVALENT FUNCTIONS

and the index $n(\Gamma, 0) = 1$. If h is a nonnegative continuous and increasing function of $R \in (0, +\infty)$, show that

$$\int_0^{2\pi} h(R(\theta)) \, d\Phi(\theta) \geq 0.$$

10.3.2. Let f be analytic and univalent in the annulus $\{z : a < |z| < b\}$ and let Γ_r be the image curve of $C(0; r)$ under f, $a < r < b$. If g is continuously differentiable and $|f(re^{i\theta})| = R(r, \theta)$, $\arg f(re^{i\theta}) = \Phi(r, \theta)$, show that

$$\int_0^{2\pi} R(r, \theta) g'(R(r, \theta)) \, d_\theta \Phi(r, \theta) = r \frac{d}{dr} \int_0^{2\pi} g(R(r, \theta)) \, d\theta.$$

10.3.3. (cont.) If $n(\Gamma_r, 0) = 1$, $a < r < b$, show that $\int_0^{2\pi} |f(re^{i\theta})|^2 d\theta$ is an increasing function of r in (a, b).

10.3.4. Let Σ be the family of all functions $F(z) = z + b_0 + b_1 z^{-1} + b_2 z^{-2} + \cdots$ analytic and univalent in $K(\infty; 1)$. If $r > 1$, show that

$$r - \sum_{n=1}^\infty n|b_n|^2 r^{-2n-1} \geq 0.$$

10.3.5. (cont.) Show that $\sum_{n=1}^\infty n|b_n|^2 \leq 1$. This inequality is sometimes called the *area theorem*.

10.3.6. Let Σ_0 be the subfamily of Σ consisting of all $F \in \Sigma$ which do not vanish in $K(\infty; 1)$ and let S be the family of functions $f(z) = z + a_2 z^2 + \cdots$ analytic and univalent in $K(0; 1)$. Find the relation between the functions of S and Σ_0. Verify that $z(1 + z^{-3})^{2/3} \in \Sigma_0$.

10.3.7. If $F(z) = z + b_0 + b_1 z^{-1} + \cdots$ and $F \in \Sigma_0$, show that $|b_0| \leq 2$. Discuss the case of equality.
Hint: Consider $G(z) = \sqrt{F(z^2)}$.

10.3.8. If $F \in \Sigma$ and does not assume the values w_1, w_2, show that $|w_1 - w_2| \leq 4$.
Hint: Consider $F(z) - w_k$, $k = 1, 2$.

10.3.9. If $f(z) = z + a_2 z^2 + \cdots$, $f \in S$, show that $|a_2| \leq 2$. Also show that equality holds only for the Koebe function $f_\alpha(z) = z(1 + e^{i\alpha}z)^{-2}$.

10.3.10. If $f \in S$ does not assume the value h, show that $|h| \geq \frac{1}{4}$. Also show that equality holds only for f_α.
Hint: Consider the development of $\varphi(z) = hf(z)/(h - f(z))$.

10.3.11. Suppose that $a, b \in K(0; 1)$ and $z = z(t)$ is the function mapping the unit disk $\{t: |t| < 1\}$ onto the domain $D(a, b)$ of Exercise 2.9.20 so that $z(0) = a$. If $f \in S$, verify that also

$$\varphi(t) = [f(z(t)) - f(a)] : \left[\frac{d}{dt} f(z(t))\right]_{t=0}$$

belongs to S. Deduce the inequality

$$|f'(a)| \left|\frac{a-b}{f(a)-f(b)}\right| \cdot \frac{|1-\bar{a}b|}{1-|b|^2} \cdot \frac{|1-\bar{a}b|-|a-b|}{|1-\bar{a}b|+|a-b|} \leq 1.$$

10.3.12. (cont.) Find the following estimates:
(i) the upper bound of logarithmic derivative $|f'(a)/f(a)|$;
(ii) the upper and lower bound of $|f(a)|$;
(iii) the lower bound of logarithmic derivative.

Hint: Consider in (iii) the mapping

$$\psi(t) = \left[f\left(\frac{t+z}{1+\bar{z}t}\right) - f(z)\right] [f'(z)(1-|z|^2)]^{-1}.$$

Verify that equality can be attained in each case.

10.3.13. If $f \in S$, show that

$$(1-|z|)(1+|z|)^{-3} \leq |f'(z)| \leq (1+|z|)(1-|z|)^{-3}.$$

Verify that in either case equality can be attained by a Koebe function.

10.3.14. Let a, b ($a \neq b$) be two fixed points of $K(0; 1)$. Find the lower and the upper bound of the ratio $|f'(a)/f'(b)|$ for f ranging over S.

Hint: Consider $\varphi(\zeta) = f\left(\dfrac{\zeta+b}{1+\bar{b}\zeta}\right)$.

10.3.15. Find for $F \in \Sigma_0$ an inequality analogous to the inequality of Exercise 10.3.11.

10.3.16. Find for $F \in \Sigma_0$ sharp estimates of: (i) $|F(z)|$; (ii) $|zF'(z)/F(z)|$.

10.3.17. If $F \in \Sigma$, $F(z) = z + b_0 + b_1 z^{-1} + \ldots$, show that
(i) $\varlimsup_{|z| \to 1+} |F(z) - b_0| \leq 2$;
(ii) $|F(z) - b_0 - z| \leq 3|z|^{-1}$.

10.3.18. If $F \in \Sigma$, show that:
(i) $|F'(z) - 1| \leq (|z|^2 - 1)^{-1}$;

Hint: Use the area theorem and Schwarz inequality;
(ii) $|F'(z)| \leq |z|^2/(|z|^2 - 1)$.

10.4. CIRCULAR AND STEINER SYMMETRIZATION

10.3.19. If $F(z) = z - (h - h^{-1})(hz - 1)$, $h > 1$, show that $F \in \Sigma$ and verify that for $z = h$ the equality in Exercise 10.3.18 (ii) actually holds. Find the image domain of $K(\infty; 1)$.

10.3.20. Suppose that $f \in S$ and there exists a constant M such that $|f(z)| \leq M$ for all $z \in K(0; 1)$.
Show that

(i) $|a_2| = \tfrac{1}{2}|f''(0)| \leq 2(1 - M^{-1})$;

(ii) $-f_M(-|z|) \leq |f(z)| \leq f_M(|z|)$,

where f_M is the Pick function which maps $K(0; 1)$ onto $K(0; M)$ slit along a radius from $-M$ to $-M[2M - 1 - 2\sqrt{M(M-1)}\,]$.

Hint: Derive first the inequality (ii) by considering $g(z) = f(z)[1 + e^{i\alpha}M^{-1}f(z)]^{-2}$; $w = f_M(z)$ satisfies the equation $z(1-z)^{-2} = w(1 - w/M)^{-2}$.

10.3.21. If $f \in S$ and $\zeta \in K(0; 1)$, show that the function

$$g(z) = f'(\zeta)(1 - |\zeta|^2)\left[f\left(\frac{1 + z\bar{\zeta}}{z + \bar{\zeta}}\right) - f(\zeta)\right]^{-1}$$

belongs to the family Σ. Find the initial coefficients in the Laurent development of g and verify that $|\{f, \zeta\}| \leq 6(1 - |\zeta|^2)^{-2}$, where $\{f, \zeta\}$ is the Schwarzian derivative of f (cf. Ex. 7.2.14).

Verify that the same inequality also holds for an arbitrary function meromorphic and univalent in $K(0; 1)$.

10.4. THE INNER RADIUS. CIRCULAR AND STEINER SYMMETRIZATION

The definition of the inner radius $r(z_0; G)$ of a simply connected domain G in the finite plane (cf. Exercise 8.1.8) can be restated in terms of Green's function $g(z, z_0; G)$ as follows:

(10.4A) $$\log r(z_0; G) = \lim_{z \to z_0}[g(z, z_0; G) + \log|z - z_0|].$$

The limit on the right-hand side exists for any bounded domain regular with respect to the Dirichlet problem.

If G is an arbitrary domain in the finite plane, there exists an increasing sequence $\{G_n\}$ of bounded domains, each regular with respect to the Dirichlet problem, such that $G = \bigcup_{n=1}^{\infty} G_n$. Each point z_0 of G belongs to G_n for all n sufficiently large and one can prove that the increasing sequence $\{r(z_0; G_n)\}$ has always a limit, finite or infinite, which does not depend on a particular choice of $\{G_n\}$. This limit may be assumed as the inner radius of G. Moreover, if $r(z_0; G)$

is finite at a certain point $z_0 \in G$, it is also finite at any other point of G (cf. Exercise 10.4.4). Circular (or Pólya) symmetrization of an open set G in the extended plane is determined by a fixed ray λ with the origin at a and associates with G a symmetrized set G^* which is defined as follows.

1° If G contains a, or ∞, or both, so does G^* and conversely.

2° If $C(a; r) \subset G$, then $C(a; r) \subset G^*$, and conversely; if $C(a; r) \cap G = \emptyset$, then $C(a; r) \cap G^* = \emptyset$, and conversely.

3° If the intersection $C(a; r) \cap G$ is neither empty, nor coincides with $C(a; r)$ and has the total angular Lebesgue measure $l(r)$, then $C(a; r) \cap G^*$ consists of a single open arc bisected by λ and subtending at a an angle equal to $l(r)$.

Steiner symmetrization is determined by a straight line which we take to be the real axis in the complex plane.

Let $\sigma(x)$ be the straight line $\{z: \operatorname{re} z = x\}$. If G is an open set in the plane, then the symmetrized set G^* is defined as follows.

1° If $\sigma(x) \cap G = \emptyset$, then $\sigma(x) \cap G^* = \emptyset$.

2° If $\sigma(x) \cap G$ has the linear Lebesgue measure $l(x)$, then $\sigma(x) \cap G^*$ is a single, open segment of length $l(x)$ bisected by the real axis.

If G is a domain, so is G^* for both Steiner and circular symmetrization. Also the area of G and G^* is the same. As shown by Pólya and Szegö, for any point z_0 situated on the line (resp. on the half-line) of symmetrization and any domain G containing z_0, we have: $r(z_0, G) \leqslant r(z_0, G^*)$.

10.4.1. Show that both Steiner and Pólya symmetrization carries a domain G into a domain G^*; if G is simply connected, so is G^*.

10.4.2. If G is starshaped with respect to the origin and the ray λ of circular symmetrization emanates from the origin, show that also G^* is starshaped with respect to the origin.

10.4.3. Show that Steiner symmetrization preserves the convexity of domains. Give an example showing that circular symmetrization does not have an analogous property.

10.4.4. If f is univalent in the unit disk $K(0; 1)$ and the area of the image domain $f[K(0; 1)]$ is finite, show that

$$\lim_{|z| \to 1} (1-|z|)|f'(z)| = 0.$$

Hint: Cf. Exercise 4.2.12.

10.4. CIRCULAR AND STEINER SYMMETRIZATION

10.4.5. If G is a bounded, simply connected domain, show that the inner radius $r(w; G)$ is a continuous function of w which tends to zero as $\operatorname{dist}(w; \operatorname{fr} G) \to 0$. *Hint*: Cf. Exercise 8.1.9.

10.4.6. Let G be a bounded domain, regular with respect to the Dirichlet problem, and let $h(z, z_0)$ be the solution of the Dirichlet problem with boundary values $\log |\zeta - z_0|$.
If $z_0, z_1 \in G$, $z_0 \neq z_1$ and $\operatorname{dist}(z_k; \operatorname{fr} G) \geq \delta$ for $k = 0, 1$, show that
(i) $|h(z, z_0) - h(z, z_1)| < |z_1 - z_0|/\delta$ for any $z \in G$;
Hint: Use the maximum principle and the inequality $\log|1+z| \leq |z|$;
(ii) $|\log r(z_0; G) - \log r(z_1; G)| < 2|z_1 - z_0|/\delta$.

10.4.7. If the inner radius $r(z'; G)$ is finite for some $z' \in G$, show that it is also finite for any $z'' \in G$. Show that $r(z; G)$ is a continuous function of z.

10.4.8. If H is an ellipse, show that $r(z; H)$ has a maximum at the center of H.

10.4.9. Let $\mathscr{G} = \mathscr{G}(r, R)$ be the class of domains G, $K(0; r) \subset G \subset K(0; R)$, regular with respect to the Dirichlet problem and such that each circle $C(0; \rho)$, $r \leq \rho \leq R$, meets $\mathbf{C} \setminus G$. Find $\sup_G r(0; G)$.

10.4.10. Let $\mathscr{G} = \mathscr{G}(R)$, $R > 1$, be the class of domains G, $0 \in G \subset K(0; R)$, such that $r(0; G) = 1$. Find the least possible distance δ of the point $w = 0$ from the component of $E \setminus G$ containing the point at infinity.

10.4.11. Let $\mathscr{G}(w_0)$ be the class of simply connected domains containing $0, w_0$ and satisfying $r(0; G) = 1$. Show that
$$\sup\{r : r = r(w_0; G), G \in \mathscr{G}(w_0)\} = 1 + 4|w_0|.$$

10.4.12. Suppose that G is a simply connected domain of hyperbolic type (i.e. G can be mapped 1:1 conformally onto a disk) and G^* is the domain obtained by circular symmetrization. Show that G^* is also of hyperbolic type. Show that Steiner symmetrization does not have this property.

10.4.13. If G is of hyperbolic type, $w_1, w_2 \in G$ ($0 \leq w_1 < w_2$) and G^* arises from G by circular symmetrization with respect to the real axis, show that $\rho(w_1, w_2; G^*) \leq \rho(w_1, w_2; G)$ where ρ denotes the hyperbolic distance (cf. Ex. 8.5.11).
Hint: Show that $\rho(w_1, w_2; G) = \int_\Gamma \dfrac{|dw|}{r(w; G)}$, where Γ is the image of the h-segment under the 1:1 conformal mapping of $K(0; 1)$ onto G.

10.4.14. (cont.) If g denotes the Green function, show that
$$g(w_1, w_2; G^*) \geq g(w_1, w_2; G).$$

10.4.15. If f maps the unit disk 1:1 conformally onto G and G is invariant under circular symmetrization with respect to the real axis, moreover, $f(0) = 0$, $f'(0) > 0$, show that $M(r, f) = f(r)$.

Hint: If $w_0 = f(z_0)$, then $g(0, w_0; G) = -\log|z_0|$.

10.4.16. Suppose that f, f^* are analytic and univalent in the unit disk and the image domain $G^* = f^*[K(0; 1)]$ contains the origin and arises from $G = f[K(0; 1)]$ by circular symmetrization with respect to the positive real axis. If $f(0) = f^*(0) = 0$, show that $M(r, f) \leq M(r, f^*)$ for all $r \in (0, 1)$.

10.4.17. Suppose that $f \in S$ and $D_f = f[K(0; 1)]$. Verify Koebe one quarter theorem (Ex. 10.3.10) by considering $r(0; D_f)$.

10.5. THE METHOD OF INNER RADIUS MAJORIZATION

10.5.1. Let D be a domain possessing the classical Green function $g(w, a_0; D) = g(w)$ and let $\mathscr{F}(a_0; D)$ be a family of functions analytic in the unit disk and such that $f(0) = a_0$, $f[K(0; 1)] = D_f \subset D$. Prove that $|f'(0)| \leq r(a_0; D)$ for any $f \in \mathscr{F}(a_0; D)$.

Hint: Consider $h(z) = g(f(z)) + \log|z|$, $z \in K(0; 1)$.

10.5.2. (cont.) Prove an analogous theorem for arbitrary D.

Hint: Consider $D(\rho) = f[K(0; \rho)]$.

10.5.3. Prove following symmetrization principle. Let f be analytic in the unit disk and let $f(0) = a_0$. If the symmetrized domain D^* of $D_f = f[K(0; 1)]$ with respect to a line (Steiner symmetrization), or a ray through a_0 (circular symmetrization) is situated in D_0, show that $|f'(0)| \leq r(a_0; D_0)$.

10.5.4. Suppose that $0 < \alpha < 2$ and that the function $f(z) = a_0 + a_1 z + \ldots$ is analytic in $K(0; 1)$ and $D_f \cap C(0; r)$ is for any $r > 0$ a system of arcs of total length $\pi \alpha r$ at most. Show that $|a_1| \leq 2\alpha |a_0|$, with equality holding for

$$f(z) = a_0 \left(\frac{1+z}{1-z}\right)^\alpha.$$

10.5.5. If $f(z) = a_0 + a_1 z + \ldots$ is analytic in $K(0; 1)$ and $R = R_f$ is the least upper bound of all $r > 0$ such that $C(0; r) \subset D_f$, show that $|a_1| \leq 4(|a_0| + R)$ with equality when $a_0' \geq 0$ and $f(z) = a_0 + 4(a_0 + R)z(1-z)^{-2}$.

10.5.6. If $f(z) = z + a_2 z^2 + \ldots$ is analytic (not necessarily univalent) in the unit disk, show that for any $\varepsilon > 0$ there exists $r > \frac{1}{4} - \varepsilon$ such that $C(0; r) \subset D_f$.

10.5.7. Deduce Koebe one quarter theorem from Exercise 10.5.6.

10.5. THE INNER RADIUS MAJORIZATION

10.5.8. Suppose that $f(z) = a_0 + a_1 z + \ldots$ is analytic in the unit disk and that there exists $h > 0$ and a real-valued function $v(u)$, $-\infty < u < +\infty$ such that on any straight line $\operatorname{re} w = u$ each point $u + i(v(u) + nh)$ ($n = 0, \mp 1, \mp 2, \ldots$) belongs to $\mathbf{C} \setminus D_f$. Show that $|a_1| \leqslant 2h/\pi$.
Hint: Consider $g(z) = \exp\{2\pi f(z)/h\}$.

10.5.9. Suppose that any circle $C(0; r)$, $r > 0$, contains a point of $\mathbf{C} \setminus D_f$ and f is analytic in $K(0; 1)$. Show that
$$(1 - |z|^2)|f'(z)| \leqslant 4|f(z)|$$
for any $z \in K(0; 1)$.
Hint: Exercise 10.5.5.

10.5.10. If f is analytic, univalent and does not vanish in the unit disk, show that
$$|f(0)| \left(\frac{1-|z|}{1+|z|}\right)^2 \leqslant |f(z)| \leqslant |f(0)| \left(\frac{1+|z|}{1-|z|}\right)^2.$$

10.5.11. If φ is analytic in the unit disk and never assumes both values w, $-w$, show that
$$(1 - |z|^2)|\varphi'(z)| \leqslant 2|\varphi(z)|.$$

10.5.12. Suppose that f is a Bieberbach–Eilenberg function, i.e. f is analytic in the unit disk, $f(0) = 0$ and $f(z_1)f(z_2) \neq 1$ for any $z_1, z_2 \in K(0; 1)$. Show that
(i) f is bounded in the unit disk;
(ii) $(1+f)/(1-f)$ never assumes both values w, $-w$.

10.5.13. Show that for any Bieberbach–Eilenberg function f we have:
(i) $(1 - |z|^2)|f'(z)| \leqslant |1 - f^2(z)|$;
(ii) $|f'(0)| \leqslant 1$.

10.5.14. If $f \in S$ and $L(r, f)$ is the angular measure of values omitted by f and situated on $C(0; r)$, then $L(r, f)$ is equal to zero for $r \leqslant \tfrac{1}{4}$ and any $f \in S$ and can be equal to 2π for $r \geqslant 1$ (e.g. for $f(z) \equiv z$).
If $r \in (\tfrac{1}{4}, 1)$, show that
$$L(r) = \sup_{f \in S} L(r, f) = 4 \arcsin(2\sqrt{r} - 1).$$
Show that the extremal function maps $K(0; 1)$ onto
$$\mathbf{C} \setminus (\{w \colon |w| = r, |\arg w| \leqslant 2\arcsin(2\sqrt{r} - 1)\} \cup [r, +\infty)).$$
Find the extremal function (cf. Ex. 2.9.22).

10.5.15. Solve an analogous problem for starlike univalent functions.

Hint: Verify first that the mapping

$$w = F(t) = \left(\frac{1+\sqrt{t}}{1-\sqrt{t}}\right)^\theta \frac{1-\theta\sqrt{\bar{t}}+t}{1+\theta\sqrt{\bar{t}}+t}, \quad 0 < \theta < 2,$$

carries $1:1$ conformally the upper half-plane $\operatorname{im} t > 0$ onto the domain $H(\theta) = \{w: |w| < 1\} \cup \{w: 0 < \arg w < \theta\pi\}$. Next find the mapping $f: K(0;1) \to H(\theta)$. Using Exercise 10.4.2 show that $f(z)/f'(0)$ is the extremal mapping.

SOLUTIONS

CHAPTER 1

Complex Numbers. Linear Transformations

1.1.1. $\frac{1}{5}+\frac{3}{5}i$; 64; i; $2+i2\sqrt{3}$.

1.1.2. 0, 1, $-\frac{1}{2}\mp i\frac{1}{2}\sqrt{3}$.

1.1.4. (i) z is real; (ii) $z = \cos\theta + i\sin\theta$, θ is real.

1.1.5. $\zeta = \dfrac{1}{\sqrt{2}}[(x+\sqrt{x^2+y^2})^{1/2}+iy(x+\sqrt{x^2+y^2})^{-1/2}]$.

1.1.6. Use the identity $|a+b|^2 = (a+b)(\bar{a}+\bar{b})$; the sum of squares of all sides in a parallelogram equals the sum of squares of both diagonals.

1.1.7. Square both sides and use Exercise 1.1.6.

1.1.8. Use the identity $|a+b|^2 = (a+b)(\bar{a}+\bar{b})$.

1.1.9. The left-hand side is equal to

$\{\exp[i(n+1)\theta]-1\}:(e^{i\theta}-1) = \{\sin\frac{1}{2}(n+1)\theta\exp[\frac{1}{2}(n+1)\theta i]\}:(\sin\frac{1}{2}\theta\exp\frac{1}{2}\theta i)$.

1.1.10. $z_1z_2+z_2z_3+z_3z_1 = z_1z_2z_3(\bar{z}_1+\bar{z}_2+\bar{z}_3) = 0$, hence $(z-z_1)(z-z_2)\times (z-z_3) = z^3-a$ with $|a|=1$.

1.1.11. We have as before $\sum x_k x_l x_m = 0$, hence

$$\prod_{k=1}^{4}(z-z_k) = (z^2-a^2)(z^2-b^2), \quad \text{where} \quad |a|=|b|=1.$$

1.1.12. $\bar{z}_1 z_2 = x_1x_2+y_1y_2+i(x_1y_2-y_1x_2)$.

1.1.13. The area is equal to

$\frac{1}{2}[z_1,z_2]\times[z_1,z_3] = \frac{1}{2}\text{im}(\bar{z}_2-\bar{z}_1)(z_3-z_1) = \frac{1}{2}(\bar{z}_2z_3+\bar{z}_3z_1+\bar{z}_1z_2)$

since $\text{im}\,a = \text{im}(-\bar{a})$.

1.1.14. Induction and Exercise 1.1.13.

1.1.15. $\cos\alpha = |z_2-z_1|^{-1}\operatorname{im}(\bar{z}_1 z_2+\bar{z}_2 z_3+\bar{z}_3 z_1)$, cf. Exercises 1.1.12, 1.1.13.

1.1.16. Take $C(0; 1)$ as the circum-circle and use Exercise 1.1.10.

1.1.17. (i) Logarithmic spiral;
(ii) the circle $C(\frac{1}{2}; \frac{1}{2})$ with the point $z = 0$ removed;
(iii) ellipse with semi-axes $|a \mp a^{-1}|$;
(iv) cycloid;
(v) involute of $C(0; a)$;
(vi) cardioid;
(vii) the right-hand half of parabola $y = x^2$;
(viii) a branch of hyperbola $xy = 1$.

1.1.18. If $|x'|+|y'| > 0$, then $z'(t)$ is the tangent vector.

1.1.19. $\alpha = \arg[z'(\theta)/z(\theta)] = \operatorname{arccot}[r'(\theta)/r(\theta)]$.

1.1.20. (i) If $x \neq 0$, then $\arg z = \arctan(y/x)$; (ii) $|z| = \sqrt{x^2+y^2}$.

1.1.21. (i) The line of symmetry of $[a, b]$;
(ii) closed ellipse with foci $\mp c$;
(iii) a part of the right half-plane outside $C(1; \sqrt{2})$;
(iv) a strip bounded by the real axis and the line $y = -1$ with the real axis included;
(v) the domain to the right of the right-hand branch of equilateral hyperbola $x^2-y^2 = \alpha$;
(vi) the left half-plane;
(vii) the closure of the domain to the left of parabola $y^2-1 = -2x$;
(viii) the interior domains of both loops of a lemniscate;
(ix) the union of a part of the exterior of $C(i; \sqrt{2})$ contained in the right half-plane and of a part of $K(i; \sqrt{2})$ contained in the left half-plane.

1.1.22. If $z = kZ+l$, $a = k+l$, $b = -k+l$, then the given set becomes $\{Z: \arg(Z^2-1) = \text{const}\}$ which is an equilateral hyperbola center at the origin.

1.1.23. $R > |b-\frac{1}{4}a^2|$.

1.1.24. Circle $C(BA^{-1}; |A|^{-1}\sqrt{|B|^2-AC})$.

1.1.25. The radius: $k|a-b||1-k^2|^{-1}$; the center: $(a-k^2b)(1-k^2)^{-1}$.

1.1.26.
$$\begin{vmatrix} |z|^2 & z & \bar{z} & 1 \\ |z_1|^2 & z_1 & \bar{z}_1 & 1 \\ |z_2|^2 & z_2 & \bar{z}_2 & 1 \\ |z_3|^2 & z_3 & \bar{z}_3 & 1 \end{vmatrix} = 0.$$

1.1.27. Cf. Exercises 1.1.26, 1.1.24.

1.1.28. (i) If $m_1 = 0$, then $z_0 \in [z_2, z_3]$; if $m_1 = 1$, then $z_0 = z_1$; if $m_1 \neq 0, 1$, then $z_0 = m_1 z_1 + (1-m_1)\zeta_1$, where $\zeta_1 = (1-m_1)^{-1}(m_2 z_2 + m_3 z_3) \in [z_2, z_3]$ and therefore $z_0 \in [z_1, \zeta_1] \in T$.

(ii) If the change of m_j by h_j gives the same point z_0, then $h_1 z_1 + h_2 z_2 + h_3 z_3 = 0$, or $h_1(z_1 - z_3) + h_2(z_2 - z_3) = 0$ since $h_1 + h_2 + h_3 = 0$. This means that z_j are collinear.

1.1.29, 30. Induction with respect to n and Exercise 1.1.28.

1.1.31. If $\mu = \sum_{k=1}^{n} |\zeta - z_k|^{-2}$, $m_k = \mu^{-1}|\zeta - z_k|^{-2}$, then $m_k > 0$, $\sum_{k=1}^{n} m_k = 1$ and $\sum_{k=1}^{n} m_k(\zeta - z_k) = 0$, or $\zeta = \sum_{k=1}^{n} m_k z_k$.

1.1.32. If $P(z) = A(z-z_1) \ldots (z-z_n)$, then
$$\frac{P'(z)}{P(z)} = \frac{1}{z-z_1} + \frac{1}{z-z_2} + \ldots + \frac{1}{z-z_n};$$
(cf. now Exercise 1.1.31).

1.1.33. $\log z_n = \dfrac{n}{2}\left[\dfrac{2x}{n} + O\left(\dfrac{1}{n^2}\right)\right] \to x$, or $|z_n| \to e^x$;
$\arg z_n = n \arctan(y/(n+x)) \to y$.

1.1.34. The sequence is divergent; $|z_n|^2 \to R^2 = \prod_{n=1}^{\infty}\left(1 + \dfrac{1}{n^2}\right)$, however, $\arg \dfrac{z_n}{z_{n-1}} = \arctan \dfrac{1}{n}$ is a term of a divergent series; every point of $C(0; R)$ is an accumulation point for $\{z_n\}$.

1.1.35. $\sum b_n \zeta_n$ is absolutely convergent, since $\{\zeta_n\}$ is bounded.

1.1.36. The convergence of $\sum z_n$ implies the convergence of $\sum x_n$; since $x_n \geq 0$, $\sum x_n^2$ is also convergent. The convergence of $\sum y_n^2$ follows now from the convergence of both $\sum x_n^2$, $\sum z_n^2$.

1.1.37. Consider first the case $\lim \zeta_n = 0$ and show using (i), (ii) that $\lim z_n = 0$.

1.1.38. Put $a_{nk} = p_k/(p_1 + p_2 + \ldots + p_n)$ for $k \leq n$, and $a_{nk} = 0$ for $k > n$; apply Exercise 1.1.37.

1.1.39. Express w_n by means of z_n and use Exercise 1.1.37.

1.1.40. Express w_n in terms of $s_n = \sum_{k=1}^{n} z_k$ and use Exercise 1.1.37.

1.1.41. Put $p_n = \mu_n^{-1}$ and cf. Exercise 1.1.40.

1.1.42. If $v \neq 0$, consider the sequence w_n/v; note that $\dfrac{1}{nv}v_n$, $\dfrac{1}{nv}v_{n-1}, \ldots$
$\ldots, \dfrac{1}{nv}v_1, 0, 0, \ldots$ are lines of a matrix of Exercise 1.1.37. If $v = 0$, replace v_n by $V_n = g + v_n$ with $g \neq 0$.

1.2.1. Use the parametric equation of the straight line NA: $X = x_1 t$, $Y = x_2 t$, $Z = 1 + (x_3 - 1)t$.

1.2.2. $(\cos\alpha; \sin\alpha; 0)$, $(-\tfrac{2}{3}; \tfrac{2}{3}; \tfrac{1}{3})$, $(\tfrac{3}{13}; -\tfrac{4}{13}; \tfrac{12}{13})$.

1.2.3. $C \setminus K(0; 1)$; $K(0; 1)$.

1.2.4. Note that all projecting rays are situated in one plane through N and the given straight line.

1.2.5. The points of S situated on a circle satisfy: $\sum_{k=1}^{3} A_k x_k + B = 0$; use now Exercises 1.2.1, 1.1.24.

1.2.6. Use the formulas of Exercise 1.2.1.

1.2.7. (i) opposite points on the same parallel;
(ii) points on the same parallel symmetric w.r.t. the plane $Ox_1 x_3$;
(iii) opposite points on the circle $x_1 = C$, $x_2^2 + x_3^2 = 1 - C^2$.

1.2.8. Put $x_1 = \cos\theta\cos\varphi$ etc. into the formulas of Exercise 1.2.1.

1.2.9. Cf. Exercise 1.2.1.

1.2.10. Note that the equation of the stereographic projection of the great circle is identically satisfied by $-\bar{z}^{-1}$ (Ex. 1.2.9).

1.2.11. The equations $|z - z_0|^2 = 1 + |z_0|^2$, $z\bar{z} - z_0\bar{z} - \bar{z}_0 z - 1 = 0$, are equivalent (cf. Ex. 1.2.10).

1.2.12. $C(z_0; R)$, $z_0 = -14 - \tfrac{27}{2} i$, $R = \tfrac{1}{2}\sqrt{1517}$.

1.2.13. Apply the formula for the distance of two points and express their coordinates by means of formulas of Exercise 1.2.1.

1.2.14. $\dfrac{d\sigma}{ds} = \lim\limits_{z_1 \to z} \dfrac{\sigma(z_1, z_2)}{|z_1 - z|}$ (cf. Ex. 1.2.13); hence $d\sigma^2 = \lambda(z)(dx^2 + dy^2)$ and this implies preservation of angles.

1. COMPLEX NUMBERS. LINEAR TRANSFORMATIONS

1.2.15. $\sigma(z, a) = \sigma(\zeta, a)$, $\sigma(z, -\bar{a}^{-1}) = \sigma(\zeta, -\bar{a}^{-1})$, hence

$$\left|\frac{\zeta-a}{1+\bar{a}\zeta}\right| = \left|\frac{z-a}{1+\bar{a}z}\right|;$$

moreover, z and ζ are situated on circles through a, $-\bar{a}^{-1}$ intersecting at an angle φ, hence

$$\arg\frac{\zeta-a}{\zeta+\bar{a}^{-1}} - \arg\frac{z-a}{z+\bar{a}^{-1}} = \varphi.$$

1.2.16. $|z-a_1||z-a_2|^{-1} = (1+|a_1|^2)^{1/2}(1+|a_2|^2)^{-1/2}$ which is Apollonius circle with limit points a_1, a_2.

1.2.17. $\frac{1}{2}\sigma(|a|-r, |a|+r)$.

1.2.18. Evaluate e.g. the length of inscribed polygonal line using Exercise 1.2.13 and consider its limit for a normal sequence of partitions.

1.2.19. The stereographic projection of a rhumb line intersects all rays $\arg z$ = const at a constant angle hence it must be a logarithmic spiral (cf. Ex. 1.1.19).

1.2.20. The equation of stereographic projection is $r = r_1 e^{k\theta}$, $k = \frac{1}{\alpha}\log\frac{r_2}{r_1}$, hence

$$l(\Gamma) = \int_{r_1}^{r_2}\frac{2\sqrt{1+k^{-2}}}{1+r^2}\,dr = 2\sqrt{1+k^{-2}}\,(\arctan r_2 - \arctan r_1).$$

The numerical example corresponds to $\arctan r_1 = \pi/6$, $\arctan r_2 = \pi/3$, hence

$$l(\Gamma) = \frac{\pi}{3}\left[1+\left(\frac{\pi}{6\log 3}\right)^2\right]^{1/2}.$$

1.2.21. $d\sigma^2 = \dfrac{4(dx^2+dy^2)}{(1+|z|^2)^2}$, hence

$$|D| = \iint_\Delta \sqrt{EG-F^2}\,dx\,dy = \iint_\Delta \frac{4}{(1+|z|^2)^2}\,dx\,dy.$$

1.2.22. Without loss of generality we may take the south pole as one of the vertices of T. The stereographic projection T_0 of T is bounded by two straight line segments emanating from the origin and an arc of a circle: $z = z_0 + \sqrt{1+|z_0|^2}e^{i\varphi}$, $\varphi_1 \leqslant \varphi \leqslant \varphi_2$ (cf. Ex. 1.2.11). Using the formula of Exercise 1.2.21 in polar coordinates we obtain:

$$|T| = \iint_{T_0}\frac{4r}{(1+r^2)^2}\,dr\,d\theta = \int_{\theta_1}^{\theta_2}\frac{2r^2}{1+r^2}\,d\theta;$$

we now introduce a new variable φ putting $re^{i\theta} = z_0 + \sqrt{1+|z_0|^2}\, e^{i\varphi}$. We have:

$$r^2 = 1 + 2|z_0|^2 + 2\sqrt{1+|z_0|^2}(x_0\cos\varphi + y_0\sin\varphi),$$

$$\frac{d\theta}{d\varphi} = \frac{d}{d\varphi}\arg(z_0 + \sqrt{1+|z_0|^2}\, e^{i\varphi})$$

$$= r^{-2}[1 + |z_0|^2 + \sqrt{1+|z_0|^2}(x_0\cos\varphi + y_0\sin\varphi),]$$

where $x_0 + iy_0 = z_0$, (cf. Ex. 1.1.20). This gives:

$$|T| = \int_{\varphi_1}^{\varphi_2} d\varphi = \varphi_2 - \varphi_1 = (\alpha+\beta+\gamma) - \pi,$$

which follows from elementary geometric considerations.

1.2.23. The stereographic projection of T is bounded by $[i, i(1+\sqrt{2})]$ and arcs of $C(0; 1)$, $C(-1+i; \sqrt{3})$; hence

$$|T| = \frac{\pi}{2} + 2\arctan\sqrt{2} - \pi = \arctan\frac{1}{2\sqrt{2}}.$$

1.3.1. The angle of rotation: $\arg a$; the ratio of homothety center at the origin: $|a|$.

1.3.3. $w = \pi i(z - \tfrac{1}{2})$.

1.3.4. (i) $w = az + b$, $a > 0$, b real;
(ii) $w = az + b$, $a < 0$, b real;
(iii) $w = -i(az + b)$, $a > 0$, b real.

1.3.5. $w = (b-a)^{-1}[(B-A)z + Ab - aB]$.

1.3.6. $w = (1+i)(1-z)$.

1.3.7. $w = \dfrac{k+i}{b_2 - b_1}(z - ib_2)$.

1.3.8. $z_0 = b/(1-a)$; the angle of rotation: $\arg a$, the ratio of homothety: $|a|$.

1.3.9. (i) $z_0 = \tfrac{1}{5}(3+i)$, $\arg a = -\tfrac{3}{4}\pi$, $|a| = \sqrt{2}$;
(ii) $z_0 = b_2(1-ik)/(b_2 - b_1 - k - i)$, $a = (k+i)/(b_2 - b_1)$.

1.3.10. It is sufficient to consider the case $W = aZ$ (cf. Ex. 1.3.8); if $Z(\theta) = A|a|^{\theta/\alpha}e^{i\theta}$, $-\infty < \theta < +\infty$, $A > 0$, arbitrary, and $\alpha = \arg a \neq 0$, then evidently $aZ(\theta) = Z(\theta + \alpha)$. Hence the spiral $Z = Z(\theta)$ remain unchanged after the transformation $W = aZ$.

1.4.1. If $c \neq 0$, then $w = \dfrac{bc-ad}{c} z_1 + \dfrac{a}{c}$, $z_1 = z_2^{-1}$, $z_2 = cz+d$.

1.4.2. If $z(\theta) = r(\theta)e^{i\theta}$ is the equation of γ, then the image curve has the equation $w(\theta) = (r(\theta))^{-1}e^{-i\theta}$. The angle α between γ and the radius vector satisfies $\cot \alpha = r'(\theta)/r(\theta)$; a corresponding value for the image curve after taking the opposite orientation is the same.

1.4.3. Use Exercises 1.4.1 and 1.4.2.

1.4.4. The equality
$$(w^{-1}-a)(\overline{w}^{-1}-\overline{a}) = r^2$$
is equivalent with the equality
$$[w-\overline{a}(|a|^2-r^2)^{-1}][\overline{w}-a(|a|^2-r^2)^{-1}] = r^2(|a|^2-r^2)^{-2};$$
the image curve of $C(a; |a|)$ is a straight line.

1.4.5. Use Exercise 1.4.4 with $a = bi$, $r^2 = b^2+1$.

1.4.6. Use Exercises 1.4.1 and 1.4.4.

1.4.7. Verify that $w = az+b$, $w = z^{-1}$ do not change the cross-ratio and cf. Exercise 1.4.1.

1.4.8. $w = \dfrac{b-c}{b-a} \cdot \dfrac{z-a}{z-c}$.

1.4.9. (i) Straight lines $\operatorname{re} w = (2a)^{-1}$;
(iii) circles $b(u^2+v^2)+u+v = 0$;
(iii) straight lines $v = -ku$;
(iv) circles through $w = 0$ and $w = z_0^{-1}$;
(v) cissoid $u^2(v+1)+v^3 = 0$.

1.4.10. Linear transformation defined by the equation $(z_1, z_2, z_3, z) = (a, b, c, w)$, where a, b, c are real and different from each other carries the circle (or possibly the straight line) C determined by z_1, z_2, z_3 into the real axis. Since the r.h.s. is real iff w is real, so is the l.h.s. iff $z \in C$.

1.4.11, 12. Verify that identity, inverse transformations and superposition are also transformations of the same type.

1.5.1. The equation satisfied by z^* is equivalent to the conditions: $|z-a| \times |z^*-a| = r^2$, $\arg(z-a) = \arg(z^*-a)$.

1.5.2. If $Z = a+re^{i\theta}$, $z = a+\rho e^{i\varphi}$, then
$$|Z-z^*||Z-z|^{-1} = \left|re^{i\theta} - \frac{r^2}{\rho}e^{i\varphi}\right| |re^{i\theta}-\rho e^{i\varphi}|^{-1} = \frac{r}{\rho}.$$

1.5.3. $\frac{9}{2}+i$.

1.5.4. (i) The straight line $\operatorname{re} z = \frac{1}{2}$;
(ii) lemniscate $(x^2+y^2)^2 = x^2-y^2$.

1.5.5. If z_0 is the center of the circle orthogonal to $C(0;1)$, its radius is $\sqrt{|z_0|^2-1}$. Its reflection satisfies $(\bar\zeta^{-1}-z_0)(\zeta^{-1}-\bar z_0) = |z_0|^2-1$, which is equivalent to $(\zeta-z_0)(\bar\zeta-\bar z_0) = |z_0|^2-1$.

1.5.6. This is an immediate consequence of the fact that symmetry is defined in terms of circles and their orthogonality and both properties are preserved under linear transformations.

1.5.7. $z^* = (\bar a_2-\bar a_1)[(a_2-a_1)\bar z+a_1\bar a_2-\bar a_1 a_2]$.

1.5.8. If $L_k = C(a_k; r_k)$, $k = 1, 2$, then
$$Z = a_2+r_2^2(z-a_1)[(\bar a_1-\bar a_2)(z-a_1)+r_1^2]^{-1}.$$
Hence $(z = a_1) \leftrightarrow (Z = a_2)$. If the same holds after interchanging the order of reflections, then $r_1^2+r_2^2 = |a_1-a_2|^2$ which shows that L_1, L_2 are orthogonal. The sufficiency is easily verified after an additional linear transformation carrying L_1, L_2 into perpendicular straight lines.

1.5.9. An immediate consequence of Exercise 1.5.1.

1.5.10. The roots of the equation:
$$\bar a + \frac{r^2}{z-a} = (a_2-a_1)^{-1}[(\bar a_2-\bar a_1)z+\bar a_1 a_2-a_1\bar a_2].$$

1.5.11. Circles through $-7\mp 4\sqrt{3}$ which are symmetric points with respect to both circles.

1.6.1. (i) The lower half-disk of $K(0;1)$;
(ii) $K(0;1) \cap (\mathbf{C}\setminus K(-\frac{5}{4}i; \frac{3}{4}))$;
(iii) $\{w\colon \operatorname{im} w < 0\} \cap \{\mathbf{C}\setminus K(\frac{1}{2}(1-i); \frac{1}{2}\sqrt{2})\}$;
(iv) $K(\frac{1}{2}; \frac{1}{2})\setminus \bar K(\frac{3}{4}; \frac{1}{4})$;
(v) $\{w\colon \operatorname{re} w > \frac{1}{2}\} \cap (\mathbf{C}\setminus \bar K(\frac{4}{3}; \frac{2}{3}))$.

1.6.2. $w = (z-4)(z-1)^{-1}$.

1.6.3. If $A = (b-a)(d-c)(c-b)^{-1}(d-a)^{-1}$, then $k = 2A+1-2\sqrt{A(A+1)}$; $(w, b, c, d) = (z, -1, 1, k^{-1})$.

1.6.4. If a, b are real and $(z = a) \leftrightarrow (w = b)$, then
$$w = (a+1)bz[(a-b)z+a(b+1)]^{-1}, \quad ab(a+1)(b+1) > 0.$$

1.6.5. $C\setminus \bar{K}(-\tfrac{1}{2}; 1)$; $\{w: \operatorname{im} w < 0\} \cap [C\setminus \bar{K}(-\tfrac{1}{2}; 1)]$.

1.6.6. $z = (1-2w)(1+w)^{-1}$.

1.6.7. $z = i(w+1)(w-1)^{-1}$; Apollonius circles $|w+1|/|w-1| = r$.

1.6.8. If $(z = a) \leftrightarrow (w = 0)$, $|a| < 1$, then
$$w = e^{i\alpha}(z-a)/(1-\bar{a}z).$$

1.6.9. $z = -(1+i)(w-i)/(w+1-2i)$.

1.6.10. The points symmetric w.r.t. either circle are ∓ 3, (cf. Ex. 1.5.9). Hence $w = 2(z+3)/(z-3)$, $R = 4$.

1.6.11. The points symmetric w.r.t. either circle are:
$z_1 = \tfrac{1}{2}(3+i)$, $z_2 = 2+i$; $w = 2^{-1/2}(2z-3+i)/(z-2+i)$, $R = \sqrt{3/2}$.

1.6.12. $w = (z-\sqrt{h^2-R^2})/(z+\sqrt{h^2-R^2})$; ρ is found by putting $z = h-R$.

1.6.13. $w = a(z-2)^{-1} + b$, a is real.

1.7.1. $w = (az-\alpha\beta)/(z+a-\alpha-\beta)$, $a \neq \alpha, \beta$.

1.7.3. $cwz+wd-az-b \equiv czw+dz-aw-b \equiv 0$, iff $a+d = 0$ since $w \neq z$.

1.7.4. If $c = 0$, then $w = -z+A$ and $\tfrac{1}{2}A$, ∞ are invariant points, if $c \neq 0$, then the discriminant of the quadratic equation for invariant points: $cz^2 - 2az - b = 0$, i.e. $a^2 + bc = bc - ad \neq 0$.

1.7.5. $W = W(Z)$, where $W = \dfrac{w-\alpha}{w-\beta}$, $Z = \dfrac{z-\alpha}{z-\beta}$ has two invariant points $0, \infty$ (cf. Ex. 1.7.2).

1.7.6. $A = \dfrac{w-\alpha}{w-\beta} \dfrac{z-\beta}{z-\alpha} = w'(\alpha) = \dfrac{a+d+\sqrt{\Delta}}{a+d-\sqrt{\Delta}}$.

1.7.7. $A = \tfrac{1}{2}(-1+i\sqrt{3})$, $\alpha = \tfrac{1}{2}(1-\sqrt{3})(1+i)$, $\beta = \tfrac{1}{2}(1+\sqrt{3})(1+i)$.

1.7.8. The equation $cz^2 + (d-a)z - b = 0$ for invariant points has no finite roots, iff $c = 0$, $d = a$, $b \neq 0$, i.e. $w = z+h$; if $W = (w-\alpha)^{-1}$, $Z = (z-\alpha)^{-1}$, then necessarily $W = Z+h$, since ∞ is the only invariant point.

1.7.9. $(R-w)^{-1} = (R-z)^{-1} + ih$, h is real.

1.7.10. The angle subtended by $[\alpha, \beta]$ at $z \in C_z$ is equal to $\mp\dfrac{\pi}{2}$, hence for $w \in C_z$ we have:

$$\varphi = \arg \frac{w-\alpha}{w-\beta} = \arg A \mp \frac{\pi}{2} = \theta \mp \frac{\pi}{2};$$

note that $2R = |\alpha-\beta| \left|\operatorname{cosec}\left(\theta \mp \frac{\pi}{2}\right)\right|$.

1.7.11. Consider the sequence $Z_n = (z_n-\alpha)^{-1}$; if the transformation is parabolic and $Z_n = (z_n-\alpha)(z_n-\beta)^{-1}$, it has two finite, invariant points α, β. In the former case $Z_n \to \infty$, or $z_n \to \alpha$ (cf. Ex. 1.7.8); in the latter case the behavior of $\{z_n\}$ depends on the constant A in the canonical representation of Exercise 1.7.5. If $|A| \neq 1$, $\{z_n\}$ is convergent to one of the invariant points, otherwise $\{z_n\}$ is divergent.

1.7.12. The sequence $\{z_n\}$ is divergent; $A = \dfrac{-1+i\sqrt{3}}{2}$; the points $0, -1, -i$ are points of accumulation of $\{z_n\}$.

1.7.13. If A is real, the circle through α, β is mapped onto itself; if $A = e^{i\varphi}$ the circles of Apollonius with limit points α, β are mapped onto themselves.

1.7.14. Put $W = (w-\alpha)^{-1}$, $Z = (z-\alpha)^{-1}$ and consider the family of straight lines parallel to the vector h (cf. Ex. 1.7.8), which remain invariant under the transformation $W = Z+h$.

1.7.15. If $\alpha \notin C$ and α^* is its reflection w.r.t. C, then α^* is invariant, too. However, there are at most two invariant points, hence $\beta = \alpha^*$.

1.7.16. If C is mapped onto itself, then either α, β are symmetric, or both lie on C. In the former case $\left|\dfrac{w-\alpha}{w-\beta}\right| = \left|\dfrac{z-\alpha}{z-\beta}\right|$ for $w, z \in C$ (cf. Ex. 1.5.2), which means that $|A| = 1$. In the latter case $\arg \dfrac{w-\alpha}{w-\beta} = \arg \dfrac{z-\alpha}{z-\beta}$ ($w, z \in C$) which means that A is real.

1.7.17. The discriminant of the quadratic equation for invariant points is equal to $(a+d)^2-4$ (cf. Ex. 1.7.6, 1.7.5).

1.7.18. Cf. Exercise 1.2.15.

1.7.19. We find h such that $\alpha+h = a$; $\beta+h = -\bar{a}^{-1}$ which is possible, when $|\beta-\alpha| = |a+\bar{a}^{-1}| = |a|+|a|^{-1} \geq 2$. The elliptic transformations with invariant points $a, -\bar{a}^{-1}$ correspond to rotations of the sphere (cf. Ex. 1.2.15).

1.7.20. If $a-\bar{a} = 2ih$, invariant points α, β are roots of the equation $\bar{b}z^2 + 2ihz+b = 0$, and consequently $\alpha\bar{\beta} = -1$; $A = (a+\bar{a}+2i\sqrt{|b|^2+h^2})/(a+\bar{a}-$

1. COMPLEX NUMBERS. LINEAR TRANSFORMATIONS

$-2i\sqrt{|b|^2+h^2}) = e^{i\varphi}$, where φ is the angle of rotation of the sphere. Hence the transformation has the form as in Exercise 1.2.15 with

$$\varphi = 2\arg(\operatorname{re} a + i\sqrt{|b^2|+(\operatorname{im} a)^2}).$$

1.7.21. $(w-1-i)/(2w+1+i) = \exp(i\pi/3)(z-1-i)/(2z+1+i)$.

1.7.22. $w = (ihz+b)/(-\bar{b}z-ih)$, h is real, a is purely imaginary (cf. Ex. 1.7.20).

1.7.23. Since w, z may be interchanged, $A^2 = 1$ and consequently $A = -1$. From $w = (az+b)(cz-a)^{-1}$ we have $\frac{1}{2}(\alpha+\beta) = \frac{a}{c}$ which is the preimage of $w = \infty$. Now, $\alpha, \beta, \frac{1}{2}(\alpha+\beta), \infty$ are situated on one straight line, their images being $\alpha, \beta, \infty, \frac{1}{2}(\alpha+\beta)$, resp., which means that the straight line considered is mapped onto itself.

1.7.24. The straight line through the invariant points $\frac{1}{2}(\mp\sqrt{3}-i)$, as well as the symmetry line of the segment whose end points are the invariant points (cf. Ex. 1.7.15).

1.8.1. If $z_1 \neq 0$, the circle through z_1, z_2 which intersects $C(0; 1)$ at a right angle, also contains $\bar{z}_1^{-1}$, hence it is uniquely determined.

1.8.2. It is determined by three different points: $z_1, \bar{z}_1^{-1}, e^{i\alpha}$.

1.8.3. L meets $C(0; 1)$ in two points $e^{i\alpha}, e^{i\beta}$; the pairs $\{z_1, e^{i\alpha}\}, \{z_1, e^{i\beta}\}$ determine two h-lines L_1, L_2 parallel to L. Each h-line through z_1 situated between L_1 and L_2 does not meet L.

1.8.4. $C(0; 1)$ remains unchanged, hence $z_0, \bar{z}_0^{-1}$ are invariant points and the corresponding linear transformation must be elliptic (cf. Ex. 1.5.2, 1.7.13, 1.7.15). Hence

$$(w-z_0)(w-\bar{z}_0^{-1})^{-1} = e^{i\theta}(z-z_0)(z-\bar{z}_0^{-1})^{-1}, \quad \theta \text{ is real.}$$

1.8.5. If z_1, z_2 are invariant points, $|z_1| = |z_2| = 1$, the linear transformation must be hyperbolic because $C(0; 1)$ is mapped onto itself (cf. Ex. 1.7.13, 1.7.15). Hence $(w-z_1)/(w-z_2) = k(z-z_1)/(z-z_2)$ with k real. If $z = \infty$, then $k = (w_0-z_1)/(w_0-z_2)$ is real and w_0 is outside $[z_1, z_2]$, hence $k > 0$. This condition is also sufficient.

1.8.6. $z_1/(w-z_1) = z_1/(z-z_1)+ih$, with real h and $z_1 = 1$ (consider $W = z_1/(z-z_1)$).

1.8.7. $w = e^{i\beta}(z-a)/(1-\bar{a}z)^{-1}$, with real β and $|a| < 1$. Hence $z = e^{-i\beta} \times (w+ae^{i\beta})/(1+\bar{a}we^{-i\beta})$, thus the inverse transformation is again an h-motion

If $w = e^{i\gamma}(Z-b)/(1-\bar{b}Z)$, $Z = e^{i\beta}(z-a)/(1-\bar{a}z)$, the superposition has the form:
$$w = \exp\{i(\beta+\gamma)\}(1+\bar{a}be^{-i\beta})(z-c)[(1+a\bar{b}e^{i\beta})(1-\bar{c}z)]^{-1},$$
where $c = (a+be^{-i\beta})/(1+\bar{a}be^{-i\beta})$. Note that $|c| < 1$ and $|1+\bar{a}be^{-i\beta}| = |1+a\bar{b}e^{i\beta}|$.

1.8.8. $z = [t(z_2-z_1)+z_1(1-\bar{z}_1 z_2)][1-\bar{z}_1 z_2+t\bar{z}_1(z_2-z_1)]^{-1}$, $0 \leq t \leq 1$; the substitution $\zeta = (z-z_1)(1-\bar{z}_1 z)^{-1}$ gives the Euclidean segment $[0, \zeta_2]$ with parametric representation $\zeta = t\zeta_2$.

1.8.9. If $w(z)$ is the given h-motion, $(w(z)-b)/(1-\bar{b}w(z))$ is a linear transformation with a zero at a and a pole at a^{-1}, hence it has the form $A(z-a)/(1-\bar{a}z)$. Moreover, for $|w| = |z| = 1$ absolute values of both expressions are equal 1, hence $|A| = 1$.

1.8.10. $\dfrac{ds}{d\sigma} = (1-|z|^2)/(1-|\zeta|^2)$ (cf. Ex. 1.8.9); if $\delta = \max_k (t_{k+1}-t_k)$, then
$$\frac{\Delta s_k}{1-|z_k|^2} = \frac{\Delta \sigma_k}{1-|\zeta_k|^2} + \varepsilon_k, \quad \text{with} \quad |\varepsilon_k| < \varepsilon = \varepsilon(\delta) \to 0 \text{ as } \delta \to 0.$$

1.8.11. If the arc-length s on γ is the parameter and $C(0; r)$ intersects γ at two points corresponding to s_1, s_2, then after removing the open arc $s_1 < s < s_2$ and rotating suitably the remaining parts so as to join these, we obtain a new, shorter regular curve. Hence we may take $\theta = \theta(r)$ as the equation of γ in polar coordinates;
$$\int_0^R (1-|z|^2)^{-1}ds = \int_0^R (1-r^2)^{-1}[1+r^2(\theta'(r))^2]^{1/2}dr$$
which is a minimum for $\theta(r) = \text{const} = 0$.

1.8.12. After a suitable h-motion $e^{i\psi}(z-z_1)(1-\bar{z}_1 z)^{-1}$ the end points z_1, z_2 become 0, $R = |z_2-z_1|/|1-\bar{z}_1 z_2|$. The h-length of C remains unchanged after the h-motion (Ex. 1.8.10), hence (Ex. 1.8.11) the geodesic lines are h-lines and
$$\rho(z_1, z_2) = \rho(0, R) = \tfrac{1}{2}\log((1+R)/(1-R)) = \operatorname{ar\,tanh} R.$$

1.8.13. The points z on the h-circle satisfy
$$|z-z_0|/|z-\bar{z}^{-1}| = |z_0|\tanh R$$
which is an Apollonius circle with limit points z_0 and its reflection w.r.t. $C(0; 1)$. After a suitable h-motion h-circle becomes $C(0; \tanh R)$, thus $l_h = \pi \sinh 2R$.

1.8.14. Only the triangle inequality is non-trivial. It follows, however easily from Exercise 1.8.12.

1. COMPLEX NUMBERS. LINEAR TRANSFORMATIONS

1.8.15. The linear transformation considered in Exercise 1.8.12 gives:
$$(z_1, z_2, e^\beta, e^{i\alpha}) = (0, R, 1, -1) = (1+R)/(1-R);$$
(cf. Ex. 1.8.12, 1.4.7).

1.8.16. Note that the Jacobian
$$\frac{\partial(\xi, \eta)}{\partial(x, y)} = \left(\frac{d\sigma}{ds}\right)^2 = \left(\frac{1-|\xi|^2}{1-|z|^2}\right)^2$$
and use the formula for the integral transformed to new variables.

1.8.17. $|\Omega|_h = \pi R^2 (1-R^2)^{-1}$.

1.8.18. Suppose the angle at $z = 0$ is equal γ. Using the polar coordinates we obtain:
$$|T|_h = \iint_T \frac{r}{(1-r^2)^2}\, dr\, d\theta = \frac{1}{2}\int_{\theta_1}^{\theta_2} \frac{r^2(\theta)}{1-r^2(\theta)}\, d\theta,$$
where $r = r(\theta)$ is the equation of the side AB in polar coordinates. Let $C(s;R)$ be the circle containing the side AB and N a variable point on AB. We now

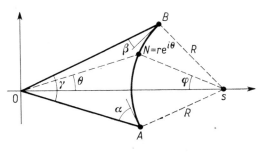

FIG. 1

introduce the new variable $\varphi = \sphericalangle OsN$; obviously $\tan\theta = \dfrac{R\sin\varphi}{d-R\cos\varphi}$, where $d = |s|$, thus
$$\sec^2\theta\, d\theta = (d-R\cos\varphi)^{-2} R(d\cos\varphi - R)\, d\varphi.$$
However,
$$\sec^2\theta = 1+\tan^2\theta = (d-R\cos\varphi)^{-2}(d^2+R^2-2dR\cos\varphi),$$
hence
$$d\theta = (d^2+R^2-2dR\cos\varphi)^{-1} R(d\cos\varphi - R)\, d\varphi.$$
Now, $r^2 = d^2+R^2-2dR\cos\varphi$, $1-r^2 = 2R(d\cos\varphi - R)$ because $1+R^2 = d^2$ (orthogonality condition).

Finally
$$(1-r^2)^{-1}r^2 d\theta = \tfrac{1}{2} d\varphi.$$
Considering the quadrangle $OAsB$ we easily verify that
$$\gamma + (\beta + \tfrac{1}{2}\pi) + (\alpha + \tfrac{1}{2}\pi) + \Delta\varphi = 2\pi$$
and this gives
$$|T|_h = \tfrac{1}{4}\int d\varphi = \tfrac{1}{4}\Delta\varphi = \tfrac{1}{4}[\pi - (\alpha+\beta+\gamma)].$$

1.8.19. The angle α at z_1 is equal to
$$\arg\{(z_3-z_1)(1-z_2\bar{z}_1)[(z_2-z_1)(1-z_3\bar{z}_1)]^{-1}\}$$
because after an h-motion considered in Exercise 1.8.12 two sides meeting at z_0 become segments.

The remaining angles are obtained by cyclic permutations. From Exercise 1.8.18 it follows that
$$|T|_h = \tfrac{1}{2}\arg(1-z_1\bar{z}_2)(1-z_2\bar{z}_3)(1-z_3\bar{z}_1)$$
if the orientation is positive.

CHAPTER 2

Regularity Conditions. Elementary Functions

2.1.1. (i), (iv) are continuous at $z = 0$; (ii), (iii) are discontinuous at $z = 0$: if $z = x(1+mi)$ and $x \to 0$, the limit depends on m.

2.1.2. Suppose that $\varepsilon > 0$ is arbitrary and $\delta > 0$ is such that $|f(z')-f(z'')| < \varepsilon$ for any $z', z'' \in K(0; 1)$ with $|z'-z''| < \delta$. If $|z_n-\zeta| < \frac{1}{2}\delta$ for all $n > N$, then $|z_m-z_n| < \delta$ and consequently $|f(z_m)-f(z_n)| < \varepsilon$ which means that $\{f(z_n)\}$ is convergent for any $\{z_n\} \to \zeta \in C(0;1)$.

2.1.3. $u_x(0) = u_y(0) = v_x(0) = v_y(0)$, however $f'(0)$ does not exist for f is discontinuous at $z = 0$. In fact, the limit of $u(x, y)$ along the line $y = mx$ depends on m.

2.1.4. Cauchy–Riemann equations are satisfied at $z = 0$ only.

2.1.5. Consider $\Delta z = \Delta x$ and $\Delta z = i\Delta y$.

2.1.6. If $\Delta z = h+ik$, then

$$\Delta u = hu_x+ku_y+o(\sqrt{h^2+k^2}), \quad \Delta v = hv_x+hv_y+o(\sqrt{h^2+k^2});$$

thus

$$\left|\frac{\Delta w}{\Delta z}\right| = \left[\frac{(hu_x+ku_y)^2+(hv_x+kv_y)^2}{h^2+k^2}\right]^{1/2}+o(1).$$

Assume that $k = 0$ and $h \to 0$, and vice versa which gives

(A) $\quad \lim\limits_{\Delta z \to 0}\left|\frac{\Delta w}{\Delta z}\right|^2 = u_x^2+v_x^2 = u_y^2+v_y^2;$

take now $k = mh$ which gives

(B) $\quad \lim\limits_{\Delta z \to 0}\left|\frac{\Delta w}{\Delta z}\right|^2 = u_x^2+v_x^2+2m(1+m^2)^{-1}(u_xv_x+u_yv_y), \quad$ or $\quad u_xv_x+u_yv_y = 0.$

It follows from (A) and (B) that $(u_x+iu_y)^2 = (v_x-iv_y)^2$ which means that either f, or $\bar{f}$ satisfies Cauchy–Riemann equations.

2.1.9. $u_x = v_y = 2uu_y$, $v_x = 2uu_x = -u_y$, thus $(1+4u^2)u_x = 0$, i.e. $u_x = 0$. Hence $v_y = v_x = u_y = 0$.

2.1.10. If $f = u+iv$, then $|f| = \sqrt{u^2+v^2}$; $f' = u_x+iv_x = v_y-iu_y$. Hence
$$\Delta|f| = |f|^{-1}(u_x^2+v_x^2+u_y^2+v_y^2)-|f|^{-3}[u^2(u_x^2+u_y^2)+v^2(v_x^2+v_y^2)-2uv(u_xv_x+u_yv_y)]$$
$$= |f|^{-1}2|f'|^2-|f|^{-3}(u^2+v^2)|f'|^2 = |f'|^2|f|^{-1},$$
because $u_xv_x+u_yv_y = 0$.

2.1.11. $\dfrac{\partial^2}{\partial x^2}|f|^2 = 2(uu_{xx}+vv_{xx}+u_x^2+v_x^2)$; $\dfrac{\partial^2}{\partial y^2}|f|^2$ has an analogous form; note that $\Delta u = \Delta v = 0$.

2.1.12. $U_r = r^{-1}V_\theta$, $V_r = -r^{-1}U_\theta$; $f'(z) = e^{-i\theta}(U_r+iV_r)$.

2.1.13. $U = r^n\cos n\theta$, $V = r^n\sin n\theta$;
$$f' = e^{-i\theta}nr^{n-1}(\cos n\theta+i\sin n\theta) = nz^{n-1}$$
(cf. Exercise 2.1.12).

2.2.1. In the right half-plane u has the form $\varphi(y/x)$ and the condition $\Delta u = 0$ gives $\dfrac{d}{dt}[\varphi'(t)(1+t^2)] = 0$, where $t = y/x$. Hence $\varphi(t) = A\arctan t+B$, or $u(z) = A\arg z+B$, where A, B are real constants.

2.2.2. u has the form $\varphi(x^2+y^2)$ and $\Delta u = 0$ gives $\dfrac{d}{dt}[t\varphi'(t)] = 0$, where $t = x^2+y^2$. Hence $\varphi(t) = A\log t+B$, or $u(z) = 2A\log z+B$, where A, B are real constants.

2.2.3. $u_x = x(x^2+y^2)^{-1} = v_y$, $u_y = y(x^2+y^2)^{-1} = -v_y$.

2.2.4. $u_x = e^x\cos y = v_y$, $u_y = -e^x\sin y = -v_x$;
$$\exp(\log|z|+i\arg z) = \exp(\log|z|)[\cos(\arg z)+i\sin(\arg z)]$$
$$= |z|\exp(i\arg z) = z;$$
$$\mathrm{Log\,exp}\,z = \log e^x+i\mathrm{Arg}(\cos y+i\sin y) = x+iy = z.$$

2.2.5. $f'(z) = u_x-iu_y$, hence
$$(\mathrm{Log}\,z)' = \frac{x-iy}{x^2+y^2} = z^{-1} \quad \text{and} \quad (\exp z)' = e^x\cos y+ie^x\sin y = \exp z.$$

2.2.6. If $f = u+iv$ and $u_x = v_y$, $u_y = -v_x$, then
$$U = \log|f| = \tfrac{1}{2}\log(u^2+v^2), \quad V = \operatorname{Arg}\frac{v}{u}.$$
Hence
$$U_x = (uu_x+vv_x)(u^2+v^2)^{-1} = (uv_y-vu_y)(u^2+v^2)^{-1} = V_y$$
and similarly $U_y = -V_x$; $F' = U_x - iU_y = f'\bar{f}|f|^{-2} = f'/f$.

2.2.7. $u_{xx} = -u_{yy}$ because $\Delta u = 0$; $u_{xy} = u_{yx}$ because the partials of second order are continuous.

2.2.8. Cauchy–Riemann equations imply
$$\frac{\partial v_1}{\partial x} = \frac{\partial v_2}{\partial x}, \quad \frac{\partial v_1}{\partial y} = \frac{\partial v_2}{\partial y}.$$

2.2.9. From Euler's formula for homogeneous functions:
$$u(x,y) = m^{-1}(xu_x+yu_y)$$
it follows by differentiation;
$$u_x = m^{-1}(u_x+xu_{xx}+yu_{yx}) = v_y = m^{-1}(u_x+yu_{xy}-xu_{yy}),$$
and similarly, $u_y = -v_x$.

2.2.10. (i) u is homogeneous of degree 2, hence by Exercise 2.2.9
$$v(x,y) = \tfrac{1}{2}[y(2x+y)-x(x-2y)] = \tfrac{1}{2}(y^2-x^2)+2xy; \quad f(z) = (1-\tfrac{1}{2}i)z^2.$$
(ii) $v(x,y) = -2x^3+3x^2y+6xy^2-y^3$, $f(z) = (1-2i)z^3$;
(iii) $v(x,y) = -y(x^2+y^2)^{-1}$, $f(z) = z^{-1}$;
(iv) $u = \operatorname{re} z^2 z^{-4} = \operatorname{re} z^{-2}$, $v = \operatorname{im} z^{-2}$, $f(z) = z^{-2}$.

2.2.11. The denominator has the form $(1+z^2)(1+\bar{z}^2)$, whereas the numerator is equal $\tfrac{1}{2}(z+\bar{z})(1+z\bar{z}) = \operatorname{re} z(1+\bar{z}^2)$. Hence
$$u = \operatorname{re} z(1+z^2)^{-1}, \quad f = z(1+z^2)^{-1},$$
$$v = \operatorname{im} z(1+z^2)^{-1} = y(1-x^2-y^2)[1+2(x^2-y^2)+(x^2+y^2)^2]^{-1}.$$

2.2.12. If v exists and $F = u+iv$, then $F' = u_x+iv_x = u_x-iu_y = f$. If a primitive $F = P+iQ$ does exist, then
$$F' = P_x+iQ_x = Q_y+iQ_x = f = u_x-iu_y.$$
Hence $u_x = Q_y$, $u_y = -Q_x$ which means that Q is a conjugate harmonic function for u.

2.2.13. $u_x-iu_y = (z+1)e^z$ has a primitive ze^z, hence
$$v = \operatorname{im} ze^z = e^x(y\cos y + x\sin y).$$

2.2.14. $r\dfrac{\partial}{\partial r}\left(r\dfrac{\partial F}{\partial r}\right)+\dfrac{\partial^2 F}{\partial \theta^2}=0.$

2.2.15. (i) $u=Axy+B,\ v=\tfrac{1}{2}A(y^2-x^2)+C$;

(ii) $u=A\sqrt{x+\sqrt{x^2+y^2}}+B,\ v=Ay(x+\sqrt{x^2+y^2})^{-1/2}+C$,
i.e. $u+iv=Az^{1/2}+B+iC$ (cf. Ex. 1.1.5);
(iii) do not exist.

2.2.16. Put $\varphi(t)=\int\exp(-\int\psi(t)dt)\,dt$ which means that $\varphi''+\varphi'\psi=0$ in (a,b). Hence
$$\Delta(\varphi\circ F)=(F_x^2+F_y^2)\varphi''\circ F+\Delta F\cdot\varphi'\circ F=(F_x^2+F_y^2)(\varphi''+\varphi'\psi)\circ F=0.$$

2.2.17. $(r^2F_r^2+F_\theta^2)^{-1}\left[r\dfrac{\partial}{\partial r}\left(r\dfrac{\partial F}{\partial r}\right)+\dfrac{\partial^2 F}{\partial \theta^2}\right]=\psi\circ F.$

2.2.18. The equation of the family of parabolas is $x+\sqrt{x^2+y^2}=p=\text{const.}$
If $F(x,y)=x+\sqrt{x^2+y^2}$, then $(F_x^2+F_y^2)^{-1}\Delta F=(2F)^{-1}$; hence $u=A\,\mathrm{re}\sqrt{z}+B$ (cf. Ex. 2.2.15 (ii)).

2.2.19. $f(z)=\exp(Az^{-1}+B)$ with A real.

2.2.20. $f(z)=\exp(Ai\,\mathrm{Log}\,z+B),\ A$ is real.

2.2.21. $u(r,\theta)=A(\log|z|-\lambda\arg z)+B;\ A,B$ are real.

2.2.22. $f(z)=\exp(-Aiz^{-2}+C),\ A$ is real.

2.3.1. $f'(-1)=-\tfrac{1}{2}$, hence $\alpha=\pi,\ \lambda=\tfrac{1}{2}$.

2.3.2. (i) The circle $|z+d/c|=|c|^{-1}|ad-bc|^{-1/2}$;
(ii) the straight line $z=-d/c+c^{-1}\sqrt{ad-bc}\,t,\ -\infty<t<+\infty$.

2.3.3. The tangent vector of $\Gamma=f(\gamma)$, i.e. $f'(z(t))z'(t)$ has a constant direction. If s is the arc length on Γ and $w=u+iv=f\circ z$, then $du/ds=a_1,\ dv/ds=a_2$ and hence $w=as+b,\ a=a_1+ia_2$.

2.3.4. $\arg f'(z)=0$ on the straight line of Exercise 2.3.2 (ii) hence on its image line: $\arg f'(z)z'(t)=\arg z'(t)=\text{const.}$

2.3.5. Infinitesimal segments are expanded (i) outside $C(0;\tfrac{1}{2})$, (ii) outside $C(-1;\tfrac{1}{2})$, (iii) inside $C(0;1)$.

2.3.6. $|f'(z_0)|=75,\ \arg f'(z_0)=\arg(7+24i)+\pi$.

2.3.7. $\dfrac{\partial(u,v)}{\partial(x,y)}=u_xv_y-u_yv_x=u_x^2+v_x^2=|u_x+iv_x|^2=|f'|^2$ which is equal to the ratio of area of infinitesimal squares corresponding to each other under f.

2. REGULARITY CONDITIONS. ELEMENTARY FUNCTIONS

2.3.8. The mapping is locally 1:1 because the Jacobian is $\neq 0$ at z_0; note that the curves $u(x, y) = \text{const}$, $v(x, y) = \text{const}$ are mapped onto perpendicular segments.

2.3.9. (i) equilateral hyperbolas $x^2-y^2 = \text{const}$, $xy = \text{const}$;
(ii) circles $C(0; r)$ and rays $\text{Arg}\, z = \text{const}$.

2.3.10. $\int_a^b |f'(z(t))|\,|z'(t)|\,dt$; $\iint_\Omega |f'(z)|^2\,dx\,dy$ (cf. Ex. 2.3.7).

2.3.11. $C(1; 1) = \{z : z = 2\cos\theta\, e^{i\theta}\}$, hence
$$w = 4\cos^2\theta\, e^{2i\theta} = 2(1+\cos\varphi)e^{i\varphi}, \quad \varphi = 2\theta;$$
$$\int_{-\pi/2}^{\pi/2} 2|z(\theta)|\,|z'(\theta)|\,d\theta = 16, \quad |\Omega| = \iint_{K(1;1)} 4|z|^2\,dx\,dy = 6\pi.$$

2.3.12. Under $w = z^2$ D is mapped onto the rectangle $\Delta: 1 < u < 4,\ 0 < v < 2$, hence
$$\iint_D |z|^2\,dx\,dy = \iint_D |w|\left|\frac{dz}{dw}\right|^2 du\,dv = \frac{1}{4}\iint_\Delta du\,dv = \frac{3}{2}.$$

2.3.13. $\frac{1}{4}\pi$.

2.4.1. $w = 1+2iz$; $|z^2-(1+2iz)| = |z-i|^2 < \frac{1}{100}$.

2.4.2. The boundary of the image domain consists of the segment $[-1, 1]$ and two arcs of parabolas symmetric w.r.t. Ov: $\mp u = 1-\frac{1}{4}v^2$, $u+iv = w$; $l = 2+\sqrt{2}+\log(1+\sqrt{2})$.

2.4.3. $w = (z^2-2i)(z^2+2i)^{-1}$.

2.4.4. $w = (1+z)^2$.

2.4.5. $w = 1-2(a/z)^2$ (consider first $Z = z^2$).

2.4.6. Superposition of two transformations:
$$w = (\sqrt{p}-\zeta)(\zeta-\sqrt{p}(\sqrt{2}-1))^{-1}, \quad z = \tfrac{1}{2}p-\zeta^2.$$

2.4.7. $w = i(2z^2-1)$.

2.4.8. If w_1, w_2 have the same image point, then
$$2a^2 w_1^2(1+w_1^2)^{-1} = 2a^2 w_2^2(1+w_2^2)^{-1}, \quad \text{i.e.} \quad w_1^2 = w_2^2.$$

Since $f(-w) = -f(w)$, we have $w_1 = w_2$ which means that $z = f(w)$ is univalent. If $|w^2| < 1$, then $\text{re}\{2a^2 w^2(1+w^2)^{-1}\} < a^2$ [consider the linear transformation $2a^2 W(1+W)^{-1}$ in $K(0;1)$].

Note that the domain between both branches of hyperbola is characterized by the inequality $\operatorname{re} z^2 < a^2$.

2.4.9. $z = \rho^{-2}(w^2 - a^2)$.

2.4.10. The mapping $W = w^2$ carries the inside of lemniscate into the double sheeted disk $K(a^2; \rho^2)$ with the branch point $W = 0$. A subsequent linear transformation $Z = \rho^2 W[\rho^4 + a^2(W-a^2)]^{-1}$ yields the double sheeted unit disk with branch point $Z = 0$. Ultimately

$$z = w\rho^{-1}[1 + a^2\rho^{-4}(w^2 - a^2)]^{-1/2}.$$

2.4.11. The transformation $W = \sqrt{z}$ (im $W > 0$) yields a strip parallel to the real axis; $w = (\sqrt{2}-1)^{-1}[2\sqrt{z} - i(1+\sqrt{2})]$.

2.5.1. The linear transformation $z = (\eta w + 1)(\eta w - 1)^{-1}$ with suitably chosen η ($|\eta| = 1$) carries C into the imaginary axis $\operatorname{re} w = 0$. If w_k are image points of z_k, $k = 1, 2$, then $z_1 z_2 = 1$ implies $w_1 + w_2 = 0$, i.e. the imaginary axis separates w_k, and consequently C separates z_k.

2.5.2. The roots z_1, z_2 of the equation $w = \frac{1}{2}(z + z^{-1})$ satisfy $z_1 z_2 = 1$ and this means (Ex. 2.5.1) that the mapping is univalent in either component of $\mathbf{C}\setminus C$. If C intersects the real axis at an angle α its image under the mapping $\zeta = (z-1)^2(z+1)^{-2}$ is a ray subtending with the real axis an angle 2α which is mapped under $\zeta = (w-1)(w+1)^{-1}$ onto a circular arc γ with end points $-1, 1$ inclined to the real axis at an angle 2α. Hence both components of $\mathbf{C}\setminus C$ are mapped 1:1 onto $\mathbf{C}\setminus\gamma$.

2.5.3, 4. Particular cases of Exercise 2.5.2, $\alpha = 0$ and $\alpha = \frac{1}{2}\pi$, resp.

2.5.5. A simple, closed Jordan curve containing the image arc C_1 of C inside and having a cusp at $w = 1$ with a common tangent with C_1.

2.5.6. (i) Ellipses $u^2[\frac{1}{2}(R+R^{-1})]^{-2} + v^2[\frac{1}{2}(R-R^{-1})]^{-2} = 1$ with foci at ∓ 1;
(ii) half-branches of hyperbolas $u^2 \cos^{-2}\theta - v^2 \sin^{-2}\theta = 1$ with foci at ∓ 1.

2.5.7. $w = z + z^{-1}$, $R = 7$ (cf. Ex. 2.5.6 (i)).

2.5.8. Take first R so that the circle $C(0; R)$ is mapped under $w = \frac{1}{2}c(Z + Z^{-1})$ onto the given ellipse and put $zR = Z$;

$$w = \frac{1}{2}[(a+b)z + (a-b)z^{-1}].$$

2.5.9. The mapping $Z = z + z^{-1}$ carries the given domain onto the outside of some segment and this is mapped by a suitable similarity onto $\mathbf{C}\setminus[-1, 1]$. Finally

$$\tfrac{1}{2}(w + w^{-1}) = [2(z + z^{-1}) - (b + b^{-1} - a - a^{-1})](a + a^{-1} - b - b^{-1})^{-1}.$$

2. REGULARITY CONDITIONS. ELEMENTARY FUNCTIONS

2.5.10. (i) $w = z+z^{-1}-2$; (ii) $W = \sqrt{w} = (1-z)z^{-1/2}$.

2.5.11. The mapping $Z_1 = \frac{1}{2}(z+z^{-1})$ yields the upper half-plane slit along $[\frac{3}{4}i, +i\infty)$ and this is mapped under $Z_2 = Z_1^2$ onto $\mathbf{C}\setminus\{(-\infty, -\frac{9}{16}] \cup [0, +\infty)\}$. Again, the linear transformation $Z_3 = 1+\frac{9}{16}Z_2^{-1}$ yields $\mathbf{C}\setminus[0, +\infty)$. Ultimately,
$$w = k\sqrt{Z_3} = (4z^4+17z^2+4)^{1/2}(1+z^2)^{-1}.$$

2.5.12. $w = z+z^{-1}$; $R_1 = \frac{1}{2}(\sqrt{5}+1)$, $R_2 = \frac{1}{2}(\sqrt{5}+3)$, $R_2/R_1 = \frac{1}{2}(\sqrt{5}+1)$.

2.5.13. The mapping $w = \frac{1}{2}(z+z^{-1})$, $\operatorname{im} z > 0$, carries the rays $\arg z = \theta$, $0 < \theta < \pi/2$ onto the right-hand branches of hyperbolas $u^2\cos^{-2}\theta - v^2\sin^{-2}\theta = 1$, hence it is just the desired mapping.

2.5.14. w is the image of $z \in C(0; r)$ under the mapping $w = z+z^{-1}$; note that $W = Z+Z^{-1}$, where $Z = z^3$.

2.5.15. The quadrant considered is swept out by arcs of confocal ellipses with foci ∓ 2 and each arc is mapped 1:1 onto an arc of confocal ellipse situated in the complementary domain of the first quadrant (cf. Ex. 2.5.14).

2.5.16. The transformation $Z = 2(z^2-1)$ carries the upper half of the domain considered into the domain of Exercise 2.5.15 which is mapped under $Z = w^3-3w$ onto the quadrant $(+; -)$ in the w-plane. One proves in an analogous manner that the same transformation carries the lower half of the domain considered into $(+; +)$. Ultimately $z^2 = 1+\frac{1}{2}(w^3-3w)$ maps $\{w: \operatorname{re} w > 0\}$ onto the given domain.

2.6.1. (i) $|e^z| = e^x\sqrt{\sin^2 y + \cos^2 y} = e^x$;
(ii) $\exp(z+2\pi i) = e^x[\cos(y+2\pi)+i\sin(y+2\pi)] = e^z$;
(iii) $(\exp z_1)(\exp z_2) = e^{x_1+x_2}[\cos y_1 \cos y_2 - \sin y_1 \sin y_2 + i(\sin y_1 \cos y_2 + \cos y_1 \sin y_2)] = e^{x_1+x_2}[\cos(y_1+y_2)+i\sin(y_1+y_2)] = \exp(z_1+z_2)$.

2.6.2. Putting $w = R(\cos\Phi + i\sin\Phi)$, $w = e^z$, and comparing absolute values, we obtain $x = \log R$; also $y = \Phi + 2k\pi$ is determined uniquely, Φ being given.

2.6.3. $\mathbf{C}\setminus(-\infty, 0]$; circles $C(0; e^{x_0})$ with deleted points $-e^{x_0}$; rays $\arg w = y_0$.

2.6.4. Logarithmic spiral $R = \exp[(\Phi-\pi)m^{-1}]$; R, Φ are polar coordinates.

2.6.5. w-plane slit along the logarithmic spiral $R = \exp[(\Phi-\pi)m^{-1}]$.

2.6.6. (i) On straight lines $y = k\pi$; (ii) on straight lines $y = (k+\frac{1}{2})\pi$, $k = 0$, $\mp 1, \mp 2, \ldots$; $\exp(2+i) = 3.992\ldots +i\,6.218\ldots$

2.6.7. The image domain of the square is the domain $|\arg w| < \varepsilon$, $e^{a-\varepsilon} < |w| < e^{a+\varepsilon}$, with area $\varepsilon(e^{2a+2\varepsilon} - e^{2a-2\varepsilon})$, hence

$$\lim_{\varepsilon \to 0} \frac{1}{4\varepsilon}(e^{2a+2\varepsilon} - e^{2a-2\varepsilon}) = e^{2a}.$$

2.6.8. The chord $[r_1 e^{i\theta}, r_2 e^{i\theta}]$, $-\arcsin\rho \leqslant \theta \leqslant \arcsin\rho$, of $\overline{K}(1;\rho)$ is mapped 1:1 under $w = \mathrm{Log}\, z$ onto the segment $[\log r_1 + i\theta, \log r_2 + i\theta]$ whose center is $\frac{1}{2}\log r_1 r_2 + i\theta = \frac{1}{2}\log(1-\rho^2) + i\theta$ which shows the symmetry of the image domain w.r.t. the straight line $\mathrm{re}\, w = \frac{1}{2}\log(1-\rho^2)$. Symmetry w.r.t. $\mathrm{im}\, w = 0$ is obvious. The parametric equation of the boundary of the image domain is: $w(\theta) = \mathrm{Log}(1+\rho e^{i\theta})$; hence

$$\arg w'(\theta) = \tfrac{1}{2}\pi + \theta - \arg(1+\rho e^{i\theta}),$$

$$\frac{d}{d\theta} \arg w'(\theta) = (1+\rho\cos\theta)/(1+2\rho\cos\theta+\rho^2) > 0$$

(cf. Ex. 1.1.20) which shows the convexity.

2.6.9. The image domain of $\mathbf{C} \setminus [\alpha, \beta]$ under $Z = (z-\alpha)(z-\beta)^{-1}$ is $\mathbf{C} \setminus (-\infty, 0]$ which is mapped under $\mathrm{Log}\, Z$ onto $\{w: |\mathrm{im}\, w| < \pi\}$; (i) straight lines $\mathrm{im}\, w = \mathrm{const}$; (ii) straight lines $\mathrm{re}\, w = \mathrm{const}$; (iii) $w = 0$.

2.6.10. The mapping is univalent as a superposition of linear transformation and the univalent mapping of Exercise 2.6.9. The mapping $W = (\beta-\alpha)w^{-1}$ carries the strip $|\mathrm{im}\, w| < \pi$ onto the outside of two circles with diameters $|\beta-\alpha|\pi^{-1}$ tangent to each other externally at the origin with the common tangent parallel to $[\alpha, \beta]$.

2.6.11. $\mathrm{Log}\, f$ is analytic in D, hence $\log\varphi(x) + \log\psi(y)$ is harmonic. This implies

$$\frac{d}{dx^2}\log\varphi(x) = -\frac{d}{dy^2}\log\psi(y) = 2a$$

(a is a real constant). Hence

$$\log\varphi(x) = ax^2 + b_1 x + c_1, \quad \log\psi(y) = -ay^2 + b_2 y + c_2;$$
$$\log|f(z)| = a(x^2-y^2) + b_1 x + b_2 y + c_1 + c_2, \quad \arg f(z) = 2axy + b_1 y - b_2 x + d_1$$

and finally

$$f(z) = \exp(az^2 + bz + c)$$

with real a and complex b, c.

2.6.12. The rays $\arg z = \theta$ are image lines of meridians, the circles $|z| = r$

2. REGULARITY CONDITIONS. ELEMENTARY FUNCTIONS

are image lines of parallels. Hence $w = u(\theta)+iv(r)$; $v(r)$ as a harmonic function constant on circles $|z| = r$ has the form $k\log|z|$ (cf. Ex. 2.2.2), thus

$$w = ik\,\text{Log}\,z = ik\,(\log|z|+i\arg z) = -k\theta+ik\log(\tan\tfrac{1}{4}\pi+\tfrac{1}{2}\varphi),$$

(cf. Ex. 1.2.8). The conformal mapping of the sphere so obtained which arises by a superposition of stereographic projection and logarithm is called Mercator projection.

2.7.1. It follows from the definition of e^z that $e^{ix} = \cos x+i\sin x$, $e^{-ix} = \cos x-i\sin x$ (x is real).

2.7.2. $\cos z = \tfrac{1}{2}(e^{-y+ix}+e^{y-ix}) = \cosh y \cos x - i\sinh y \sin x$,
$\sin z = \sin x \cosh y + i\cos x \sinh y$;
$\tan z = (\sin 2x + i\sinh 2y)(\cos 2x + \cosh 2y)^{-1}$.

2.7.3. $|\cos z|^2 = \cosh^2 y \cos^2 x + \sinh^2 y \sin^2 x = \sinh^2 y(1-\cos^2 x)+\cosh^2 y \cos^2 x = \cos^2 x + \sinh^2 y$; similarly $|\sin z|^2 = \sin^2 x + \sinh^2 y$; $\cos z = 0$ means $|\cos z|^2 = 0$, hence $y = 0$, $x = (k+\tfrac{1}{2}\pi)$, similarly $\sin z = 0$ for $y = 0$, $x = k\pi$ (k is an integer).

2.7.4. On vertical sides $|\sin z|^2 = 1+\sinh^2 y \geqslant 1$; on horizontal sides $|\sin z|^2 \geqslant \sinh^2 y \geqslant \sinh^2 \tfrac{1}{2}\pi > (2.29)^2 > 1$.

2.7.5. Cf. Exercises 2.7.4, 2.7.3.

2.7.6. E.g. $|\sin z| = \tfrac{1}{2}|e^{iz}-e^{-iz}| \leqslant \tfrac{1}{2}(|e^{iz}|+|e^{-iz}|) = \cosh y \leqslant \cosh R$ $(z = x+iy)$.

2.7.7. E.g. $\left|\dfrac{\sin z}{\cos z}\right|^2 \leqslant \dfrac{1+\sinh^2 y}{\sinh^2 y} \leqslant 1+(\sinh\delta)^{-2}$.

2.7.8. $\sin^2 z+\cos^2 z = \tfrac{1}{4}(e^{iz}+e^{-iz})^2-\tfrac{1}{4}(e^{iz}-e^{-iz})^2 = 1$.

2.7.9. E.g. $\sin(x-iy) = \cosh y \sin x + i\sinh y \cos x = \overline{\sin(x+iy)}$.

2.7.10. (i) sin, cos, tan are real on the real axis; moreover, sine is real on lines $x = (k+\tfrac{1}{2})\pi$ and cosine on lines $x = k\pi$.
(ii) Sine is purely imaginary on lines $x = k\pi$ and cosine is purely imaginary on lines $x = (k+\tfrac{1}{2})\pi$, whereas tangent is purely imaginary on lines $x = \tfrac{1}{2}k\pi$ ($k = 0, \mp 1, \mp 2, \ldots$).

2.7.11. $\cos(5-i) = \cos 5 \cosh 1 - i\sinh 1 \sin 5 \approx 0.438-1.127\,i$;
$\sin(1-5i) \approx 62.45-40.99i$;
$\sin z_0 = \sin \tfrac{1}{2}\pi \cosh[\log(4+\sqrt{15})] = \tfrac{1}{2}[4+\sqrt{15}+(4+\sqrt{15})^{-1}] = 4$.

2.7.12. Use the formulas of Exercise 2.7.1 and the property proved in Exercise 2.6.1 (iii).

2.7.13. $\cosh z = \cosh x \cos y + i \sinh x \sin y$, $\sinh z = \sinh x \cos y + i \cosh x \sin y$.

2.7.14. $|\sinh z|^2 = \sinh^2 x + \sin^2 y$, $|\cosh z|^2 = \sinh^2 x + \cos^2 y$.

2.7.15. $\cos z = \cosh iz$, $i \sin z = \sinh iz$, $i \tan z = \tanh iz$; hence cos arises from cosh by rotation of z-plane by an angle $\frac{1}{2}\pi$; similarly sine and tangent arise from corresponding hyperbolic functions by rotations of z-plane and w-plane by $\frac{1}{2}\pi$ and $-\frac{1}{2}\pi$, resp.

2.7.16. $\cot(\alpha+i\beta)+\cot(\alpha-i\beta) = \dfrac{\sin[(\alpha+i\beta)+(\alpha-i\beta)]}{[\sin(\alpha+i\beta)\sin(\alpha-i\beta)]} = \dfrac{2\sin 2\alpha}{\cos 2i\beta - \cos 2\alpha}$

$$= \dfrac{2\sin 2\alpha}{\cosh 2\beta - \cos 2\alpha};$$

$$i[\cot(\alpha+i\beta)-\cot(\alpha-i\beta)] = \dfrac{i\sin[(\alpha-i\beta)-(\alpha+i\beta)]}{\sin(\alpha+i\beta)\sin(\alpha-i\beta)} = \dfrac{2\sinh 2\beta}{\cosh 2\beta - \cos 2\alpha}.$$

2.7.17. The image line of the segment $(-\frac{1}{2}\pi+y_0, \frac{1}{2}\pi+y_0)$ is a half-ellipse $u = \sin x \cosh y_0$, $v = \cos x \sinh y_0$ with foci ∓ 1 for $y_0 \neq 0$, or the segment $(-1, 1)$ for $y_0 = 0$, resp. The correspondence between half-ellipses and segments is $1:1$, because u is a strictly increasing function of x. Moreover, the segments with y_0 varying sweep out the whole strip $|\mathrm{re}\, z| < \frac{1}{2}\pi$, whereas their images sweep out the domain $\mathbf{C} \setminus \{(-\infty, -1] \cup [1, +\infty)\}$. This implies univalence of sine in $\{z: |\mathrm{re}\, z| < \frac{1}{2}\pi\}$. The image lines of lines $x = x_0$ are branches of hyperbolas $u = \sin x_0 \cosh y$, $v = \cos x_0 \sinh y$ for $x_0 \neq 0$, and the imaginary axis for $x_0 = 0$.

2.7.18. A quarter of the ellipse with semi-axes $\sinh a$, $\cosh a$ and foci ∓ 1; the right angle at $z = \frac{1}{2}\pi$ corresponds to the angle π at $w = 1$ (note that the derivative is $= 0$ at this point).

2.7.19. Note that $\cos z = \sin(\frac{1}{2}\pi - z)$ and cf. Exercise 2.7.17.

2.7.20. $W = (1+w)/(1-w)$ maps $1:1$ the right half-plane onto $K(0; 1)$ in the W-plane and $[1, +\infty) \leftrightarrow (-1, 0]$. Hence $W = (1-\cos z)/(1+\cos z) = \tan^2 \frac{1}{2} z$.

2.7.21. The mapping $Z = \frac{1}{2}\pi \sqrt{z}$ carries $D \setminus (-\infty, 0]$ onto $\{Z: 0 < \mathrm{re}\, Z < \frac{1}{2}\pi\}$ which is mapped under $w = \tan^2 \frac{1}{2} Z = \tan^2 \frac{1}{4}\pi \sqrt{z}$ onto $K(0; 1) \setminus (-1, 0]$ (cf. Ex. 2.7.20). However, $\tan^2 \frac{1}{4}\pi \sqrt{z}$ is analytic on $(-\infty, 0]$ and maps it $1:1$ onto $(-1, 0]$ so that both slits can be removed.

2.8.1. $w = \mp i \log(z + \sqrt{z^2 - 1})$.

2.8.2. Cf. Exercise 2.7.19.

2. REGULARITY CONDITIONS. ELEMENTARY FUNCTIONS

2.8.3. $-i\operatorname{Log}(1+i) = \frac{1}{4}\pi - \frac{1}{2}i\log 2$.

2.8.4. Evaluate w from the equation $iz = (e^{2iw}-1)/(e^{2iw}+1)$; the mapping $Z = (1+iz)/(1-iz)$ carries $\mathbf{C}\setminus\{(-i\infty, -i]\cup[i, +i\infty)\}$ into $\mathbf{C}\setminus(-\infty, 0]$ and the latter domain is mapped under $w = \dfrac{1}{2i}\operatorname{Log} z$ onto $|\operatorname{re} w| < \frac{1}{2}\pi$.

2.8.5. The image lines of Apollonius circles $|(z-i)/(z+i)| = \text{const}$ are segments $\operatorname{im} w = \text{const}$; their orthogonal trajectories, i.e. circles through $i, -i$ are straight lines $\operatorname{re} w = \text{const}$.

2.8.6. (i) $\frac{1}{2}\operatorname{Arg}(-2+i)+\frac{1}{4}i\log 5$;
(ii) $\frac{1}{2}\pi+\frac{1}{2}i\log|\tan(\frac{1}{4}\pi+\frac{1}{2}\theta)|$.

2.8.7. The strip $|\operatorname{re} w| < \frac{1}{4}\pi$ (cf. Ex. 2.8.6 (ii)).

2.8.8. $w = \frac{1}{2}\operatorname{Log}((1+z)/(1-z))$; the strip $|\operatorname{im} w| < \frac{1}{4}\pi$, (cf. Ex. 2.8.7).

2.8.9. The half-strip: $u > 0$, $0 < v < \frac{1}{2}v$.

2.8.10. The mapping $Z = e^z$ carries the strip $0 < y < \pi$ into the upper half-plane which is mapped by $w = \operatorname{Arcsin} Z$ onto the half-strip $|\operatorname{re} w| < \frac{1}{2}\pi$, $\operatorname{im} w > 0$. The vertical boundary rays correspond to the rays: $x \geq 0$, $y = 0$ and $x \geq 0$, $y = \pi$ $(x+iy = z)$. Since the mapping function has symmetry property w.r.t. the boundary rays, we can propagate the mapping onto the whole z-plane with removed rays $y = k\pi$, $x \leq 0$ $(k = 0, \mp 1, \mp 2, \ldots)$ and the values will cover the upper half-plane $\operatorname{im} w > 0$.

2.9.1. The linear transformation $Z = (z-a)/(z-b)$ carries the wedge into an angular domain with vertex $Z = 0$ and a subsequent mapping $w = Z^{\pi/\alpha}$ gives a half-plane.

2.9.2. After the transformation $Z = z\exp[-\frac{1}{2}i(\alpha+\beta)]$ we obtain an angle symmetric w.r.t. the real axis; hence

$$w = Z^{\pi/(\beta-\alpha)} = z^{\pi/(\beta-\alpha)} \exp[-\frac{1}{2}\pi i(\alpha+\beta)/(\alpha-\beta)].$$

2.9.3. The linear transformation $Z = (z+1)/(z-1)$ carries the semi-disk into the quadrant $(-;-)$ and a subsequent mapping $w = Z^2$ gives the upper half-plane. Hence $w = [(z+1)/(z-1)]^2$.

2.9.4. The mapping $W = z^3$ carries the given sector into the upper semi-disk which is mapped under $Z = [(W+1)/(W-1)]^2$ onto the upper half-plane. On

the other hand, $Z = i(1+w)/(1-w)$ maps $K(0;1)$ onto the upper half-plane $\operatorname{im} Z > 0$. Hence
$$i(1+w)/(1-w) = [(z^3+1)/(z^3-1)]^2.$$

2.9.5. $w = i(z+1)^3(z-1)^{-3}$ (cf. Ex. 2.9.1).

2.9.6. The mapping $Z = (z+i)/(z-i)$ carries the wedge into the angle $\frac{3}{4}\pi < \arg Z < \frac{5}{4}\pi$ which is mapped under $W = Z^2$ onto the right half-plane. Hence
$$w = i(W-1)/(W+1) = 2z(1-z^2)^{-1}.$$

2.9.7. $w = \tanh\frac{1}{2}\pi z$ (cf. Ex. 2.8.7); Apollonius circles with limit points $-1, 1$; circular arcs joining 1 to -1.

2.9.8. The linear transformation $Z = (1-z)^{-1}$ maps the given domain onto the strip $\frac{1}{2} < \operatorname{re} Z < 1$ which is carried under similarity $W = 2i(Z-\frac{3}{4})$ into the strip $|\operatorname{im} W| < \frac{1}{2}$; we now apply Exercise 2.9.7; finally
$$w = \tanh\pi i((1-z)^{-1}-\tfrac{3}{4}) = i\tan\pi((1-z)^{-1}-\tfrac{3}{4}).$$

2.9.9. The mapping $Z = z^{-1}$ carries the given domain into the strip $|\operatorname{im} Z| < \frac{1}{2}$ which is mapped under $w = \coth\frac{1}{2}\pi Z$ onto $\mathbf{C}\setminus \overline{K}(0;1)$ (cf. Ex. 2.9.7). Hence $w = \coth(\pi/2z)$.

2.9.10. Two circles intersect at an angle $\frac{1}{4}\pi$ at $0, 2$ and the third circle being an Apollonius circle with limit points $0, 2$ is orthogonal to former ones. The mapping $Z = z/(z-2)$ yields the circular sector: $0 < |Z| < 2$, $\frac{5}{4}\pi < \arg Z < \frac{3}{2}\pi$ which can be mapped onto $\operatorname{im} w > 0$ similarly as the sector of Exercise 2.9.4. Hence
$$w = [16(z-2)^4-z^4]^2[16(z-2)^4+z^4]^{-2}.$$

2.9.11. The mapping $Z = \sqrt{w+\frac{1}{4}}$ carries the slit w-plane into the right half-plane $\operatorname{re} Z > 0$ so that $(w=0) \leftrightarrow (Z=\frac{1}{2})$; the mapping $Z = \frac{1}{2}(1+z)/(1-z)$ carries $K(0;1)$ into the right half-plane, hence $w = z(1-z)^{-2}$ is the desired mapping. This is the Koebe function which plays an important role in various extremal problems in conformal mapping.

2.9.12. The linear mapping $Z = (1+2w)/(1-2w)$ carries the given domain into $\mathbf{C}\setminus(-\infty, 0]$ and a subsequent mapping $W = \sqrt{Z}$ gives $\{W: \operatorname{re} W > 0\}$. Hence $w = z(1+z^2)^{-1}$.

2.9.13. $w = (1+z)^2(1-z)^{-2}$.

2.9.14. The linear transformation $Z = (z-ih)/(z+ih)$ carries the slit half-plane into $K(0;1)\setminus(-1, 0]$ which is mapped under $W = Z(1-Z)^{-2}$ onto $\mathbf{C}\setminus(-\infty, 0]$; finally $w = 2ih\sqrt{W} = \sqrt{z^2+h^2}$.

2. REGULARITY CONDITIONS. ELEMENTARY FUNCTIONS

2.9.15. After the inversion $Z = z^{-1}$ we obtain the domain of Exercise 2.9.13, thus $w = (hz)^{-1}(z^2+h^2)^{1/2}$.

2.9.16. By Exercise 2.9.11 the given mapping carries $K(0; 1)$ into $\mathbf{C}\setminus(-\infty, -\frac{1}{4}\rho^{-1}(1-\rho)^2]$. Note that $(-1, 0]$ corresponds to $(-\frac{1}{4}\rho^{-1}(1-\rho)^2, 0]$.

2.9.17. $w = (1+\eta z)^2(1-\eta z)^{-2}$, $|\eta| = 1$.

2.9.18. $\left(\dfrac{1+t}{1-t}\right)^2 = \dfrac{(1-\rho)^2}{\rho} \dfrac{z}{(1-z)^2}$ (cf. Ex. 2.9.17, 2.9.16).

2.9.19. The mappings $Z = \dfrac{(1+\rho)^2}{4\rho} \dfrac{z}{(1-z)^2}$, $Z = \dfrac{t}{(1-t)^2}$ carry the slit disk and the full disk, resp. into the Z-plane slit along $(-\infty, -\frac{1}{4}]$ so that the desired mapping has the following implicit form:

$$\tfrac{1}{4}(1+\rho)^2\rho^{-1}z(1-z)^{-2} = t(1-t)^{-2}.$$

2.9.20. The linear transformation

$$z = -e^{-i\alpha}(\zeta-\alpha)/(1-\bar{\alpha}\zeta), \quad \text{where} \quad (b-a)/(1-\bar{a}b) = e^{i\alpha}|b-a|/|1-\bar{a}b|,$$

maps $D(a, b)$ onto $K(0; 1)\setminus(-1, -\rho]$ with $\rho = |b-a|/|1-\bar{a}b|$; moreover $(\zeta = a) \leftrightarrow (z = 0)$. The mapping function is obtained by substituting the above given values of z, ρ into the formula obtained in Exercise 2.9.19.

2.9.21. $\left(\dfrac{d\zeta}{dt}\right)_{t=0} = \left(\dfrac{dz}{dt}\right)_{t=0} : \left(\dfrac{dz}{d\zeta}\right)_{\zeta=a} = -4e^{i\alpha}\dfrac{|b-a||1-\bar{a}b|(|1-\bar{a}b|-|b-a|)}{(1-|b|^2)(|1-\bar{a}b|+|b-a|)}$.

2.9.22. If z_1, z_2 are roots of the equation $z^2+(\rho-w)z-w\rho^{-1} = 0$, then $z_1 z_2 = -w\rho^{-1}$; moreover,

$$z_1(z_1+\rho)(1+\rho z_1)^{-1} = z_2(z_2+\rho)(1+\rho z_2)^{-1} = w\rho^{-1}.$$

Hence $z_2 = -(z_1+\rho)(1+\rho z_1)^{-1}$ and this means that $|z_1| < 1$ implies $|z_2| > 1$, i.e. the mapping is univalent in $K(0; 1)$, as well as in $K(\infty; 1)$. If $|z| = 1$, then $|w| = \rho$. The end points of circular slit on $C(0; \rho)$ are found by solving the equation $dw/dz = 0$.

2.9.23. Put $\rho = \sqrt{2}$ in Exercise 2.9.22; $w = z(z+\sqrt{2})(1+z\sqrt{2})^{-1}$.

2.9.24. $w = Rz(z+\rho)(1+\rho z)^{-1}$, $\rho = \operatorname{cosec}\tfrac{1}{2}\alpha$.

CHAPTER 3

Complex Integration

In this chapter I denotes the integral to be evaluated

3.1.1. (i) $z(t) = (1+i)t$, $\operatorname{re} z(t) = t$, $t \in [0, 1]$;

$$I = \int_0^1 t(1+i)\,dt = \tfrac{1}{2}(1+i);$$

(ii) $z(\theta) = re^{i\theta}$, $\operatorname{re} z(t) = r\cos\theta$, $\theta \in [-\pi, \pi]$;

$$I = \int_{-\pi}^{\pi} r\cos\theta \cdot ire^{i\theta}\,d\theta = i\pi r^2.$$

3.1.2. For example

$$\int_\gamma u(x,y)\,ds = \int_a^b u(x(t),y(t))\sqrt{x'^2(t)+y'^2(t)}\,dt = \int_a^b u(x(t),y(t))|z'(t)|\,dt.$$

The same holds for $v = \operatorname{im} f$ and the result follows.

3.1.3. $z(\theta) = \cos\theta + i\sin\theta$, $|z-1| = 2\sin\tfrac{1}{2}\theta$, $0 \leqslant \theta \leqslant 2\pi$;

$$I = \int_0^{2\pi} 2\sin\tfrac{1}{2}\theta\,d\theta = 8.$$

3.1.4. (i) $z(t) = it$, $-1 \leqslant t \leqslant 1$;

$$I = \int_{-1}^{1} i|t|\,dt = i;$$

(ii) $z(\theta) = -\sin\theta - i\cos\theta$, $0 \leqslant \theta \leqslant \pi$;

$$I = \int_0^{\pi} (-\cos\theta + i\sin\theta)\,d\theta = 2i;$$

(iii) $z(\theta) = \sin\theta - i\cos\theta$, $0 \leqslant \theta \leqslant \pi$;

$$I = \int_0^{\pi} (\cos\theta + i\sin\theta)\,d\theta = 2i.$$

3. COMPLEX INTEGRATION

3.1.5. (i) If $\int_a^b f(t)dt = 0$, the inequality is obvious; it is easy to verify that $\int_a^b cf(t)dt = c\int_a^b f(t)dt$ for any complex c, hence for real θ we have:

$$\mathrm{re}\left[e^{-i\theta}\int_a^b f(t)dt\right] = \int_a^b \mathrm{re}[e^{-i\theta}f(t)]dt \leq \int_a^b |f(t)|dt.$$

Choose now θ such that $e^{-i\theta}\int_a^b f(t)dt = \left|\int_a^b f(t)dt\right|$;

(ii) $\left|\int_\gamma f(z)dz\right| = \left|\int_a^b f(z(t))z'(t)dt\right| \leq \int_a^b |f(z(t))||z'(t)|dt = \int_\gamma |f(z)||dz|$

(cf. Ex. 3.1.2).

3.1.6. By Exercise 3.1.5 (i):

$$\left|\int_\gamma f(z)dz\right| \leq \int_a^b |f(z(t))||z'(t)|dt \leq M\int_a^b |z'(t)|dt = ML.$$

3.1.7. On the lower side of ∂Q we have: $z(t) = z_0 - a(1+i) + 2at$ $(0 \leq t \leq 1)$, whereas on the upper side $z(t) = z_0 + a(1+i) - 2at$ $(0 \leq t \leq 1)$, thus the sum of both integrals is equal to $4i\int_0^1 \frac{dt}{1+(2t-1)^2} = \pi i$. The sum of integrals over two remaining sides is also equal πi.

3.1.8. The existence of a primitive in D would imply $\int_C (z-z_0)^{-1}dz = 0$ for any closed curve $C \subset D$ which contradicts Exercise 3.1.7.

3.1.9. z has in E a primitive $\frac{1}{2}z^2$, hence $\int_\Gamma z dz = 0$, and consequently $\int_\Gamma xdz = -i\int_\Gamma y dz$, $\int_\Gamma xdx+y dy = 0$; $\frac{1}{2}\int_\Gamma \bar{z}dz = \frac{1}{2}\int_\Gamma xdx+y dy + \frac{1}{2}i\int_\Gamma xdy-y dx = iA$; $\int_\Gamma xdz = \frac{1}{2}\int_\Gamma (z+\bar{z})dz = \frac{1}{2}\int_\Gamma \bar{z}dz$.

3.1.10. $\frac{4}{3}$.

3.1.11. πi.

3.1.12. From the existence of the total differential at z_0 it follows that

$$f(z_0+re^{i\theta}) = f(z_0)+f'_x(z_0)r\cos\theta + if'_y(z_0)r\sin\theta + ro(1)$$

for small $r > 0$. Hence

$$\int_{C(z_0,r)} f(z)dz = [f'_x(z_0)+if'_y(z_0)]i\pi r^2 + 2\pi r^2 o(1).$$

Note that Cauchy–Riemann equations at z_0 are equivalent to the equation $f'_x(z_0)+if'_y(z_0)=0$.

3.1.13. $\int_{[a,b]} f(z)dz = F(b)-F(a) = (b-a)\int_0^1 f[a+(b-a)t]dt$; consider now a partition $t_0=0<t_1<\ldots<t_m=1$ of the interval $[0,1]$ and put $f_k=f[a+\Delta t_k(b-a)]$, $\Delta t_k = t_k - t_{k-1}$. The point $\zeta = \sum_{k=1}^m f_k \Delta t_k$, as the center of mass of a system of particles $f_1,\ldots,f_m$ with masses $\Delta t_1,\ldots,\Delta t_m$ belongs to $\operatorname{conv}[f_1,\ldots,f_m] \subset \operatorname{conv}\Gamma$ (cf. Ex. 1.1.29). Consider now a normal sequence of partitions and a corresponding sequence $\{\zeta_n\}$. Then $\zeta_n \to \zeta = \int_a^b f(z)dz$, and on the other hand $\zeta_n \in \operatorname{conv}\Gamma$ which is a closed set. Hence $\zeta \in \operatorname{conv}\Gamma$.

3.1.14. Suppose that $z_1 \neq z_2$ and $F(z_1)=F(z_2)$. It follows from convexity of Δ that $[z_1,z_2] \subset \Delta$. If Γ is the curve: $w = F'[z_1+(z_2-z_1)t]$ $(0\leqslant t\leqslant 1)$, then $F(z_2)-F(z_1) = (z_2-z_1)\zeta$, where $\zeta = 0 \in \operatorname{conv}\Gamma$. However, $\operatorname{re} F'(z)>0$ implies $\operatorname{re}\zeta > 0$ for any $\zeta \in \operatorname{conv}\Gamma$, and this contradicts $\zeta = 0$.

3.1.15. $K(0;1)$ is a convex domain; $\operatorname{re} F'(z) = n\operatorname{re}(1+z^{n-1})>0$ for $z\in K(0;1)$.

3.1.16. The half-plane $\operatorname{re} z < 0$ is a convex domain; $\operatorname{re} F'(z) = 1+e^x\cos y > 0$ for $x<0$ $(z=x+iy)$.

3.1.17. A similar proof as in Exercise 3.1.16.

3.1.18. Continuity. If $\operatorname{dist}(z_0;\gamma) = \delta > 0$, then for any $z\in K(z_0;\tfrac{1}{2}\delta)$ we have:

$$|F(z)-F(z_0)| = |z-z_0|\left|\int_\gamma (\zeta-z)^{-1}(\zeta-z_0)^{-1}\varphi(\zeta)d\zeta\right| < |z-z_0|2ML\delta^{-2},$$

where $M = \sup|\varphi(\zeta)|$ on γ and L is the length of γ. This implies continuity.

Differentiability. We have:

$$\frac{F(z)-F(z_0)}{z-z_0} = \int_\gamma \frac{\varphi(\zeta)}{(\zeta-z)(\zeta-z_0)}d\zeta = \int_\gamma \frac{\varphi_1(\zeta)}{\zeta-z}d\zeta = F_1(z),$$

where $\varphi_1(\zeta) = (\zeta-z_0)^{-1}\varphi(\zeta)$. Since F_1 is continuous at any point not on γ, the limit $\lim_{z\to z_0} F_1(z)$ exists and is equal to $\int_\gamma (\zeta-z_0)^{-1}\varphi_1(\zeta)d\zeta$.

3.1.19. We have: $h'(t) = [z(t)-a]^{-1}z'(t)$ at continuity points of $z'(t)$. Consequently, $u'(t) = 0$ except for a finite set of values t, and from the continuity of $u(t)$ it follows that $u(t) = \operatorname{const} = u(\alpha) = z(\alpha)-a = \exp(-h(\beta))[z(\beta)-a]$, hence, by $z(\alpha) = z(\beta)$, it follows that $h(\beta) = 2k\pi i$.

3. COMPLEX INTEGRATION

3.1.20. $n(\gamma; a) =$ const as a continuous function of $a \in D$ (Ex. 3.1.18) whose all values are integers (Ex. 3.1.19).

3.1.21. $\hat{C} \setminus K$ is a connected set disjoint with γ hence $n(\gamma; a) =$ const $= 0$ in $\hat{C} \setminus K$, since $\infty \in \hat{C} \setminus K$.

3.1.22. $n(\gamma; a) = n(\gamma; \infty) = 0$.

3.1.23. If $z_0 = re^{i\theta}$ and $r < r(\theta)$, the segment $[0, z_0]$ does not contain any points of γ, hence

$$n(\gamma; z) = \text{const} = n(\gamma; 0) = (2\pi i)^{-1} \int_0^{2\pi} [i + r'(\theta)/r(\theta)] d\theta = 1$$

for any $z \in [0, z_0]$. On the other hand, for $z_1 = re^{i\theta}$ with $r > r(\theta)$ the ray: $\arg z = \theta$, $|z| \geqslant r$, does not meet γ and $n(\gamma; z) =$ const $= n(\gamma; \infty) = 0$ for any z on this ray. The corresponding sets E_0, E_∞ are obviously domains being open and arc-wise connected. $E_0 \cap E_\infty = \emptyset$, since $n(\gamma; z)$ takes different values in either set.

3.1.24. If $|z| < \delta < \delta_k$ for all $k > k_0$, then the points $0, z$ are situated in $K(0; \delta)$ which does not contain any points of γ_k. Hence $n(\gamma_k; z) = n(\gamma_k; 0) = 1$ (Ex. 3.1.20).

3.1.25. (i) 2, 1, 0; (ii) 0, 1, 0.

3.1.26. If $a \neq b$, then $(z-a)^{-1}(z-b)^{-1} = (a-b)^{-1}[(z-a)^{-1} - (z-b)^{-1}]$ and, consequently,

$$I = 2\pi i (a-b)^{-1}[n(\gamma; a) - n(\gamma; b)] = 0.$$

If $a = b$, again $I = 0$ by continuity of I as a function of a (Ex. 3.1.18).

3.1.27. Differentiate both sides of the formula of Exercise 3.1.26 $(m-1)$ times w.r.t. a and then $(n-1)$ times w.r.t. b (cf. Ex. 3.1.18).

3.1.28. I has the form $\sum_{k=0}^{n} a_k \int_{C(0;r)} z^{-n-1+k}(z-a)^{-1}dz$. Since $n(C(0;r); 0) = n(C(a;r); 0)$ (cf. Ex. 3.1.20), we have $I = 0$ by Exercise 3.1.27.

3.1.29. Use the equality $I = 2\pi i(a-b)^{-1}[n(\gamma; a) - n(\gamma; b)]$.

3.1.30. Differentiate both sides of the formula of Exercise 3.1.29 $(m-1)$ times w.r.t. a and $(n-1)$ times w.r.t. b. This gives

$$T = (-1)^n 2\pi i \binom{m+n-2}{m-1}(b-a)^{-m-n+1}.$$

3.1.31. $n(C(0; 2); i) = n(C(0; 2); -i) = 1$ (cf. Ex. 3.1.26).

3.1.32. Any polynomial has a primitive in the open plane, hence $\int_\gamma W(z)dz = 0$ for any closed curve γ.

3.1.33. $(z-a)^{-n}$ has a primitive in $\mathbf{C}\setminus a$ for any integer $n \geqslant 2$.

3.1.34. After developing R in partial fractions we obtain a polynomial plus terms of the form $A_{jn}(z-a_j)^{-n}$ and the integration cancels the polynomial, as well as all terms $A_{jn}(z-a_j)^{-n}$ with $n \geqslant 2$ (cf. Ex. 3.1.32, 3.1.33).

3.1.35. If $z \in C$, $\bar{z} = r^2 z^{-1}$, hence

$$|z-a|^2 = z\bar{z} - a\bar{z} - \bar{a}z + |a|^2 = r^2 - ar^2 z^{-1} - \bar{a}z + |a|^2,$$

or

$$I = \frac{ir}{\bar{a}} \int_C \frac{dz}{(z-a)(z-r^2/\bar{a})}.$$

Now, one of the points a, $r^2/\bar{a}$ is situated inside C, another one is outside C, hence $I = 2r||a|^2 - r^2|^{-1}$ (cf. Ex. 3.1.29).

3.1.36. $(2\pi)^{-1} \int_{C(0;r)} |z-1|^{-2} |dz| = r/(1-r^2)$ (cf. Ex. 3.1.35).

3.1.37. If γ is a curve consisting of $[1, r]$ and of a circular arc: $z = re^{i\theta}$ ($0 \leqslant \theta \leqslant \varphi$), then $\int_\gamma z^{-1} dz = \log r + i\varphi$; note that $\Gamma - \gamma$ is a cycle, hence

$$(2\pi i)^{-1} \int_{\Gamma-\gamma} \frac{dz}{z} = n(\Gamma-\gamma; 0) = k, \quad \text{i.e.} \quad \int_\Gamma z^{-1} dz = 2\pi i k + \int_\gamma z^{-1} dz.$$

3.2.1. $I = 0$ since Γ is contained in the rectangle $|x| < \frac{3}{2}$, $|y| < \frac{3}{4}$ where $(1+z^2)^{-1}$ is analytic.

3.2.2. $I = 0$ since $n(C; a) = 0$ for all a being the zeros of the denominator.

3.2.3. If $0 < r_1 < r_2 < R$, then $\Gamma = C(0; r_1) - C(0; r_2) \sim 0 \pmod{D}$, $D = \{z: 0 < |z| < R\}$; hence

$$\int_\Gamma z^{-1} f(z) dz = 0, \quad \text{or} \quad \int_{C(0;r_1)} z^{-1} f(z) dz = \int_{C(0;r_2)} z^{-1} f(z) dz.$$

3.2.4. $f(0) = (2\pi i)^{-1} \int_{C(0;r)} z^{-1} f(z) dz = (2\pi)^{-1} \int_0^{2\pi} f(re^{i\theta}) d\theta$, hence $I = 2\pi f(0)$.

3.2.5. Given u, find its complex conjugate v in $K(0; R)$, apply the formula of Exercise 2.3.4 for $u+iv$ and compare real parts of both sides.

3.2.6. $\log|re^{i\theta} - a| = \operatorname{re} \log(z-a)$ is harmonic in $C(0; |a|)$, hence $I = 2\pi \log|a|$.

3.2.7. Put $f(z) = (z-a)^{-1}(A+\varepsilon(z))$ and $z = a+re^{i\theta}$ ($\varepsilon(z) \to 0$ for $r \to 0$); hence
$$\int_{C(a;r)} f(z)\,dz = 2\pi i A + o(1) = 2\pi i A$$
by Exercise 3.2.3.

3.2.8. Let $C = C(0; R)$ be a circle containing γ inside and leaving all a_n with $n \geqslant N$ outside. If C_k is a circle center at a_k such that all $a_n, n \neq k$, are situated outside C_k, then
$$\Gamma = \gamma - \sum_{k=1}^{N} n_k C_k \sim 0 \pmod{D} \quad \text{for} \quad n_k = n(\gamma; a_k).$$
Hence
$$\int_{\Gamma} f(z)\,dz = 0, \quad \text{or} \quad \int_{\gamma} = \sum_{k=1}^{N} n_k \int_{C_k} = 2\pi i \sum_{k=1}^{N} n_k A_k,$$
(cf. Ex. 3.2.7).

3.2.9. $2\pi i [\tfrac{1}{2} n(C; 1) + \tfrac{1}{2} n(C; -1) - n(C; 0)]$.

3.2.10. $\tfrac{1}{2}\pi i$.

3.2.11. $2\pi i a^{-1} \sin a$.

3.2.12. $\tfrac{1}{2}(ze^z)''_{z=a} = e^a(1+\tfrac{1}{2}a)$.

3.2.13. (i) 1; (ii) $-\tfrac{1}{2}(z^{-1}e^z)''_{z=1} = -\tfrac{1}{2}e$.

3.2.14. $I = 2\pi i(a-b)^{-1}[f(a)-f(b)]$; on the other hand
$$|I| \leqslant 2\pi R M(R-|a|)^{-1}(R-|b|)^{-1} \to 0 \quad \text{as} \quad R \to \infty$$
since f is bounded: $|f(z)| \leqslant M$. Hence $I = 0$, or $f(a) = f(b)$.

3.2.15. $I = \dfrac{2\pi i}{(m-1)!} \dfrac{d^{m-1}}{dz^{m-1}} [(z-b)^{-1}]_{z=a} = -2\pi i(b-a)^{-m}$.

3.2.16. By Cauchy integral formula for the cycle $\Gamma = C - C(0; R)$ which is $\sim 0 \bmod (\mathbf{C} \setminus 0)$, and for the function $z^{-1}f(z)$ we have:
$$n(\Gamma; z) z^{-1} f(z) = (2\pi i)^{-1} \int_{\Gamma} [\zeta(\zeta-z)]^{-1} f(\zeta)\,d\zeta = (2\pi i)^{-1} \int_{C} [\zeta(\zeta-z)]^{-1} f(\zeta)\,d\zeta$$
since $\int_{C(0;R)} \to 0$ as $R \to +\infty$.

Now, $n(\Gamma; z) = 0$ for z situated inside C and $\neq 0$, while $n(\Gamma; z) = 1$ for z situated outside C.

3.2.17. If $A_k = \lim_{z \to z_k} (z-z_k)/P_n(z) = [(z_k-z_1) \ldots (z_k-z_{k-1})(z_k-z_{k+1}) \ldots (z_k-z_n)]^{-1}$, then $I = 2\pi i(\varepsilon_1 A_1 + \ldots + \varepsilon_n A_n)$ where $\varepsilon_k = n(C; z_k) = 0, 1$. Now, $A_1 + A_2 + \ldots + A_n = 0$, which is verified by taking $C = C(0; R)$ and $R \to +\infty$, hence I can take at most $2^n - 1$ different values (all $\varepsilon_k = 0$ and all $\varepsilon_k = 1$ give the same value 0).

3.3.1. $\Gamma = \gamma - n(\gamma, a)C(a; r) \sim 0 \pmod{K(a; R) \setminus a}$ for any $0 < r < R$, hence

$$\int_\Gamma f(z)dz = 0 = \int_\gamma f(z)dz - n(\gamma; a) \int_{C(a;r)} f(z)dz.$$

Note that $\int_{C(a;r)} f(z)dz = 0$ (Ex. 3.2.7).

3.3.2. Suppose that $0 < r_1 < |\zeta - a|$; then $\Gamma = C(a; r) - C(a; r_1) \sim 0 \pmod{K(a; R) \setminus a}$, hence

$$f(\zeta) = (2\pi i)^{-1} \int_\Gamma (z-\zeta)^{-1} f(z)dz = (2\pi i)^{-1} \int_{C(a;r)} (z-\zeta)^{-1} f(z)dz,$$

because $\int_{C(a;r_1)} = 0$ by Exercise 3.3.1.

3.3.3. The integral $(2\pi i)^{-1} \int_{C(a;r)} (z-\zeta)^{-1} f(z)dz = \varphi(\zeta)$ is an analytic function of $\zeta \in K(a; r)$ (cf. Ex. 3.1.18).

3.3.4. $\lim_{z \to 0} z^{-1} f(z) = \lim_{z \to 0} [f(z) - f(0)]/(z-0) = f'(0)$.

3.3.5. For any positive integer p and real $x \to +\infty$ we have $x^{-p} e^x \to +\infty$, hence $\lim_{z \to 0} z^p \exp(1/z)$ does not exist.

3.3.6. Put $z = 1/iy$; if $y \to +\infty$, then $z^p \sin(1/z) = (2i^{p+1})^{-1} y^{-p}(e^{-y} - e^y)$ has no finite limit for any positive integer p.

3.3.7. $(\sin z)' = \cos z$, hence $\lim_{z \to k\pi} \sin z/(z-k\pi) = \cos k\pi = (-1)^k$ and consequently $\lim_{z \to k\pi} (z-k\pi)^2 (\sin z)^{-2} = 1$.

3.3.8. (i) $0, \mp 1$ are simple poles, $z = \infty$ is a removable singularity;

(ii) $z = 1$ is a pole of second order, $z = \infty$ is a pole of order 3;

(iii) $z = \mp i$ are simple poles, $z = \infty$ is an essential singularity;

(iv) $z = (2k+1)\pi i$, $k = 0, \mp 1, \mp 2, \ldots$ are simple poles, $z = \infty$ is a point of accumulation of poles;

(v) $z = 1$ is an essential singularity, $z = \infty$ is a removable singularity;

3. COMPLEX INTEGRATION

(vi) $z = 1$ is an essential singularity, $z = 2k\pi i$ are simple poles, $z = \infty$ is an accumulation point of poles;

(vii) $z = 2[(2k+1)\pi]^{-1}$ are essential singularities, $z = 0$ is their accumulation point;

(viii) $z = 2[(2k+1)\pi]^{-1}$ are essential singularities, $z = 0$ is their accumulation point, $z = \infty$ is a removable singularity.

3.3.9. f is analytic in $K(0; 2R)$, hence $|f(z)| \leqslant M_1$ for all $z \in K(0; R)$; $f(\zeta^{-1})$ is analytic in $K(0; 2R^{-1})$, hence $|f(\zeta^{-1})| \leqslant M_2$ for $|\zeta| \leqslant R^{-1}$, and consequently $|f(z)| \leqslant \max(M_1, M_2)$ (cf. Ex. 3.2.14).

3.3.10. If a_k are poles of order n_k, $k = 1, 2, \ldots, N$, then

$$\varphi(z) = f(z) \prod_{k=1}^{N} (z-a_k)^{n_k}$$

is analytic except for a pole at ∞ and consequently φ is a polynomial.

3.3.11. We have: $f(z) = f(a) + (z-a)f_1(z)$, f_1 having a removable singularity at $z = a$; similarly $f_1(z) = f_1(a) + (z-a)f_2(z), \ldots, f_{n-1}(z) = f_{n-1}(a) + (z-a)f_n(z)$ and this implies:

$$f(z) = f(a) + (z-a)f_1(a) + (z-a)^2 f_2(a) + \ldots + (z-a)^n f_n(z).$$

Differentiating and putting $z = a$ we obtain $f^{(k)}(a) = k! f_k(a)$.

3.3.12. $\lim_{z \to a} (z-a)^{-m} f(z) = (m!)^{-1} f^{(m)}(a)$ (cf. Ex. 3.3.11).

3.3.13. $\lim_{z \to a} (z-a)^m f(z)^{-1} = m! [f^{(m)}(a)]^{-1} \neq 0$.

3.3.14. By Exercise 3.3.10 $f(z) = P_m(z)/P_n(z)$, where P_m, P_n are polynomials of order m and n, resp.; if e.g. $m > n$, then f has $q = m$ zeros, n finite poles and $z = \infty$ is a pole of order $m-n$.

3.3.15. If $0 < r < R$ and $M = \sup_{\varphi} |f(re^{i\varphi})|$ then

$$f_n(z) = (2\pi i)^{-1} \int_{C(a;r)} (\zeta-a)^{-n}(\zeta-z)^{-1} f(\zeta) d\zeta, \quad |z-a| < r$$

therefore $|f_n(z)| \leqslant r^{-n}(r-|z-a|)^{-1} Mr$. By Exercise 3.3.11 and $f^{(k)}(a) = 0$ we have:

$$|f(z) - f(a)| \leqslant r^{-n}(r-|z-a|)^{-1} Mr |z-a|^n \to 0$$

for fixed $z \in K(a; r)$.

3.3.16. In case no such m exists, we should have $f = \text{const}$ by Exercise 3.3.15.

3.3.17. If $f(a) \neq 0$, then $1/f$ is analytic in some neighborhood of a; if $f(a) = 0$, then f has a zero of order m at $z = a$ which means that $1/f$ has a pole of order m at a (cf. Ex. 3.3.16, 3.3.13).

3.3.18. We have $|f(z)-w_0| \geq \delta$, hence $\lim_{z \to a}[f(z)-w_0]^{-1}(z-a) = 0$ which means that $\varphi(z) = [f(z)-w_0]^{-1}$ is analytic in $K(a; R)$; by Exercise 3.3.17, $z = a$ is either a pole, or a removable singularity for both $1/\varphi$ and f.

3.3.19. By Exercise 3.3.18 the values taken by f in any annular neighborhood of a (i.e. in $\{z: 0 < |z-a| < \delta\}$) form a set dense in $\mathbf{C}$ which is mapped under any non-constant polynomial P onto a set dense in $\mathbf{C}$. This implies that $P \circ f$ cannot have a as a pole, or regularity point.

3.3.20. The assumption $|f(z)-w_0| \geq \delta > 0$ for all $z \in K(a;r) \setminus a$ gives a contradiction, similarly as in Ex. 3.3.18.

3.4.1. If $n(\gamma; a) = m$ and $0 < \delta < r$, then the cycle $\Gamma = \gamma - mC(a; \delta)$ $\sim 0 \pmod{K(a;r) \setminus a}$ and by Cauchy's theorem $\int_\Gamma f(z)dz = 0$. Now, $\omega = \int_{C(a;\delta)} f(z)dz$ does not depend on δ (Ex. 3.2.3), hence

$$\int_\gamma f(z)dz - m\omega = 0, \quad \text{or} \quad \int_\gamma [f(z)-\omega(2\pi i)^{-1}(z-a)^{-1}]dz = 0.$$

Hence $A = (2\pi i)^{-1}\omega = (2\pi i)^{-1} \int_{C(a;\delta)} f(z)dz$.

3.4.2. Note that $f - A_1(z-a)^{-1}$ has a primitive in $K(a; R)$ and consequently

$$\int_\gamma [f(z) - A_1(z-a)^{-1}]dz = 0$$

for any closed curve $\gamma \subset K(a; \delta) \setminus a$.

3.4.3. $\operatorname{res}(0; f) = 2$, $\operatorname{res}(1; f) = -\frac{3}{4}$, $\operatorname{res}(-1; f) = -\frac{5}{4}$.

3.4.4. Cf. Exercise 3.2.7.

3.4.5. Obviously $\operatorname{res}(0; \Gamma) = 1$; if for some nonnegative integer k we have $h\Gamma(-k+h) \to (k!)^{-1}(-1)^k$ as $h \to 0$, then

$$h\Gamma(-k-1+h) = (-k-1+h)^{-1}h\Gamma(-k+h) \to [(k+1)!]^{-1}(-1)^{k+1}.$$

3.4.6. (i) $\operatorname{res}(0; f) = -1$, $\operatorname{res}(1; f) = e$;
(ii) $\operatorname{res}(i; f) = \frac{1}{2}i$, $\operatorname{res}(-i; f) = -\frac{1}{2}i$.

3.4.7. $\operatorname{res}(-a; f) = (2\pi i)^{-1} \int_{C(-a;r)} f(z)dz = (2\pi i)^{-1} \int_{C(a;r)} f(\zeta)d\zeta = \operatorname{res}(a; f)$ for odd f. Similarly $\operatorname{res}(-a; f) = -\operatorname{res}(-a; f)$ for even f.

3.4.8. Use the representation:
$$f(z) = (z-a)^m[A+(z-a)\varphi(z)], \quad g(z) = (z-a)^n[B+(z-a)\psi(z)],$$
where $A \neq 0$, $B \neq 0$, and φ, ψ are analytic in some neighborhood of a; m, n are positive for a being a zero and negative for a being a pole.

3.4.9. $\lim_{z \to z_v} (z-z_v)(z^4+a^4)^{-1} = \lim_{z \to z_v} 1/4z^3 = -\frac{1}{4}z_v a^{-4}$ (cf. Ex. 3.4.4, 3.4.8).

3.4.10. $\text{res}(z_v; f) = \lim_{z \to z_v} z^{n-1}(z-z_v)(z^n+a^n)^{-1} = \lim_{z \to z_v} [nz^{n-1}-(n-1)z_v z^{n-2}]/nz^{n-1}$
$= 1/n$.

3.4.11. By Exercise 3.3.11:
$$(z-a)^{-k}f(z) = (z-a)^{-k}f(a)+(z-a)^{-k+1}f'(a)+ \ldots +(z-a)^{-1}\frac{f^{(k-1)}(a)}{(k-1)!}+f_k(z)$$
where f_k is analytic in some neighborhood of a (cf. Ex. 3.4.2).

3.4.12. The point $z = z_k$ is a removable singularity for $(z-z_k)^2 f(z)(\cos z)^{-2}$, hence by Exercise 3.4.11,
$$\text{res}(z_k; f(z)(\cos z)^{-2}) = \frac{d}{dz}[f(z)(z-z_k)^2(\cos z)^{-2}]_{z=z_k} = f'(z_k),$$
the corresponding limit can be evaluated by using Exercise 3.4.8.

3.4.13. $\text{res}(0; f) = \lim_{z \to 0} \frac{1}{2}\left(\frac{\sin \alpha z}{\sin \beta z}\right)'' = \alpha\beta^{-1}(\beta^2-\alpha^2)/6$.

3.4.15. $\int_{C(0;r)} P(z)dz = 0$, $\int_{C(0;r)} \varphi(z)z^{-2}dz = O(r^{-1})$ and having a constant value, the latter integral must vanish.

3.4.16. (i) $\text{Log}[(1-a\zeta)/(1-b\zeta)] = (b-a)\zeta+\zeta^2\varphi(\zeta)$, where φ is analytic in some neighborhood of $\zeta = 0$ (cf. Ex. 3.3.11); hence $\text{res}(\infty; f) = a-b$;
(ii) $\sqrt{(z-a)(z-b)} = \mp[z-\frac{1}{2}(a+b)-\frac{1}{8}(a-b)^2 z^{-1}+z^{-2}\varphi(z)]$, where $\varphi(z) = O(1)$ as $z \to \infty$. Hence $\text{res}(\infty; f) = \mp\frac{1}{8}(a-b)^2$. The above representation is obtained by putting $z = \zeta^{-1}$ and applying Exercise 3.3.11.

3.4.17. If $K(0;r)$ contains all finite singularities $a_1, a_2, \ldots, a_n$ and each $K(a_k; r_k)$ contains only one singularity, then
$$C - \sum_{k=1}^{n} C_k \sim 0 \pmod{\mathbf{C}\setminus\bigcup\{a_k\}}, \quad \text{where} \quad C = C(0;r), \; C_k = C(a_k; r).$$
Hence
$$\int_C f(z)dz = \sum_{k=1}^{n} \int_{C_k} f(z)dz = -2\pi i\,\text{res}(\infty; f) = 2\pi i \sum_{k=1}^{n} \text{res}(a_k; f).$$

3.4.18. (i) $\text{res}(-1;f) = 2\sin 2 = -\text{res}(\infty;f)$;
(ii) $\text{res}(2;f) = -\text{res}(\infty;f) = -143/24$.

3.4.19. (i) $\text{res}(0;f) = 1$, $\text{res}(1;f) = \text{res}(-1;f) = -\frac{1}{2}$;
(ii) $\text{res}(0;f) = \frac{1}{4}$, $\text{res}(2i;f) = \frac{1}{16}i(\cos 2 + i\sin 2)$,
$\text{res}(-2i;f) = -\frac{1}{16}i(\cos 2 - i\sin 2)$, $\text{res}(\infty;f) = \frac{1}{8}(\sin 2 - 2)$;
(iii) $\text{res}(k\pi;f) = 0$, $k = 0, \mp 1, \mp 2, \ldots$;
(iv) $\text{res}(k\pi;f) = -1$, $k = 0, \mp 1, \mp 2, \ldots$;
(v) $\text{res}(-1;f) = -\text{res}(\infty;f) = -\cos 1$;
(vi) $\text{res}(0;f) = 1/2$, $\text{res}(2k\pi i/n;f) = 1/2k\pi i$, $k = \mp 1, \mp 2, \ldots$,
(vii) $\text{res}(0;f) = 0$ for negative and odd positive n, $\text{res}(0;f) = (-1)^{n/2}[(n+1)!]^{-1}$ for even positive n, $\text{res}(\infty;f) = -\text{res}(0;f)$.

3.4.20. $\lim\limits_{z \to a} (z-a) f \circ \varphi(z) = \lim\limits_{\zeta \to \varphi(a)} f(\zeta)[\zeta - \varphi(a)] \left[\dfrac{\varphi(z) - \varphi(a)}{z-a} \right]^{-1} = A/\varphi'(a)$.

3.4.21. (i) $\left(z^{-2} + \dfrac{1}{1!} z^{-3} + \dfrac{1}{2!} z^{-4} + \ldots \right) \left[(\beta - \alpha) z + \dfrac{1}{2}(\beta^2 - \alpha^2)z^2 + \dfrac{1}{3}(\beta^3 - \alpha^3)z^3 + \ldots \right] = \ldots + \left[(\beta - \alpha) + \dfrac{1}{2!}(\beta^2 - \alpha^2) + \dfrac{1}{3!}(\beta^3 - \alpha^3) + \ldots \right] z^{-1} + \ldots$; hence $\text{res}(0;f) = e^\beta - e^\alpha$;

(ii) $\text{Log}[1 + (z-1)] \cos(z-1)^{-1} = \left[(z-1) - \dfrac{1}{2}(z-1)^2 + \dfrac{1}{3}(z-1)^3 - \ldots \right] \times$

$$\times \left[1 - \dfrac{1}{2!}(z-1)^{-2} + \dfrac{1}{3!}(z-1)^{-4} - \ldots \right]$$

$$= \ldots + \left(-\dfrac{1}{1}\dfrac{1}{2!} + \dfrac{1}{3}\dfrac{1}{4!} - \dfrac{1}{5}\dfrac{1}{6!} + \ldots \right) z^{-1} + \ldots;$$

$$\text{res}(1;f) = -\dfrac{1}{1}\dfrac{1}{2!} + \dfrac{1}{3}\dfrac{1}{4!} - \dfrac{1}{5}\dfrac{1}{6!} + \ldots$$

3.5.1. C is a contour containing inside one simple pole $-2^{-1/2}(1+i)$ with residue $(1+i)/4\sqrt{2}$, hence $I = (i-1)\pi/2\sqrt{2}$.

3.5.2. (i) $-\pi i/\sqrt{2}$; (ii) $-2\pi i/3$; (iii) $3\pi i/64$; (iv) 0 (show that $\text{res}(\infty;f) = 0$).

3.5.3. (i) There are two simple poles inside C: $a_{1,2} = \frac{1}{2} \mp i\frac{1}{2}\sqrt{3}$ with residues $-\frac{1}{3}a_k$;
(ii) $C(0;R)$ contains inside a simple pole $z = 0$, $\text{res}(0;f) = 1/2\pi i$, and 6

simple poles $a_\nu = \sqrt[3]{k}\varepsilon^\nu$, $\varepsilon = \exp(\pi i/3)$, on the circles $C(0;\sqrt[3]{k})$; $\mathrm{res}(a_\nu;f) = 1/6\pi i$;
(iii) there is one simple pole $\tfrac{1}{2}(1+i)$ inside γ, with residue $\tfrac{1}{4}(1+i)e^{\pi/2}$.

3.5.4. $\mathrm{res}(\infty;f) = \tfrac{1}{3}$, hence $I = -2\pi i/3$.

3.5.5. There is exactly one root of the equation $e^z - w = 0$, i.e. $z = \mathrm{Log}\,w$, situated in Q. Thus the integrand f has inside Q a simple pole $z = \mathrm{Log}\,w$ with

$$\mathrm{res}(\mathrm{Log}\,w;f) = \lim_{z \to \mathrm{Log}\,w} (z - \mathrm{Log}\,w)ze^z(e^z - w)^{-1} = \mathrm{Log}\,w.$$

3.5.6. $\mathrm{res}(\infty;f) = -(n+1)^{-1}(b^{n+1} - a^{n+1})$, hence $I = -2\pi i\,\mathrm{res}(\infty;f)$.

3.5.7. $\mathrm{res}(\infty;f) = \tfrac{1}{8}(a-b)^2$, $I = -2\pi i\,\mathrm{res}(\infty;f)$.

3.5.8. We have

$$[(z-a)(z-b)]^{-1/2} = z^{-1}\left(1 - \frac{a+b}{z} + \frac{ab}{z^2}\right)^{-1/2} = z^{-1} + \tfrac{1}{2}(a+b)z^{-2} + O(z^{-3}),$$

hence: (i) $\mathrm{res}(\infty;f) = -\tfrac{1}{2}(a+b)$; (ii) $\mathrm{res}(\infty;f) = -1$.

3.5.9. (i) $\int_{C(0;2)} [z(z+1)]^{-1/2} z\,dz = -\pi i$ (cf. Ex. 3.5.8 (i));
(ii) πi (cf. Ex. 3.5.8 (ii)).

3.5.10. We have $z^{n-1}(1+z^{-2})^{-1/2} = z^{n-1} - \dfrac{1}{2}z^{n-3} + \dfrac{1\cdot 3}{2\cdot 4}z^{n-5} - \ldots$, hence $\mathrm{res}(\infty;f) = 0$ for odd n and

$$\mathrm{res}(\infty;f) = (-1)^{k+1}\frac{1\cdot 3\cdot\ldots\cdot(2k-1)}{2\cdot 4\cdot\ldots\cdot 2k} \quad \text{for } n = 2k.$$

3.5.11. Let $H(R)$ be a contour consisting of an arc of $C(0;R)$, $R > 1$, cut off by the parabola and situated in the right half-plane and of an arc of Γ contained in $K(0;R)$. There are two simple poles of f inside $H(R)$: $\mp 2^{-1/2}(1+i)$ with residues $(8\sqrt{1+\sqrt{2}})^{-1}(-1-\sqrt{2}\mp i)$ and hence $\int_{H(R)} = \tfrac{1}{2}\pi i\sqrt{1+\sqrt{2}}$ since $H(R)$ is described in the negative sense. Note that the integral over the circular arc does not exceed in absolute value $R(R^4-1)^{-1}(R^2-1)^{-1/2}$ which tends to 0 as $R \to +\infty$, hence $\int_{H(R)} = \int_\Gamma$.

3.6.1. If $z = e^{i\theta}$, then $\cos\theta = \dfrac{1}{2}(z+z^{-1})$, $\sin\theta = \dfrac{1}{2i}(z-z^{-1})$, $d\theta = -iz^{-1}dz$; hence

$$I = -i\int_{C(0;1)} R\left[\frac{1}{2}(z+z^{-1}), \frac{1}{2i}(z-z^{-1})\right]z^{-1}dz.$$

3.6.2. We have
$$I = \frac{i}{2b} \int_{C(0;1)} [z^2(z-\alpha)(z-\beta)]^{-1}(z^2-1)^2 dz,$$
where α, β are roots of the polynominal $z^2+2ab^{-1}z+1$; if f is the integrand, then $\text{res}(0;f) = -2ab^{-1}$, $\text{res}(\alpha;f) = 2b^{-1}(a^2-b^2)^{1/2}$, β is situated outside $C(0;1)$ since $\alpha\beta = 1$;
$$I = -\frac{\pi}{b}[\text{res}(0;f)+\text{res}(\alpha;f)].$$

3.6.3. (ii) $\cos n\theta = \frac{1}{2}(z^n+z^{-n})$ for $z = e^{i\theta}$.

3.6.4. We have $\cos n\theta = \text{re}\, z^n$ for $z = e^{i\theta}$, hence
$$I = \text{re}\,\frac{1}{i}\int_{C(0;1)} \frac{(z^2+z+1)^n}{z^3+3z+1}\,dz;$$
the only singularity inside the unit disk is a simple pole $z_1 = \frac{1}{2}(-3+\sqrt{5})$ with residue $5^{-1/2}(3-\sqrt{5})^n$.

3.6.5. Let J_ν denote the integrals taken over the consecutive sides ∂R_n; then
$$J_1 = \int_{-\pi}^{\pi} \frac{x}{a-\cos x+i\sin x}\,dx = -i\int_{-\pi}^{\pi} \frac{x\sin x}{1+a^2-2a\cos x}\,dx$$
the real part of integrand being an odd function;
$$J_2+J_4 = 2\pi i \int_0^n \frac{dy}{a+e^y} \to 2\pi i \int_0^{+\infty} \frac{dy}{a+e^y} = \frac{2\pi i}{a}\log(1+a) \quad \text{as } n \to \infty;$$
$$J_3 = -\int_{-\pi}^{\pi} \frac{(in+x)}{a-e^n e^{-ix}}\,dx \to 0.$$

If $n > \log a$, there is only one simple pole inside R_n with residue $a^{-1}\log a$. Hence
$$J_1+2\pi i a^{-1}\log(1+a)+o(1) = 2\pi i a^{-1}\log a,$$
or
$$I = 2\pi a^{-1}\log[(1+a)/a], \quad a > 1.$$
If $0 < a < 1$, then $1/a > 1$ and this gives
$$\int_{-\pi}^{\pi} \frac{x\sin x}{1+a^{-2}-2a^{-1}\cos x}\,dx = 2\pi a\log(1+a), \quad \text{i.e.} \quad I = 2\pi a^{-1}\log(1+a).$$

3. COMPLEX INTEGRATION

3.6.7. We have

$$\int_{C(0;1)} e^z z^{-n-1} dz = i \int_0^{2\pi} \exp(\cos\theta) \exp[i(\sin\theta - n\theta)] d\theta$$

$$= \int_0^{2\pi} \exp(\cos\theta) \sin(n\theta - \sin\theta) d\theta + i \int_0^{2\pi} \exp(\cos\theta) \cos(n\theta - \sin\theta) d\theta$$

$$= 2\pi i/n!$$

since the integrand has a pole of order $n+1$ at the origin, the residue being equal to $1/n!$. Compare now the real and imaginary parts of both sides.

3.7.2. The inequality $\sin\theta > \dfrac{2}{\pi}\theta$ is a consequence of convexity of sine in $(0, \tfrac{1}{2}\pi)$;

$$\int_0^\pi \exp(-A\sin\theta) d\theta = 2\int_0^{\pi/2} \exp(-A\sin\theta) d\theta < 2\int_0^{\pi/2} \exp(-2A\theta/\pi) d\theta < \pi A^{-1}.$$

3.7.3. E.g. (iii) can be solved as follows. The only singularity of $e^{imz}(a^2+z^2)^{-2}$ in the upper half-plane is a double pole $z = ai$ with residue $-i\tfrac{1}{4}a^{-3}e^{-am}(1+am)$; if $R > a$, then

$$\int_{-R}^{R} e^{imx}(a^2+x^2)^{-2} dx + \int_{\Gamma(R)} e^{imz}(a^2+z^2)^{-2} dz = \tfrac{1}{2}\pi a^{-3}e^{-am}(1+am);$$

$$\left|\int_{\Gamma(R)}\right| < R(R^2-a^2)^{-2} \int_0^\pi e^{-mR\sin\theta} d\theta < \pi m^{-1}(R^2-a^2)^{-2} \to 0 \quad \text{as } R \to +\infty$$

(cf. Ex. 3.7.2).

Equating the real parts and making $R \to +\infty$, we obtain

$$\int_{-\infty}^{+\infty} (a^2+x^2)^{-2} \cos mx\, dx = \tfrac{1}{2}\pi a^{-3}e^{-am}(1+am);$$

(ii) integrate $ze^{iz}(z^2+a^2)^{-2}$ and compare imaginary parts of both sides after making $R \to +\infty$.

3.7.4. $\int_{\gamma_r} f(z) dz = i\int_{\theta_1}^{\theta_2} [b + \varepsilon(a+re^{i\theta})] d\theta = ib(\theta_2 - \theta_1) + o(1).$

3.7.5. We have: $\int_{[-R,-r]} = \int_{[r,R]}$; moreover, $\lim_{z\to 0} zf(z) = -2i$, hence $\int_{-\Gamma(r)} \to -2\pi$ as $r \to 0$ (cf. Ex. 3.7.4);

$$\left|\int_{\Gamma(R)}\right| \leqslant 2\pi R^{-1} \to 0 \quad \text{as} \quad R \to +\infty;$$

this implies
$$2\int_0^{+\infty} x^{-2}(1-e^{2ix})dx - 2\pi = 0.$$
Equating real parts we obtain $I = \tfrac{1}{2}\pi$.

3.7.6. Integrate $z^{-2}(z^2+a^2)^{-1}(1-e^{2miz})$ over the contour of Exercise 3.7.5 and compare after making $R \to +\infty$, $r \to 0$, the real parts of both sides. The integrand has a simple pole at $z = ai$ with residue $(2ia^3)^{-1}(e^{-2am}-1)$, $\int_{-\Gamma(r)} \to -2\pi m a^{-2}$ as $r \to 0$ (cf. Ex. 3.7.4), hence
$$2\int_0^{+\infty} x^{-2}(x^2+a^2)^{-1}(1-\cos 2mx)\,dx = \pi a^{-3}(2am+e^{-am}-1) = 4I.$$

3.7.7. We have
$$(\sin x)^3 = (2i)^{-3}(e^{ix}-e^{-ix})^3 = -(8i)^{-1}[e^{3ix}-e^{-3ix}-3(e^{ix}-e^{-ix})]$$
$$= -\tfrac{1}{4}\sin 3x + \tfrac{3}{4}\sin x = \operatorname{im}[\tfrac{1}{4}(1-e^{3ix})-\tfrac{3}{4}(1-e^{ix})].$$
Integrate now $f(z) = [\tfrac{1}{4}(1-e^{3iz})-\tfrac{3}{4}(1-e^{iz})]z^{-3}$ round the contour of Exercise 3.7.5. We have: $\lim_{z\to 0} zf(z) = \tfrac{3}{4}$ (de l'Hospital rule) which implies $\int_{-\Gamma(r)} \to -\tfrac{3}{4}\pi i$ as $r \to 0$ (Ex. 3.7.4), moreover, $\int_{\Gamma(R)} \to 0$. Hence
$$\int_{-[R,-r]} + \int_{[r,R]} = 2i\int_r^R \operatorname{im} f(x)\,dx = 2i\int_r^R (\sin x/x)^3 dx$$
and using the fact that the integral round the contour vanishes, we obtain
$$2i\int_0^{+\infty}(\sin x/x)^3 dx - \tfrac{3}{4}\pi i + o(1) = 0, \quad \text{i.e.} \quad I = 3\pi/8.$$

3.7.8. Integrate f round the contour of Exercise 3.7.5. If $0 < r < a < R$, then
$$\int_{[-R,-r]} + \int_{[r,R]} = 2\int_r^R x^{-3}(x^2+a^2)^{-1}(x-\sin x)\,dx,$$
$\int_{-\Gamma(r)} \to -\pi/2a^2$ as $r \to 0$ (Ex. 3.7.4), $\int_{\Gamma(R)} \to 0$ as $R \to \infty$. Moreover, $2\pi i \operatorname{res}(ai;f) = -\pi a^{-4}(a+e^{-a}-1)$, hence
$$2I - \pi/2a^2 + o(1) = -\pi a^{-4}(a+e^{-a}-1).$$

3.7.9. f has inside the rectangle a simple pole $z = \pi i$ with residue $-e^{a\pi i}$. Now,

$f(z+2\pi i) = e^{2\pi i a}f(z)$, hence the sum of integrals over the horizontal sides is equal to

$$(1-e^{2\pi i a}) \int_{-R}^{R} (1+e^x)^{-1}e^{ax}dx,$$

whereas a corresponding expression for vertical sides tends to 0 as $R \to +\infty$. Thus

$$(1-e^{2\pi i a}) \int_{-\infty}^{+\infty} (1+e^x)^{-1}e^{ax}dx + o(1) = -2\pi i e^{a\pi i},$$

i.e. $I = \pi/\sin a\pi$.
   (i) introduce the new variable $e^x = t$;
   (ii) take $t = x^n$, $a = m/n$ in (i).

3.7.10. $f(z) = e^{az}(1+e^z+e^{2z})^{-1}$ has inside the rectangle considered in Exercise 3.7.9 two poles $z_1 = 2\pi i/3$, $z_2 = 4\pi i/3$ with residues $(i\sqrt{3})^{-1}\exp[2\pi i(a-1)/3]$, $-(i\sqrt{3})^{-1}[\exp 4\pi i(a-1)/3]$. Similarly as before

$$(1-e^{2\pi i a})I + o(1) = 2\pi 3^{-1/2}[\exp 2\pi i(a-1)/3 - \exp 4\pi i(a-1)/3].$$

3.7.11. The only singularities of f situated inside Q_N are poles $z_k = (2k+1)i$, $k = 0, 1, \ldots, N-1$, z_0 being a double pole, all remaining poles being simple. We have: $\text{res}(z_0;f) = (2\pi i)^{-1}$, $\text{res}(z_k;f) = (2\pi i)^{-1}(-1)^{k+1}(1/k - 1/(k+1))$; $|f(z)| \leqslant [(N^2-1)(\cosh^2\frac{1}{2}\pi N - 1)]^{-1}$ on vertical sides of Q_N, whereas $|f(z)| \leqslant (4N^2-1)^{-1}$ on the upper horizontal side. Therefore the corresponding integrals tend to 0 as $N \to +\infty$ and, consequently,

$$\int_{-\infty}^{+\infty} f(x)dx = 1 + (1-\tfrac{1}{2}) - (\tfrac{1}{2}-\tfrac{1}{3}) + \ldots = 2\log 2.$$

3.7.12. Suppose that $0 < r < \tfrac{1}{2}\pi < R$ and J_ν is the integral of f taken over an arc indexed by ν as in Fig. 2, $\nu = 1$ to 6. If $\Gamma(r, R)$ is the contour of Fig. 2, then $\int_{\Gamma(r,R)} = 0 = \sum_{\nu=1}^{6} J_\nu$.

We have:

$$\text{re}(J_1+J_5) = (1-e^{a\pi})\int_r^R (e^{2y}-1)^{-1}\sin ay\, dy;$$

$\lim_{z \to 0} zf(z) = -\dfrac{1}{2i}$, hence $\lim_{r \to 0} J_2 = \dfrac{\pi}{4}$ (cf. Ex. 3.7.4); similarly $\lim J_4 = \pi e^{a\pi}/4$;

$$J_3 = \dfrac{i}{2}\int_r^{\pi-R} (e^{ix}-e^{-ix})^{-1} 2ie^{ix}e^{ax}dx,$$

hence $\operatorname{re} J_3 = (1-e^{a\pi})/2a + o(1)$; moreover, $\operatorname{re} J_6 \to 0$ as $R \to +\infty$. Therefore,
$$(1-e^{a\pi})I + \tfrac{1}{4}\pi(1+e^{a\pi}) - (e^{a\pi}-1)/2a + o(1) = 0$$
which gives I.

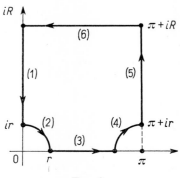

Fig. 2

3.7.15. (i), (ii): integrate $[z^{-1} - (\sinh z)^{-1}]z^{-1}$ and $(1-e^{aiz})(z\sinh z)^{-1}$, resp., round the boundary of the upper half-disk with radius $\pi(n+\tfrac{1}{2})$; the origin is a removable singularity in (i), in (ii) it should be omitted round $-\Gamma(r)$; (iii) put $x = e^t$ (cf. Ex. 3.7.9).

3.8.1. We have
$$\left| \int_{-\Gamma(r)} \right| \leqslant (a^2 - r^2)^{-1} (\log r^{-1} + \pi)^2 \pi r \to 0 \quad \text{as} \quad r \to 0:$$
$$\left| \int_{\Gamma(R)} \right| \leqslant (R^2 - a^2)^{-1} (\log R + \pi)^2 \pi R \to 0 \quad \text{as} \quad R \to +\infty.$$
Hence
$$\int_r^R (a^2 + x^2)^{-1} (\log x)^2 dx + \int_r^R (a^2 + x^2)^{-1} (\log x + \pi i)^2 dx + o(1)$$
$$= 2\pi i \operatorname{res}[ai; (a^2+z^2)^{-1}(\operatorname{Log} z)^2] = \pi a^{-1} (\log a + \tfrac{1}{2}\pi i)^2.$$
Making $r \to 0$, $R \to +\infty$ and separating real and imaginary parts, we find both integrals.

3.8.2. $\int_r^R (1+x^2)^{-2} \log x \, dx + \int_r^R (1+x^2)^{-2} (\log x + \pi i) dx + o(1)$
$$= 2\pi i \operatorname{res}[i; (1+z^2)^{-2} \operatorname{Log} z] = \tfrac{1}{4}\pi(-2 + \pi i).$$
Making $r \to 0$, $R \to +\infty$ and separating real and imaginary parts, we find I.

3.8.3. If $x > 0$, then $f(x) = x^a(1+x^2)^{-2}$; if $x < 0$, then $f(x) = (-x)^a(1+x^2)^{-2}\exp(a\pi i)$; $\left|\int_{-\Gamma(r)}\right| < (1-r^2)^{-2}r^{a+1}\pi \to 0$ for $-1 < a$ as $r \to 0$; $\left|\int_{\Gamma(r)}\right| < (R^2-1)^{-2}R^{a+1} \to 0$ for $a < 3$ as $R \to +\infty$. Hence for $-1 < a < 3$ we have:

$$[1+\exp(a\pi i)]I + o(1) = 2\pi i \operatorname{res}(i; f) = \tfrac{1}{2}\pi(1-a)\exp(\tfrac{1}{2}a\pi i),$$

or

$$I = \tfrac{1}{4}\pi(1-a)[\cosh(\tfrac{1}{2}a\pi i)]^{-1} = \tfrac{1}{4}\pi(1-a)[\cos(\tfrac{1}{2}a\pi)]^{-1}.$$

3.8.4. We have: $\int_{\Gamma(r)} \to 0$ as $r \to 0$, $\int_{\Gamma(R)} \to 0$ as $R \to +\infty$; moreover, $\int_{[r,R]} \to$

$$\to \int_0^{+\infty} (1+x^2)^{-1}\log x\, dx = 0$$

by Exercise 3.8.1. On the other hand the integral over the segment $[2^{-1/2}R(1+i), 2^{-1/2}r(1+i)]$ tends to $-2^{-1/2}(1+i)\int_0^{+\infty}(1+it^2)^{-1} \times$

$\times (\log t + \tfrac{1}{4}i\pi)\, dt$ as $r \to 0$, $R \to +\infty$. This implies $\int_0^{+\infty}(1+it^2)^{-1}(\log t + \tfrac{1}{4}i\pi)\, dt = 0$.
Separating real and imaginary parts and using Exercise 3.7.9 (ii) we obtain both integrals.

3.8.5. (i) Cf. Exercise 3.8.1;

(ii) $\int_{-R}^R (1+x^2)^{-1}\log[\sqrt{1+x^2} + i\operatorname{Arg}(x+i)]\, dx + o(1)$

$$= 2\pi i \operatorname{res}[i; (1+z^2)^{-1}\operatorname{Log}(z+i)] = \pi \operatorname{Log} 2i;$$

make $R \to +\infty$ and separate real and imaginary parts;

(iii) by Exercise 3.8.1 ($a=1$),

$$\int_0^{+\infty}(1+x^2)^{-1}\log(x+x^{-1})\, dx = \pi\log 2 = \int_0^1 + \int_1^{+\infty} = 2\int_0^1 \text{ (put } x = t^{-1}).$$

3.8.7. Integrate the branch of $\log(z-a)$ for which $0 < \arg(z-a) < 2\pi$;

$$\int_{C(0;1)} = \int_0^{2\pi} [\log|e^{i\theta}-a| + i\arg(e^{i\theta}-a)]\, d(e^{i\theta});$$

the integral over the lower edge of the slit is equal to $-\int_{a+r}^1 (\log|x-a| + 2\pi i)\, dx$

whereas an analogous integral for the upper edge is equal to $\int_{a+r}^1 \log|x-a|\, dx$;

$$\int_{-C(a;r)} = -\int_0^{2\pi}(\log r + i\theta)ire^{i\theta}\, d\theta = o(1) \quad \text{as} \quad r \to 0.$$

The sum of all four integrals is equal to 0 and the result follows by separating the real part.

3.8.9. (i) Suppose that $0 < r < \frac{1}{2}$ and $R > 1$ and integrate
$$f(z) = \{(1+z)[z^2(1-z)^{1/3}]\}^{-1},$$
which is analytic in $\mathbf{C}\setminus[0, 1]$, round the boundary of $\bar{K}(0; R)\setminus\{\bar{K}(0; r) \cup \cup \bar{K}(1; r) \cup [0, 1]\}$. Obviously $\int_{C(0;R)} = o(1)$ as $R \to +\infty$; if f is the branch taking positive values on the negative real axis, then on the upper edge of $[r, 1-r]$ we have:
$$f(z) = \exp(2\pi i/3)\{(1+x)[x^2(1-x)]^{1/3}\}^{-1}$$
and after describing $-C(1; r)$ the value of f on the lower edge is obtained by multiplying the former value by $\exp(2\pi i/3)$. Hence
$$[\exp(2\pi i/3) - \exp(4\pi i/3)]I + o(1) = 2\pi i \operatorname{res}(-1; f) = 2\pi i 2^{-1/3}$$
as $r \to 0$, $R \to +\infty$;

(ii) integrate $f(z) = z^{2n}[z(1-z^2)]^{-1/3}$ which is analytic in $\mathbf{C}\setminus[-1, 1]$ round the boundary of $K(0; R)\setminus\{[-1, 1] \cup \bar{K}(0; r) \cup \bar{K}(-1; r) \cup \bar{K}(1; r)\}$, where $R > \frac{3}{2}$ and $0 < r < \frac{1}{2}$; f takes opposite values on the upper edge of $[r, 1-r]$ and lower edge of $[-1+r, r]$ and vice versa, moreover, the integrals round the circles of radius r tend to 0 as $r \to 0$. Hence
$$2I[\exp(\pi i/3) - \exp(-\pi i/3)] + o(1) - 2\pi i \operatorname{res}(\infty; f) = 0.$$
Now,
$$f(z) = -z^{2n-1}(1-z^{-2})^{-1/3} = -z^{2n-1}\left(1 + \frac{1}{3}z^{-2} + \frac{1 \cdot 4}{3 \cdot 6}z^{-6} + \ldots\right)$$
and finally
$$2I\sin\frac{\pi}{3} = \pi\frac{1 \cdot 3 \cdot \ldots \cdot (3n-2)}{3 \cdot 6 \cdot \ldots \cdot 3n}.$$

3.8.10. After describing $-C(1; r)$ f is multiplied by $\exp(-2\pi pi)$; moreover, the integrals round $-C(0; r)$, $-C(1; r)$ and $C(0; R)$ tend to 0 as $r \to 0$, $R \to +\infty$. Hence
$$I[1 - \exp(-2\pi pi)] + o(1) = 2\pi i \operatorname{res}(-1; f) = 2\pi i e^{-\pi pi} 2^{p-3} p(p-1).$$

3.8.11. Integrate $f(z) = (1-z^n)^{-1/n}$ round the boundary of
$$K(0; R)\setminus\bigcup_{k=0}^{n-1}[0, \exp(2k\pi i/n)], \quad R > 1.$$
After describing the point $\exp(2k\pi i/n)$ the values on the opposite side of a corresponding slit are multiplied by $\exp(2\pi i/n)$; moreover, the integrals over

both edges of any slit are equal to $[1-\exp(2\pi i/n)]I$. Hence
$$n[1-\exp(2\pi i/n)]I = 2\pi i \operatorname{res}(\infty;f) = -2\pi i \exp(i\pi/n).$$

3.9.1. We have: $|-z| = 1 > |F(z)|$; hence by Rouché's theorem both equations: $-z = 0$, $-z+F(z) = 0$ have the same number of roots in $K(0; 1)$, i.e. exactly one.

3.9.2. Consider the variation of $\arg P(z)$ as z describes in the positive sense the boundary γ of $K(0; R) \cap (+, +)$: $\Delta_{[0,R]}\arg P(z) = 0$; $\Delta_{[iR,0]}\arg P(z) = -\arg[1+i(y^8+5)^{-1}(-3y^2+7y)] = O(R^{-5})$; on the circular arc $\Delta\arg P(z) = \Delta\arg[z^8(1+O(R^{-5}))] = 4\pi+o(1)$. Hence $\Delta_\gamma\arg P(z) = 4\pi+o(1) = 4\pi$ for all R sufficiently large. Now, $n(\Gamma, 0) = (2\pi)^{-1}\Delta_\gamma\arg P(z) = 2$.

3.9.3. P has no roots on coordinate axes; it is real on the real axis and attains a positive minimum at $x = \tfrac{1}{2}$; on the other hand, $\operatorname{re}P(iy) = y^4+10 \geq 10$. Suppose now that z describes the boundary γ of $K(0; R)\cap(+;+)$. We have: $\Delta_{[0,R]}\arg P(z) = 0$ because P is real and positive on $[0, R]$; on the arc of $C(0; R)$ $\Delta\arg P(z) = \Delta\arg z^4 + \Delta\arg[1+z^{-4}(2z^3-2z+10)] = 4\cdot\tfrac{1}{2}\pi + o(1) = 2\pi+o(1)$ as $R \to +\infty$, moreover, on $[iR, 0]$ the initial value of $\arg P(z)$ is $\arg[1-2i(R^4 + +10)^{-1}(R+R^3)] = o(1)$ as $R \to +\infty$, and all the time $\operatorname{re}P(iy) \geq 10$, the final value $\arg P(z)$ being 0 which means that $\Delta_{[iR,0]}\arg P(z) = o(1)$. Finally, $\Delta_\gamma\arg P(z) = 2\pi+o(1)$ for large R and being a multiple of 2π it is equal to 2π for all R sufficiently large. A similar reasoning can be made for remaining quadrants.

3.9.4. $P(iy) = (-1)^n[y^{2n}-ia^2y^{2n-1}]+b^2$, hence
$$w = P(iy) = y^{2n}+b^2-ia^2y^{2n-1} \quad \text{for even } n.$$
Suppose that z describes $\gamma = \partial[K(0; R)\cap\{z: \operatorname{re}z > 0\}]$ and R is large. For $z = iy$ moving on $[iR, -iR]$ the point $P(iy)$ describes an arc with end points $w_1 = R^{2n}+b^2-ia^2R^{2n-1}$, $w_2 = R^{2n}+b^2+ia^2R^{2n-1}$ situated in the half-plane $\{w: \operatorname{re}w > b^2\}$, hence
$$\Delta_{[iR,-iR]}\arg P(z) = \arg w_2 - \arg w_1 = 2\arccot a^{-2}(R+b^2R^{-2n+1}) = o(1)$$
as $R \to +\infty$. If z is moving on the arc of $C(0; R)$, then
$$\Delta\arg P(z) = \Delta\arg z^{2n}+\Delta\arg(1+a^2z^{-1}+b^2z^{-2n}) = 2\pi n+o(1).$$
Consequently, $\Delta_\gamma\arg P(z) = 2\pi n$ for all R sufficiently large. If n is odd, then $\Delta\arg P(z) = 2\pi n+o(1)$ on the arc of $C(0; R)$. If $z = iy$ moves on $[iR, -iR]$, then $w = P(iy) = b^2-R^{2n}+ia^2R^{2n-1}$ describes an arc Γ_1 with end points $w_1 = b^2-R^{2n}+ia^2R^{2n-1}$, $w_2 = b^2-R^{2n}-ia^2R^{2n-1}$, the equations of Γ_1 being: $u = b^2-y^{2n}$, $v = a^2y^{2n-1}$. We have: $d\arg w/dy = a^2|w|^{-2}(b^2+y^{4n-2}) > 0$ which

means that $\arg P(iy)$ decreases as z moves on $[iR, -iR]$; if $y > \sqrt[n]{b}$, then $w = u + iv \in (-; +)$; if $0 < y < \sqrt[n]{b}$, then $w \in (+; +)$; if $-\sqrt[n]{b} < y < 0$, then $w \in (+; -)$; if $y < -\sqrt[n]{b}$, then $w \in (-; -)$. Moreover, $\lim_{R \to +\infty} \arg w_1 = \pi$, $\lim_{R \to +\infty} \arg w_2 = -\pi$, hence $\Delta_{[iR, -iR]} P(z) = -2\pi + o(1)$ and finally, $n(\Gamma, 0) = n-1$.

3.9.5. Similar proof as in Exercise 3.9.4.

3.9.6. $|z^5| = 32 > |-z+16|$ on $C(0; 2)$, hence the polynomials z^5, $z^5 - z + 16$ have exactly 5 roots in $K(0; 2)$. Moreover, $|-z+16| > 14 > |z^5|$ on $C(0; 1)$, hence the polynomials $-z+16$, $z^5 - z + 16$ have no roots in $K(0; 1)$, and consequently, all the roots are situated in $\{z: 1 < |z| < 2\}$. If z moves on $[iR, -iR]$, then $P(iy) = 16 + iy(y^4 - 1)$, i.e. $\Delta_{[iR, -iR]} \arg P(z) = -\pi + o(1)$, whereas on the right-hand half of $C(0; R)$

$$\Delta \arg P(z) = 5\Delta \arg z + o(1) = 5\pi + o(1).$$

Hence $\Delta \arg P(z) = 4\pi + o(1) = 4\pi$ for all R sufficiently large which means that there are two roots of positive real part.

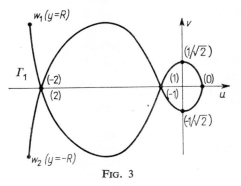

FIG. 3

3.9.7. If $z = iy$ moves on $[iR, -iR]$, then $w = u + iv = P(iy) = 1 - 2y^2 + iy(y^2 - 1)(y^2 - 4)$ describes an arc Γ_1 which can be outlined by means of the following table indicating the signs of u, v as depending on y (Fig. 3):

y	R	2		1	$2^{-1/2}$	0	$-2^{-1/2}$	-1		-2		$-R$
u	$-$	$-$	$-$	$-$	0	$+$	0	$-$	$-$	$-$	$-$	$-$
v	$+$	0	$-$	0	$+$	0	$-$	0	$+$	0	$-$

We have: $\arg w_1 = \frac{1}{2}\pi + o(1)$, $\arg w_2 = -\frac{1}{2}\pi + o(1)$ and $\Delta_{\Gamma_1} \arg w = -\pi + o(1)$ as $R \to +\infty$; moreover, $\Delta \arg w = 5\pi + o(1)$ on the right-hand half of $C(0; R)$, hence $n(\Gamma, 0) = 2 + o(1) = 2$ for R sufficiently large.

3.9.8. We find the variation of $\arg P(z)$ as z moves on the contour γ consisting of $[-R, R]$, two arcs of $C(0; R)$ contained in D and the segment $l_R = [\sqrt{R^2-1}+i, -\sqrt{R^2-1}+i]$. We have: $P(x) > 0$ on the real axis and $\Delta \arg P(z) = o(1)$ on both circular arcs, hence $\Delta \arg P(z) = o(1)$ on all three arcs. If $z = x+i$, then $P(z) = x^4 - 6x^2 + 3x + 4 + i(4x^3 - 4x + 3)$. The polynomial $4x^3 - 4x + 3$ has a unique root $x_0 \in (-2, -1)$ and $x_0^4 - 6x_0^2 + 3x_0 + 4 < 0$ which follows from the inequality: $(x^2-2)^2 < x(2x-3)$, $x \in [-2, -1]$. Hence, if R is large and z moves on l_R, then at the beginning $\arg P(z) = o(1)$ and $\operatorname{im} P(z) > 0$, afterwards $P(z)$ meets the negative real axis for $z = x_0 + i$ and then for x decreasing we have $\operatorname{im} P(z) < 0$ and at the end point again $\arg P(z) = o(1)$. This means that $\Delta'_{l_R} \arg P(z) = 2\pi + o(1)$ and also

$$\Delta_\gamma \arg P(z) = 2\pi + o(1) = 2\pi$$

for all sufficiently large R.

3.9.9. $|iz^3| = 27/8 < 4 < 81/16 - 1 \leqslant |z^4+1|$ on $C(0; \frac{3}{2})$ and by Rouché's theorem both polynomials z^4+1, z^4+iz^3+1 have four roots in $K(0; \frac{3}{2})$. The number of roots in $(+; +)$ can be found similarly as in Exercise 3.9.2.

3.9.10. (i) $|-5z+1| > 3 > |z^4|$ on $C(0; 1)$, hence $q = 1$;
(ii) $|-4z^5-1| > |z^8+z^2|$ on $C(0; 1)$, hence $q = 5$.

3.9.11. $|az^n| = a > e > e^x = |-e^z|$, $z = x+iy \in C(0; 1)$; hence both equations: $az^n = 0$, $az^n - e^z = 0$ have n solutions in $K(0; 1)$ by Rouché's theorem.

3.9.12. If $\operatorname{re} z > -\frac{3}{2}$, then $|z+2|/|z+1| > 1$, whereas $|-e^{-z}| = e^{-x} \leqslant 1$ for $z = x+iy$ in the right half-plane. Hence on the boundary of $K(0; R) \cap \{z: \operatorname{re} z > 0\}$ we have $|z+2|/|z+1| > |1-e^{-z}|$. Since $(z+2)(z+1)^{-1}$ has no roots in the right half-plane, so does $(z+2)(z+1)^{-1} - e^{-z}$.

3.9.13. Compare $|\lambda - z|$, $|-e^{-z}|$ on the boundary of $K(0; R) \cap \{z: \operatorname{re} z > 0\}$ for $R > \lambda + 1$. If $z = iy \in [-iR, iR]$, then $|\lambda - iy| \geqslant \lambda > |-e^{-iy}| = 1$; if $|z| = R$, $\operatorname{re} z \geqslant 0$, then $|\lambda - z| \geqslant |z| - \lambda = R - \lambda > 1 \geqslant |-e^{-z}|$. Now apply Rouché's theorem.

3.9.14. The case $a = 0$ is trivial. If $|a|^{-1} \leqslant 2^n$ and $z_1, \ldots, z_n$ are roots of the polynomial, then $|z_1 z_2 \ldots z_n| = |a|^{-1} \leqslant 2^n$, hence $|z_k| \leqslant 2$ for some k.
If $|a|^{-1} > 2^n$, then $|a|^{-1} > (2+\delta)^n$ for $\delta > 0$ sufficiently small and we have: $|a^{-1}(1+z)| > |a|^{-1} > (2+\delta)^n \geqslant |z|^n$ on $C(-1; 1+\delta)$ which means that the polynomials $a^{-1}(1+z)$, $a^{-1}(1+z) + z^n$ have the same number of roots in $K(-1; 1+\delta)$ for all sufficiently small $\delta > 0$ the number being equal to 1. Note that

$$\bigcap_{\delta > 0} K(-1; 1+\delta) \subset \overline{K}(0; 2).$$

3.9.15. $|(z-1)^p| = 1 > a \geqslant |-ae^{-z}| = |a|e^{-x}$ for $z = x+iy \in C(1;1)$, hence by Rouché's theorem the equation $(z-1)^p - ae^{-z}$ has p solutions in $K(1;1)$. If z_0 is a root of order $\geqslant 2$, then $(z_0-1)^p - ae^{-z_0} = 0$, $p(z_0-1)^{p-1} + ae^{-z_0} = 0$ which means that either $z_0 = 1$, or $z_0 = -p+1$, i.e. either $e^{z_0} = 0$, or $\operatorname{re} z_0 \leqslant 0$ and this is a contradiction.

3.9.16. $\tan z$ is uniformly bounded on the boundaries of squares Q_n with corners $n\pi(\mp 1 \mp i)$, hence for z moving round ∂Q_n: $\Delta \arg(z - \tan z) = \Delta \arg z + \Delta \arg(1 - z^{-1}\tan z) = 2\pi + o(1) = 2\pi$ for n sufficiently large. On the other hand, if N is the number of roots and P is the number of poles of $z - \tan z$ inside Q_n, and n is sufficiently large, then $N - P = 1$. Obviously $P = 2n$ and this implies $N = 2n+1$ and after rejecting a triple zero at the origin there are $2n-2$ non-trivial zeros left. The existence of $2n-2$ real, non-trivial zeros inside Q_n is an immediate consequence of the Darboux property of real continuous functions.

3.9.17. If $|z| = 1$, then $|(z-a_m)/(1-\bar{a}_m z)| = 1$, hence $|F(z)| = 1 > |b|$ on $C(0;1)$; by Rouché's theorem with $f = F$, $g = -b$ the equations $F = 0$, $F - b = 0$ have n roots in $K(0;1)$.

3.9.18. Apply Rouché's theorem with $f = F$, $g = -F(0)$.

3.9.19. Suppose that $r < R_1 < R_2 < R$. By the argument principle $n(\gamma_{R_2}, a) - n(\gamma_{R_1}, a) = N_a - P$, where N_a is the number of roots of the equation $f(z) - a = 0$ and P is the number of poles of f in the annulus $\{z\colon R_1 < |z| < R_2\}$. We have $N_a = P = 0$ by our assumptions.

3.9.20. Suppose that $D_\delta = \{z\colon \log|f(z)| \geqslant \delta\}$, $\delta > 0$. If δ is small enough, then a, z_0 are interior points of D_δ. Since the possible multiple points of ∂D_δ correspond to zeros of f', we can take δ so that $\partial D_\delta = \gamma_1 + \gamma_2 + \cdots + \gamma_n$, γ_k being closed analytic curves with positive orientation. If Γ is the image cycle of ∂D_δ under f, then $n(\Gamma, a) = N_a - P = 0$ because $\Gamma \subset \bar{K}(0; e^\delta)$ for $|a| > e^\delta$. Now, $P = 1$ and, consequently $N_a = 1$.

CHAPTER 4

Sequences and Series of Analytic Functions

4.1.1. $f_n(x) = x\exp(-\tfrac{1}{2}n^2x^2) \rightrightarrows 0$ on the real axis because $\sup_x |f_n(x)| = \sqrt{e}/n$. Suppose $f_n(z) \rightrightarrows f(z)$ in some disk $K(0;r)$. Since $f(z) = 0$ for all $z \in (-r, r)$, we have also $f(z) = 0$ for all $z \in K(0;r)$ and, moreover, $f_n'(z) \rightrightarrows f'(z) = 0$ in $K(0;r)$. In particular $f_n'(0) \to 0$ which is obviously false.

4.1.2. If $|z| \leqslant r < 1$, then $|u_n(z)| \leqslant r^n(1-r)^{-1}(1-r^2)^{-1} = A_n$ and $\sum A_n$ is obviously convergent. If $|z| \geqslant R > 1$, and n is sufficiently large, then $|u_n(z)| \leqslant 2/R^{n+1}$. This proves a.u. convergence. Now, $u_n(z) = (1-z^n)^{-1} - (1-z^{n+1})^{-1} + zu_n(z)$ and hence putting $s_n(z) = \sum_{k=1}^{n} u_k(z)$ we obtain $s_n(z) = (1-z)^{-1} - (1-z^{n+2})^{-1} + zs_n(z)$. If $|z| < 1$ and $n \to +\infty$, then $s(z) = \lim_n s_n(z) = (1-z)^{-1} - 1 + zs(z)$; if $|z| > 1$, then $s(z) = (1-z)^{-1} + zs(z)$ and $s(z)$ can be evaluated in each case.

4.1.3. $|\sin z|^2 = \sin^2 x + \sinh^2 y$ $(z = x+iy)$, hence $|y| \leqslant \theta \log 3$, $0 \leqslant \theta \leqslant 1$, implies:

$$3^{-n}|\sin nz| \leqslant 3^{-n}(1+\exp(n|y|)) \leqslant 3^{-n} + 3^{-(1-\theta)n}$$

which proves the uniform convergence in the strip $|\mathrm{im}\, z| \leqslant \theta \log 3$;

$$f'(0) = \sum_{n=1}^{\infty} n3^{-n} = \tfrac{3}{4}.$$

4.1.4. Suppose that $z \in \bar{K}(0;r)$, $r_1 = \tfrac{1}{2}(1+r)$ and $0 < r < 1$. We have

$$f(z) = \frac{1}{2\pi i} \int\limits_{C(0;r_1)} \frac{f(\zeta)}{\zeta - z}\, d\zeta = f(z) - f(0) = \frac{1}{2\pi i} \int\limits_{C(0;r_1)} \frac{zf(\zeta)}{\zeta(\zeta - z)}\, d\zeta$$

hence $|f(z)| \leqslant M(r)|z|$, where $M(r) = 2(1-r)^{-1}\sup_\theta |f(r_1 e^{i\theta})|$. This implies that $\sum f(z^n)$ has a majorant $\sum M(r) r^n$ in $K(0;r)$.

4.1.5. $\mathrm{im}(z+n)^{-1} = -y[(x+n)^2+y^2]^{-1}$, hence the series
$$\sum(-1)^{n+1}\mathrm{im}(z+n)^{-1}$$
is absolutely and uniformly convergent in $\overline{K}(z_0;r)$, since it has a convergent majorant $\sum An^{-2}$; $\mathrm{re}(z+n)^{-1} = (x+n)[(x+n)^2+y^2]^{-1}$ and from the fact that $t(t^2+y^2)^{-1}$ strictly decreases for $t \geqslant |y|$ it follows by Leibniz test of convergence that $\sum(-1)^{n+1}\mathrm{re}(z+n)^{-1}$ converges. Now, the rest in an alternating series is bounded in absolute value by the first term omitted, i.e. by $|x+n+1|[(x+n+1)^2+ +y^2]^{-1} \leqslant |x+n+1|^{-1} \leqslant (n+1-|x_0|-r)^{-1}$ for large n which proves the uniform convergence in $\overline{K}(z_0;r)$.

4.1.6. The points $-n$ $(n=1, 2, ...)$ are simple poles of f and $\mathrm{res}(-n;f) = (-1)^{n+1}$, hence
$$\int_\gamma f(z)dz = 2\pi i \sum(-1)^{k+1}n(\gamma,-k),$$
the sum being finite.

4.1.7. We have $\tau(n) \leqslant n$ and this implies that the power series on the right-hand side is convergent (and also a.u. convergent in $K(0; 1)$. If $0 < r < 1$ and $z \in K(0;r)$ then
$$|z^n(1-z^n)^{-1}| \leqslant r^n(1-r)^{-1}$$
and this proves the a.u. convergence of the series on the left-hand side. Moreover, both sums represent analytic functions in $K(0;1)$. The identity of both functions follows from the fact that the double series $\sum_{n=1}^{\infty}\sum_{m=1}^{\infty} z^{mn}$ is absolutely convergent and its sum does not depend on order of summation.

4.1.8. Differentiate k times the identity $(1-z)^{-1} = 1+z+z^2+\ldots$

4.1.9. (i) $K(0; 1)$, $K(\infty; 1)$;
(ii) the real axis;
(iii) $\mathbf{C}\setminus N_1$, where N_1 is the set of negative integers;
(iv) the annulus $\{z: q < |z| < q^{-1}\}$; no domain of analyticity does exist in (ii).

4.1.10. The finite limit $H(R) = \lim\limits_m \prod\limits_{n=1}^{m} [1+(R/n)^2]$ exists and consequently the given series has a convergent majorant $\sum R^2 H(R)(n+1)^{-2}$ in $\overline{K}(0;R)$.

4.1.11. The series is a.u. convergent in $\mathbf{C}\setminus N$ where N is the set of all integers, hence it can be differentiated term by term; the sum to be found is equal to $\pi^3 \cos\pi z(\sin\pi z)^{-3}$.

4.1.12. Suppose that G_0 is a subdomain of G such that $\bar{G}_0 \subset G$ and ∂G_0 is a cycle consisting of a finite number of contours. If w does not lie on $f(\partial G_0)$, the same is true for $f_n(\partial G_0)$ for all $n \geq N$, hence

$$\int_{\partial G_0} (f_n(z)-w)^{-1} f'_n(z)\,dz \to \int_{\partial G_0} (f(z)-w)^{-1} f'(z)\,dz = 0, 1$$

since f_n are univalent. This proves the univalence of f, unless it is a constant.

4.2.1. If $q_1 = 1/a_1$, $q_2 = a_1/a_2$, ..., $q_n = a_{n-1}/a_n$, then

$$|a_n|^{-1/n} = (|q_1||q_2| \cdots |q_n|)^{1/n} \to |q|.$$

4.2.2. (i) $\tfrac{1}{4}$; (ii) e; (iii) 1; (iv) $\min(1, |a|^{-1})$.

4.2.3. (i) For any $\varepsilon > 0$ there exists an integer k such that

$$|a_n|^{1/n} < R_1^{-1}+\varepsilon, \quad |b_n|^{1/n} < R_2^{-1}+\varepsilon \quad \text{for all } n \geq k;$$

hence

$$|a_n b_n|^{1/n} < (R_1 R_2)^{-1} + \varepsilon(R_1^{-1}+R_2^{-1}) + \varepsilon^2 \text{ for all } n \geq k$$

and consequently

$$R^{-1} = \overline{\lim} |a_n b_n|^{1/n} \leq (R_1 R_2)^{-1};$$

(ii) $a_n = b_n(a_n/b_n)$, cf. (i);

(iii) both series $\sum a_n z^n, \sum b_n z^n$ are absolutely convergent in $K(0; R_0)$, hence after multiplication and rearrangement according to increasing powers of z we obtain a convergent series for any z, $|z| < R_0$.

4.2.4. Take real increments of z and verify by induction that $\operatorname{im} f^{(k)}(z) = 0$ for all $z \in (-\delta, \delta)$. Note that $a_k = f^{(k)}(0)/k!$

4.2.5. $f(re^{i\theta}) = \sum_{n=0}^{\infty} a_n r^n e^{in\theta}$, $\overline{f(re^{i\theta})} = \sum_{n=0}^{\infty} \bar{a}_n r^n e^{-in\theta}$. The product $f\bar{f}$ can be arranged and written in the following form

$$\sum_{n=0}^{\infty} |a_n|^2 r^{2n} + A_1(r) e^{i\theta} + B_1(r) e^{-i\theta} + \ldots$$

which follows from its absolute convergence. The sum is uniformly convergent in $[0, 2\pi]$ for any fixed $r \in (0, R)$ and can be integrated term by term w.r.t. θ which yields the desired result because $\int_0^{2\pi} e^{in\theta} d\theta = 0$ for any integer $n \neq 0$.

4.2.6. We have

$$(2\pi)^{-1} \int_0^{2\pi} |f(re^{i\theta})|^2 d\theta = \sum_{n=0}^{\infty} |a_n|^2 r^{2n} \leq M^2,$$

hence
$$\sum_{n=0}^{m}|a_n|^2 r^{2n} \leqslant M^2 \quad \text{for any} \quad r\in[0,R), \quad \text{i.e.} \quad \sum_{n=0}^{m}|a_n|^2 R^{2n} \leqslant M^2,$$
m being arbitrary and this implies $\sum_{n=0}^{m}|a_n|^2 R^{2n} \leqslant M^2$.

4.2.7. $|a_n|^2 r^{2n} \leqslant \sum_{n=0}^{m}|a_n|^2 r^{2n} \leqslant [M(r)]^2$ by Exercise 4.2.6.

4.2.8. $\sum_{n=0}^{\infty}|a_n|^2 \leqslant M^2$ by Exercise 4.2.6, hence $a_n \to 0$.

4.2.9. Suppose that $f(z) \not\equiv z$ and $f(z_1) = f(z_2)$, i.e.
$$\sum_{n=1}^{\infty} a_n(z_1^n - z_2^n) = 0 \quad \text{for} \quad z_1 \neq z_2.$$
After dividing by $z_1 - z_2$ we obtain
$$\sum_{n=2}^{\infty} a_n(z_1^{n-1} + z_1^{n-2}z_2 + \ldots + z_2^{n-1}) = -a_1$$
and hence
$$|a_1| = \left|\sum_{n=2}^{\infty} a_n(z_1^{n-1} + z_1^{n-2}z_2 + \ldots + z_2^{n-1})\right| < \sum_{n=2}^{\infty} n|a_n|r^{n-1}$$
which is a contradiction.

4.2.10. $1 \geqslant \sum_{n=2}^{\infty} nr^{n-1}/n! = e^r - 1$ holds for $r \leqslant \log 2$. However, from the properties of exponential function it follows that $\log 2$ can be replaced by π.

4.2.11. Put in Exercise 4.2.3 (i): $a_n = c_n$, $b_n = 1/n!$; this gives $R_1 = \infty$. The sequence $\{c_n(\theta R)^n\}$ is bounded, hence $|c_n|(\theta R)^n \leqslant M(\theta)$ and also
$$|f(z)| \leqslant M(\theta) \sum_{n=0}^{\infty} (n!)^{-1}(|z|/\theta R)^n = M(\theta)\exp(|z|/\theta R).$$

4.2.12. We have
$$A(r) = \iint_{K(0;r)} |f'(\rho e^{i\theta})|^2 \rho\, d\rho\, d\theta = \int_0^r \rho\, d\rho \left(\int_0^{2\pi} |f'(\rho e^{i\theta})|^2 d\theta\right);$$
now, $f'(z) = a_1 + 2a_2 z + 3a_3 z^2 + \ldots$ and by Exercise 4.2.6
$$A(r) = 2\pi \int_0^r \left(\sum_{n=1}^{\infty} n^2 |a_n|^2 \rho^{2n-1}\right) d\rho$$
which can be integrated term by term.

4. SEQUENCES AND SERIES

4.3.1. The power series expansions of elementary functions known from the real analysis also hold in complex case. The cases (i)–(iv) can be settled by the method of undetermined coefficients.

(i) $z \equiv (z - \frac{1}{2}z^2 + \frac{1}{3}z^3 - \ldots)(1 + c_1 z + c_2 z^2 + \ldots)$, hence $c_1 - \frac{1}{2} = 0$, $c_2 - \frac{1}{2}c_1 + \frac{1}{3} = 0$, etc. and $z/\mathrm{Log}(1+z) = 1 + \frac{1}{2}z - \frac{1}{12}z^2 + \frac{1}{24}z^3 + \ldots$, $R = 1$ which is the distance between $z = 0$ and a singular point $z = -1$;

(ii) $1 + \frac{1}{3}z^2 + \frac{4}{45}z^4 - \frac{44}{7 \cdot 135}z^6 + \ldots$, $\quad R = 1$;

(iii) $1 - \frac{1}{4}z^2 - \frac{1}{96}z^4 - \frac{23}{15 \cdot 48}z^6 - \ldots$, $\quad R = \frac{\pi}{2}$;

(iv) $1 + \frac{1}{2}z^2 + \frac{5}{24}z^4 + \frac{61}{720}z^6 + \ldots$, $\quad R = \frac{\pi}{2}$;

(v) $\log 2 + \frac{1}{2}z + \frac{1}{8}z^2 - \frac{1}{192}z^4 + \ldots$, $\quad R = \pi$;

(vi) $e\left[1 + z + z^2 + \frac{5}{6}z^3 + \ldots\right]$, $\quad R = +\infty$.

4.3.2. It follows from Weierstrass theorem that
$$f^{(n)}(z) = u_0^{(n)}(z) + u_1^{(n)}(z) + \ldots, \quad |z| < R$$
and putting $z = 0$ we obtain: $a_n = a_{0n} + a_{1n} + \ldots$

4.3.3. (i) $1 + z + \frac{3}{2}z^2 + \frac{13}{6}z^3 + \ldots$, $R = 1$;
(ii) $z + z^2 + \frac{5}{6}z^3 + \frac{1}{2}z^4 + \ldots$, $\quad R = 1$;
(we can apply e.g. Exercise 4.3.2 with $u_n(z) = (n!)^{-1}z^n(1-z)^{-n}$ in (i)).

4.3.4. (i) $a_n = \frac{1}{2}\left[\frac{1}{1(n-1)} + \frac{1}{2(n-2)} + \ldots + \frac{1}{(n-1)1}\right] = \frac{1}{n}\left(1 + \frac{1}{2} + \ldots$
$\ldots + \frac{1}{n-1}\right)$ because $\frac{1}{k(n-k)} = \frac{1}{n}\left(\frac{1}{k} + \frac{1}{n-k}\right)$, $R = 1$;

(ii) $a_n = (-1)^{n-1}n^{-1}\left(1 + \frac{1}{3} + \ldots + \frac{1}{2n-1}\right)$, cf. (i), $R = 1$;

(iii) $a_n = (-1)^{n-1}2(2n+1)^{-1}\left(1 + \frac{1}{2} + \ldots + \frac{1}{2n}\right)$, $R = 1$;

(iv) $(\cos z)^2 = \frac{1}{2}(1 + \cos 2z)$, hence $a_{2n} = (-1)^n \frac{2^{2n-1}}{(2n)!}$, $a_{2n+1} = 0$ $(n \geq 1)$, $a_0 = 1$, $R = +\infty$;

(v) $f'(z) = z^{-1}[1-(1+z^2)^{-1/2}] = -\sum_{n=1}^{\infty} \binom{-1/2}{n} z^{2n-1}$ and hence $f(z) = \log 2 -$
$-\sum_{n=1}^{\infty} (2n)^{-1} \binom{-1/2}{n} z^{2n}$, $R = 1$;

(vi) $f(z) = \frac{1}{2}\text{Log}(\sqrt{1+z}+\sqrt{1-z})^2 = \frac{1}{2}\log 2 + \frac{1}{2}\text{Log}(1+\sqrt{1-z^2})$; the second term expansion can be obtained from (v) by putting iz instead of z and therefore

$$f(z) = \frac{1}{2}\log 2 + \frac{1}{2}\sum_{n=1}^{\infty} (-1)^n (2n)^{-1} \binom{-1/2}{n} z^{2n}, \quad R = 1.$$

4.3.5. $\lim_{z \to n} \sin \pi z^2 / \sin \pi z = 2n$ (cf. Ex. 3.4.8) and therefore all singularities are removable; $R = +\infty$.

4.3.6. If $t \in (-1, 1)$ then

$$\sqrt{2}\,\text{re}\,\sqrt{1+it} = (1+\sqrt{1+t^2})^{1/2} = \sqrt{2}\,\text{re}\left(1 + \frac{1}{2}it + \frac{1}{2 \cdot 4}t^2 - \frac{1 \cdot 3}{2 \cdot 4 \cdot 6}it^3 + \ldots\right).$$

Hence

(i) $(1+\sqrt{1+t^2})^{1/2} = \sqrt{2}\left(1 + \frac{1}{2 \cdot 4}t^2 - \frac{1 \cdot 3 \cdot 5}{2 \cdot 4 \cdot 6 \cdot 8}t^4 + \ldots\right)$ for real $t \in (-1,1)$

and also for complex $t \in K(0; 1)$; put next $z = t^2$;

(ii) $t(1+\sqrt{1+t^2})^{-1/2} = \sqrt{2}\,\text{im}(1+it)^{1/2}$ for $t \in (-1, 1)$ and similarly as in (i) we obtain:

$$(1+\sqrt{1+z})^{-1/2} = \sqrt{2}\left(\frac{1}{2} - \frac{1 \cdot 3}{2 \cdot 4 \cdot 6}z + \frac{1 \cdot 3 \cdot 5 \cdot 7}{2 \cdot 4 \cdot 6 \cdot 8 \cdot 10}z^2 - \ldots\right).$$

4.3.7. $(c_0 + c_1 z + c_2 z^2 + \ldots)(1-z-z^2) \equiv 1$ and by comparing the coefficients we obtain: $c_0 = 1$, $-c_0 + c_1 = 0$, $c_{n+2} - c_{n+1} - c_n = 0$;

$$(1-z-z^2)^{-1} = 5^{-1/2}\{[z + \frac{1}{2}(\sqrt{5}+1)]^{-1} + [\frac{1}{2}(\sqrt{5}-1) - z]^{-1}\},$$

hence

$$c_n = \frac{f^{(n)}(0)}{n!} = 5^{-1/2}\left[\left(\frac{\sqrt{5}+1}{2}\right)^{n+1} + (-1)^n \left(\frac{\sqrt{5}-1}{2}\right)^{n+1}\right]$$

$n = 0, 1, 2, \ldots$; $R = \frac{1}{2}(\sqrt{5}-1)$.

4.3.8. We have

$$f(z) = \frac{1+z-e^z}{z(e^z-1)} + \frac{1}{2} = \left(-\frac{z^2}{2!} - \frac{z^3}{3!} - \ldots\right) : \left(z^2 + \frac{z^3}{3!} + \ldots\right) + \frac{1}{2} \to 0 \text{ as } z \to 0,$$

hence the singularity at the origin is removable;

$$f(-z) = \frac{-e^z+1-1}{e^z-1} + z^{-1} + \frac{1}{2} = -\frac{1}{e^z-1} + z^{-1} - \frac{1}{2} = -f(z).$$

4. SEQUENCES AND SERIES

4.3.9. (i) If f is the function of Exercise 4.3.8, then $\frac{1}{2}z\coth\frac{1}{2}z = 1+f(z)$;
(ii) replace z by $2iz$ in (i).

4.3.10. (i) $z[\text{Log}(\sin z/z)]' = z\cot z - 1 = -\sum_{k=1}^{\infty}[2^{2k}B_k/(2k)!]z^{2k}$ by Exercise 4.3.9 (ii) and after integrating

$$\text{Log}(\sin z/z) = -\sum_{k=1}^{\infty}[2^{2k-1}B_k/k(2k)!]z^{2k}, \quad |z| < \pi;$$

(ii) use (i) and the identity $\text{Log}(\sin 2z/2z) = \text{Log}(\sin z/z) + \text{Log}\cos z$, $2|z| < \pi$; hence

$$\text{Log}\cos z = -\sum_{n=1}^{\infty}[2^{2n-1}(2^{2n}-1)B_n/n(2n)!]z^{2n}, \quad |z| < \tfrac{1}{2}\pi;$$

(iii) differentiate (ii):

$$\tan z = \sum_{n=1}^{\infty}[2^{2n}(2^{2n}-1)B_n/(2n)!]z^{2n-1}, \quad |z| < \tfrac{1}{2}\pi;$$

(iv) $\sec^2 z = \sum_{k=0}^{\infty}[2^{2k+1}(2^{2k+2}-1)B_{k+1}/(k+1)(2k)!]z^{2k}$, which is obtained by differentiating (iii); $|z| < \tfrac{1}{2}\pi$;
(v) $\tan^2 z = \sec^2 z - 1$;
(vi) subtract both sides in (i) and (ii):

$$\text{Log}[\tan z/z] = \sum_{n=1}^{\infty}[2^{2n}(2^{2n-1}-1)B_n/n(2n)!]z^{2n}, \quad |z| < \tfrac{1}{2}\pi;$$

(vii) differentiate both sides of (vi) and replace $2z$ by z:

$$z/\sin z = 1 + \sum_{n=1}^{\infty}[(2^{2n}-2)B_n/(2n)!]z^{2n}, \quad |z| < \pi.$$

4.3.11. We have

$$\text{Log}\left[1 - \left(\frac{z^2}{3!} - \frac{z^4}{5!} + \frac{z^6}{7!} + O(z^8)\right)\right]$$

$$= -\frac{z^2}{3!} + \frac{z^4}{5!} - \frac{z^6}{7!} + O(z^8) - \frac{1}{2}\left[\frac{z^4}{3!3!} - \frac{2z^6}{3!5!} + O(z^8)\right] +$$

$$+ \frac{1}{3}\left[\left(\frac{1}{3!}\right)^3 z^6 + O(z^8)\right] + O(z^8)$$

$$= -\frac{1}{6}z^2 - \frac{1}{180}z^4 - \frac{1}{3^4 \cdot 5 \cdot 7}z^6 + O(z^8)$$

$$= -B_1 z^2 - \frac{B_2}{3!}z^4 - \frac{2^5 B_3}{3 \cdot 6!}z^6 + O(z^8).$$

4.3.12. $\sqrt{u}\operatorname{Arctan}\sqrt{u} = u - \frac{1}{3}u^2 + \frac{1}{5}u^3 - \ldots$, $|u| < 1$. If $|z| < \frac{1}{2}$, then $|z/(1-z)| < 1$, hence the radius of convergence $R \geq \frac{1}{2}$. Now, the shortest distance from the origin to the points where φ ceases to be analytic ($z = 1$) is equal 1, hence $R = 1$.

4.3.13. By equating Taylor's coefficients of both sides we obtain $a_1 = 1$, $3a_2 = 2a_1, \ldots, (2n+1)a_{n+1} = 2na_n, \ldots$, hence

$$a_1 = 1, \qquad a_{n+1} = \frac{2n}{2n+1} a_n.$$

4.3.14. Putting $z = u^2$ in the formula of Exercise 4.3.13 we obtain

$$(1-u^2)^{-1/2}\operatorname{Arctan} u(1-u^2)^{-1/2} = (1-u^2)^{-1/2}\operatorname{Arcsin} u$$

$$= u + \frac{2}{3}u^3 + \frac{2 \cdot 4}{3 \cdot 5}u^5 + \ldots;$$

integrate now both sides.

4.3.15. $R_n(z) = f(z) - s_n(z) = z^{n+1}\varphi_n(z)$, where s_n is a polynomial of degree at most n and φ_n is analytic in $K(0; R)$. We have:

$$\varphi_n(z) = z^{-n-1}[f(z) - s_n(z)] = (2\pi i)^{-1}\int_{C(0;r_1)}(\zeta-z)^{-1}\varphi_n(\zeta)d\zeta$$

$$= (2\pi i)^{-1}\int_{C(0;r)}\zeta^{-n-1}(\zeta-z)^{-1}f(\zeta)d\zeta$$

by Exercise 3.1.28.

4.3.16. We have

$$s_n(z) = f(z) - R_n(z)$$

$$= (2\pi i)^{-1}\int_{C(0;r)}(\zeta-z)^{-1}f(\zeta)d\zeta - (2\pi i)^{-1}\int_{C(0;r)}(z/\zeta)^{n+1}(\zeta-z)^{-1}f(\zeta)d\zeta$$

$$= (2\pi i)^{-1}\int_{C(0;r)}\zeta^{-n-1}(\zeta-z)^{-1}(\zeta^{n+1}-z^{n+1})f(\zeta)d\zeta.$$

4.3.17. If $|z-a| < R$, then f is analytic in $K = K(\frac{1}{2}(a+z); R-\frac{1}{2}|z-a|)$ and the Taylor series with center $\frac{1}{2}(z+a)$ has the following form:

$$f(\zeta) = f(\tfrac{1}{2}(a+z)) + (\zeta - \tfrac{1}{2}(a+z))f'(\tfrac{1}{2}(a+z)) + \tfrac{1}{2}(\zeta - \tfrac{1}{2}(a+z))^2 f''(\tfrac{1}{2}(a+z)) + \ldots$$

Put first $\zeta = a$ and then $\zeta = z$ and subtract both sides.

4.3.18. If $a_n = \alpha_n + i\beta_n$, then

$$\sum_{n=1}^{\infty}(\alpha_n + i\beta_n)(\cos n\theta + i\sin n\theta)r^n = P(\theta) + iQ(\theta)$$

and consequently

$$P(\theta) = \sum_{n=0}^{\infty} (\alpha_n \cos n\theta - \beta_n \sin n\theta) r^n.$$

Thus by Euler–Fourier formulas

$$\pi \alpha_n r^n = \int_0^{2\pi} P(\theta)\cos n\theta \, d\theta, \qquad -\pi \beta_n r^n = \int_0^{2\pi} P(\theta)\sin n\theta \, d\theta,$$

or

$$\pi a_n r^n = \pi(\alpha_n + i\beta_n) = \int_0^{2\pi} P(\theta) e^{-in\theta} d\theta$$

and similarly

$$\pi a_n r^n = \int_0^{2\pi} iQ(\theta) e^{-in\theta} d\theta.$$

4.3.19. $|a_n| \leq (\pi r^n)^{-1} \int_0^{2\pi} P(\theta) d\theta$ by Exercise 4.3.18 and the equation $P(0) = (2\pi)^{-1} \int_0^{2\pi} P(\theta) d\theta$ (cf. Ex. 3.2.5) yields $|a_n| \leq 2r^{-n}$. Since $r \in (0, 1)$ can be arbitrary, we have $|a_n| \leq 2$.

4.3.20. $|f(z)| \leq 1 + |a_1| \cdot |z| + |a_2| \cdot |z|^2 + \ldots \leq 1 + 2|z|/(1-|z|) = (1+|z|)/(1-|z|)$. In order to obtain a lower estimate, note that $1/f$ is analytic in $K(0; 1)$ and satisfies the assumptions of Exercise 4.3.19. Equality holds for real z and $f(z) = (1+z)/(1-z)$.

4.4.1. (i) divergent everywhere on $C(0; 1)$;
 (ii) divergent only at $z = 1$;
 (iii) absolutely convergent on $C(0; 1)$;
 (iv) conditionally convergent everywhere on $C(0; 1)$ except at $z = -1$ and $z = \frac{1}{2}(1 \mp i\sqrt{3})$.

4.4.2. $a_n = (-1)^n$; $\lim_{r \to 1-} \sum_{n=0}^{\infty} a_n r^n = \lim_{r \to 1-} (1+r)^{-1} = \frac{1}{2}$.

4.4.3. $z_n = e^{i\theta}(1 - r_n \exp i\alpha_n)$, where $|\alpha_n| \leq \frac{1}{2}\pi - \delta$, $\delta > 0$; we have:

$$|z_n| = (1 - 2r_n \cos \alpha_n + r_n^2)^{1/2} = 1 - r_n \cos \alpha_n + O(r_n^2), \qquad r_n = |e^{i\theta} - z_n|.$$

Note that a necessary and sufficient condition for z_n to be situated inside a Stolz angle is: $(\cos \alpha_n)^{-1} = r_n/r_n \cos \alpha_n = O(1)$.

4.4.4. We may assume $e^{i\theta} = 1$, hence the radius of convergence of $\sum a_k z^k$ is ≥ 1.

By Exercise 4.2.3 (iii) the esries $\sum s_k z^k$ is convergent in $K(0;1)$, moreover

$$\sum_{k=0}^{\infty} s_k z^k = \sum_{k=0}^{\infty} a_k z^k \sum_{k=0}^{\infty} z^k, \quad \text{i.e.} \quad (1-z)\sum_{k=0}^{\infty} s_k z^k = \sum_{k=0}^{\infty} a_k z^k.$$

Thus $\sum_{k=0}^{\infty} a_k z_n^k$ can be considered as a transform of a convergent sequence $\{s_n\}$ by means of a Toeplitz matrix whose nth row has the following form: $1-z_n$, $z_n(1-z_n)$, $z_n^2(1-z_n)$, ... Now, cf. Exercise 1.1.37 and Exercise 4.4.3.

4.4.5. (i), (ii) The series $z - \frac{1}{2}z^2 + \frac{1}{3}z^3 - \ldots$ whose sum in $K(0;1)$ is equal to $\mathrm{Log}(1+z)$, is also convergent for all $z \in C(0;1)$ except for $z = -1$, which follows from Abel's test of convergence. If $z = \cos\theta + i\sin\theta$ and $|\theta| < \pi$, then by Abel's limit theorem:

$$\mathrm{Log}(1+e^{i\theta}) = \log(2\cos\theta/2) + i\theta/2 = \sum_{n=1}^{\infty} (-1)^{n+1} n^{-1}(\cos n\theta + i\sin n\theta).$$

Separate now real and imaginary parts.

4.4.6. The series $z + \frac{1}{3}z^3 + \frac{1}{5}z^5 + \ldots$ is convergent on $C(0;1)$ except at $z = \mp 1$ and its sum is equal to $\frac{1}{2}\mathrm{Log}[(1+z)/(1-z)]$ in $K(0;1)$. By Abel's limit theorem:

$$\tfrac{1}{2}\mathrm{Log}[(1+e^{i\theta})/(1-e^{i\theta})] = \tfrac{1}{2}\log|\cot(\theta/2)| + \tfrac{1}{4}i\pi = e^{i\theta} + \tfrac{1}{3}e^{3i\theta} + \tfrac{1}{5}e^{5i\theta} + \ldots$$
$$= \cos\theta + \tfrac{1}{3}\cos 3\theta + \tfrac{1}{5}\cos 5\theta + \ldots + i(\sin\theta + \tfrac{1}{3}\sin 3\theta +$$
$$+ \tfrac{1}{5}\sin 5\theta + \ldots), \quad 0 < \theta < \pi.$$

4.4.7. The series is an alternating series of the form $\sum(-1)^n u_n$ with $u_n = 2^{2n}(n!)^2/(2n+1)!$. From Stirling formula for asymptotic value of $n!$ it follows that $u_n = O(n^{-1/2})$, hence $u_n \to 0$. Obviously u_n decreases, hence the series is convergent. Use now Abel's limit theorem and Exercise 4.3.13 with $z = -1$.

4.4.8. The series is a convergent alternating series; put $z = i$ and apply Abel's limit theorem for $-(\mathrm{Arc}\sin z)^2$.

4.4.10. Use the expansion

$$\mathrm{Arsinh}\, z = \mathrm{Log}(z + \sqrt{1+z^2}) = z - \frac{1}{3}\frac{1}{2}z^3 + \frac{1}{5}\frac{1\cdot 3}{2\cdot 4}z^5 - \ldots$$

and Abel's limit theorem for $z = 1$.

4.5.1. For example $\sum_{n=1}^{\infty} n^{-2}z^n + \sum_{n=1}^{\infty} n^{-2}z^{-n}$; the regular part is convergent in $\overline{K}(0;1)$ and the principal part is convergent in $\overline{K}(\infty;1)$.

4.5.2. The principal part is convergent for $|z| > \overline{\lim}|a_{-n}|^{1/n} = r$ and divergent for $|z| < r$, whereas the regular part is convergent for $|z| < (\overline{\lim}|a_n|^{1/n})^{-1} = R$ and divergent for $|z| > R$. Hence the annulus of convergence is not empty iff $r < R$.

4.5.3. (i) $a_n = (-\frac{1}{3})^{n+1}$, $a_0 = -\frac{1}{3}$, $a_{-n} = (-2)^{n-1}$;
(ii) $a_n = a_0 = a_{-1} = 0$, $a_{-n} = (-1)^n(3^{n-1}-2^{n-1})$; $n = 1, 2, 3, \ldots$

4.5.4. $[z-\frac{1}{2}(a+b)]-\frac{1}{12}(a-b)^2[z-\frac{1}{2}(a+b)]^{-1}+\ldots$

4.5.5. (i) $(a-b)^{-1}\{a^{-1}[a/z+(a/z)^2+\ldots]+b^{-1}[1+z/b+(z/b)^2+\ldots]\}$;
(ii) $z^{-2}+(a+b)z^{-3}+(a^2+ab+b^2)z^{-4}+\ldots$

4.5.6. (i) $z^{-2}-z^{-4}+z^{-6}-\ldots-\frac{1}{2}+\frac{1}{4}z^2-\frac{1}{8}z^4-\ldots$;
(ii) $\sum_{n=2}^{\infty}(-1)^n(2^{n-1}-1)z^{-2n}$.

4.5.7. $z^{-2}+\frac{1}{2}z^{-4}+\frac{1}{3}z^{-6}+\ldots$

4.5.8. (i) $a_n = (\pi)^{-1}\int_0^{\pi}\exp(2\cos\theta)\cos n\theta\, d\theta$;

(ii) $a_n = \sum_{k=0}^{\infty}[k!(n+k)!]^{-1}$, $n \geq 0$; $a_{-n} = a_n$.

4.5.9. We have

$$a_n = (2\pi i)^{-1}\int_{C(0;1)}\exp[\tfrac{1}{2}z(\zeta-\zeta^{-1})]\zeta^{-n-1}d\zeta = (2\pi)^{-1}\int_{-\pi}^{\pi}\exp[i(z\sin\theta-n\theta)]d\theta$$

$$= (2\pi)^{-1}\int_{-\pi}^{\pi}\cos(z\sin\theta-n\theta)d\theta+i(2\pi)^{-1}\int_{-\pi}^{\pi}\sin(z\sin\theta-n\theta)d\theta$$

$$= (2\pi)^{-1}\int_{-\pi}^{\pi}\cos(n\theta-z\cos\theta)d\theta;$$

$a_n = \sum_{k=0}^{\infty}(-1)^k[k!(n+k)!]^{-1}(z/2)^{2k+n}$ which is the coefficient of ζ^n in the product $\exp(z\zeta/2)\exp(-z\zeta^{-1}/2)$.

4.5.10. $f(z) = 1+(z+1)^{-1}+2(z-2)^{-1}+3(z-2)^{-2}$; $a_0 = \frac{3}{4}$, $a_n = (3n-1)2^{-n-2}$.

4.5.11. (i) $z^{-1}+\sum_{n=0}^{\infty}z^n$;

(ii) $-(z-1)^{-1}+\sum(-1)^n(z-1)^n$;

(iii) $-\sum_{n=2}^{\infty}z^{-n}$.

4.5.12. $\sum_{n=1}^{\infty} c_{-n} z^{-n} + \sum_{n=0}^{\infty} c_n z^n$, where

$$c_{-n} = 2^{-1/2} i (-1)^n \Big[\binom{-1/2}{n} + \sum_{m=1}^{\infty} \binom{-1/2}{m+n} \binom{-1/2}{m} 2^{-m} \Big],$$

$$c_n = -2^n c_{-n}, \quad c_0 = 2^{-1/2} i \Big[1 + \sum_{m=1}^{\infty} \binom{-1/2}{m}^2 2^{-m} \Big].$$

4.5.13. $\mp [z - \tfrac{1}{2}(a+b) + \sum_{n=1}^{\infty} c_{-n} z^{-n}]$ for $|z| > \max(|a|, |b|)$, with

$$c_{-(n-1)} = (-1)^n \Big[\binom{1/2}{n} b^n + \binom{1/2}{n-1} \binom{1/2}{1} b^{n-1} a + \ldots + \binom{1/2}{n} a^n \Big].$$

4.5.14. The sum of the series $\sum_{n=-\infty}^{+\infty} (z-n)^{-2}$ is an analytic function h in $\mathbf{C} \setminus N$. Suppose F is a compact set such that $F \cap N = \emptyset$. If $\delta_n = \inf_{z \in F} |1 - z/n|$ then $\delta_n \to 1$ as $n \to \mp\infty$ and consequently $|z \mp n|^{-2} \leqslant \tfrac{1}{4} n^{-2}$ for all n sufficiently large which implies a.u. convergence and analyticity of h. Obviously $z = n$ is a pole of second order with principal part $(z-n)^{-2}$. If $t = z - n$, then

$$\pi^2 \operatorname{cosec}^2 \pi z - (z-n)^{-2} = \pi^2 \operatorname{cosec}^2 \pi t - t^{-2}$$
$$= [\pi^2 t^2 - (\pi t - (\pi t)^3/3! + \ldots)][t^2(\pi t - (\pi t)^3/3! + \ldots)^2]^{-1} = O(1) \quad \text{as } t \to 0$$

which means that $\pi^2 / \sin^2 \pi z - h(z)$ has a removable singularity at $z = n$.

4.5.15. Evidently both $h(z)$ and $\pi^2/\sin^2 \pi z$ are periodic with period 1. Since $g(z) = \pi^2/\sin^2 \pi z - h(z)$ has removable singularites at $z \in N$, and has period 1, it is bounded in the strip $|\operatorname{im} z| \leqslant 1$. Suppose now that $|\operatorname{im} z| \geqslant 1$ and $0 \leqslant \operatorname{re} z \leqslant 1$. We have:

$$\sum_{n=-\infty}^{\infty} |n-z|^{-2} = \sum_{n=-\infty}^{\infty} \frac{1}{(x-n)^2 + y^2} \leqslant 2 \sum_{n=0}^{\infty} (n^2 + y^2)^{-1} \leqslant 2 \sum_{n=0}^{\infty} (n^2 + 1)^{-1}.$$

Hence h is bounded in $\{z: |\operatorname{im} z| \geqslant 1\}$. The same is true for $\pi^2/\sin^2 \pi z$ and consequently g is bounded in $\mathbf{C}$, i.e. $g = \operatorname{const} = 0$ which follows from Exercise 4.5.14.

4.5.16. The series is a.u. convergent, hence it can be differentiated term by term and this gives the equality of Exercise 4.5.15. Note that $\pi \cot \pi z - z^{-1} \to 0$ as $z \to 0$.

4.5.17. (i) $2z/(z^2 - n^2) = (-2z/n^2)(1 + (z/n)^2 + (z/n)^4 + \ldots)$ and consequently $\pi \cot \pi z = z^{-1} - 2s_1 z - 2s_2 z^3 - 2s_3 z^5 - \ldots$ Hence $(2\pi)^{2k} B_k = 2(2k)! s_k$;
(ii) $3z^{-1} + 2z^{-3} + 2z^{-5} + \ldots - 2(s_1 - 1)z - 2(s_2 - 1)z^3 - 2(s_3 - 1)z^5 - \ldots$

4. SEQUENCES AND SERIES

4.6.1. We have $|\cot \pi z| \leq M$ for each $z \in \text{fr} Q_N$ and each N, where M does not depend on N, (cf. Ex. 2.7.3, 2.7.7). Hence

$$\left| \int_{\partial Q_N} f(z) \cot \pi z \, dz \right| \leq M \int_{\partial Q_N} |f(z)| \, |dz| \leq 8M |\zeta_N| |f(\zeta_N)|,$$

where $\sup_{z \in \text{fr} Q_N} |f(z)| = |f(\zeta_N)|$. Hence $\int_{\partial Q_N} f(z) \cot \pi z \, dz \to 0$ as $N \to +\infty$. If $a_1, \ldots, a_m$ are inside Q_N, then

$$\int_{\partial Q_N} f(z) \pi \cot \pi z \, dz$$

$$= 2\pi i \left\{ \sum_{n=-N}^{N} \text{res}[n; \pi f(z) \cot \pi z] + \sum_{k=1}^{m} \text{res}[a_k; \pi f(z) \cot \pi z] \right\} \to 0 \quad \text{as } N \to +\infty.$$

Note that

$$\text{res}[n; \pi f(z) \cot \pi z] = \lim_{z \to n} f(z) \cos \pi z (\pi z - \pi n)/(\sin \pi z - \sin \pi n) = f(n).$$

4.6.2. The function $f(z) = (z^2 + z + 1)^{-1}$ satisfies the assumptions of Exercise 4.6.1, hence $s = -\pi(b_1 \cot \pi a_1 + b_2 \cot \pi a_2)$ where $a_1 = -(1 + i\sqrt{3})/2$, $a_2 = \bar{a}_1$ are poles of f and $b_1 = \text{res}[a_1; f] = i/\sqrt{3}$, $b_2 = \text{res}[a_2; f] = -i/\sqrt{3}$. Hence

$$s = 3^{-1/2} \pi i [\cot \pi (1 + i\sqrt{3})/2 - \cot \pi (1 - i\sqrt{3})/2]$$
$$= -3^{-1/2} 2\pi i \tan i\pi \sqrt{3}/2 = 2\pi 3^{-1/2} \tanh \pi \sqrt{3}/2,$$

because $\cot \left(\dfrac{\pi}{2} + \alpha \right) = -\tan \alpha$.

4.6.3. (i) If $a \neq 0, \mp i, \mp 2i, \ldots$ then $f(z) = (z^2 + a^2)^{-1}$ satisfies the assumptions of Exercise 4.6.1 and therefore

$$\sum_{n=-\infty}^{\infty} (n^2 + a^2)^{-1} = -(2ai)^{-1} \pi [\cot \pi ai - \cot(-\pi ai)] = a^{-1} \pi \coth \pi a;$$

(ii) if $a^4 \neq 0, -1^4, -2^4, \ldots$ then $f(z) = (z^4 + a^4)^{-1}$ satisfies the assumptions of Exercise 4.6.1; it has 4 simple poles $a_\nu = 2^{-1/2} a (\mp 1 \mp i)$ with residues $b_\nu = -\tfrac{1}{4} a^{-3} a_\nu$ and therefore

$$\sum_{n=-\infty}^{\infty} (n^4 + a^4)^{-1} = \tfrac{1}{4} \pi a^{-4} \sum_{\nu=1}^{4} a_\nu \cot \pi a_\nu$$

$$= 2^{-3/2} a^{-3} \pi [(1+i) \cot \pi a (1+i)/\sqrt{2} + (1-i) \cot \pi a (1-i)/\sqrt{2}]$$

$$= \pi a^{-3} 2^{-1/2} (\sin \pi a \sqrt{2} + \sinh \pi a \sqrt{2})/(\cosh \pi a \sqrt{2} - \cos \pi a \sqrt{2})$$

by Exercise 2.7.16 and the identity $z \cot z = -z \cot(-z)$;

(iii) if $a^4 \neq 0, -1^4, -2^4, \ldots$ then

$$2s = \sum_{n=-\infty}^{\infty} n^2/(n^4+a^4) = -\tfrac{1}{4}\pi \sum_{\nu=1}^{4} a_\nu^{-1} \cot \pi a_\nu$$

where a_ν are the same as in (ii); hence by $a_1 a_2 = -a^2$ we obtain:

$$2s = (2a^2)^{-1}\pi(a_2 \cot \pi a_1 + a_1 \cot \pi a_2)$$
$$= 2^{-3/2} a^{-1} \pi[(1+i)\cot(-1+i)a\pi/\sqrt{2} + (-1+i)\cot(1+i)a\pi/\sqrt{2}]$$

and the result follows similarly as in (ii);

(iv) if $a \neq ni^\nu$ ($\nu = 1, \ldots, 4$; $n = 0, 1, 2, \ldots$), then

$$\sum_{n=-\infty}^{\infty} (n^4-a^4)^{-1} = -\tfrac{1}{4} a^{-4} \pi \sum_{\nu=1}^{4} a_\nu \cot \pi a_\nu,$$

where $a_\nu = ai^\nu$, $\nu = 1, \ldots 4$;

$$-\tfrac{1}{2} a^{-4} \pi(a_1 \cot \pi a_1 + a_2 \cot \pi a_2) = -\tfrac{1}{2} a^{-3} \pi(\cot \pi a + i \cot \pi ai)$$
$$= -\tfrac{1}{2} a^{-3} \pi(\cot \pi a + \coth \pi a).$$

4.6.4. The function $f(z) = (z-a)^{-2}$ has a double pole at $z = a$; moreover, $\text{res}[a; \pi(z-a)^{-2}\cot \pi z] = -\pi^2/\sin^2 \pi a$, hence by Exercise 4.6.1

$$\sum_{n=-\infty}^{\infty} (n-a)^{-2} = \pi^2/\sin^2 \pi a.$$

4.6.5. $f(z) = (z-a)^{-1}(z-b)^{-1}$ satisfies the assumptions of Exercise 4.6.1. We have: $\text{res}[a; \pi f(z) \cot \pi z] = \pi(a-b)^{-1} \cot \pi a$ and consequently $s = -\pi(a-b)^{-1} \times (\cot \pi a - \cot \pi b)$.

4.6.6. Putting $a = i$, $b = -i$ in Exercise 4.6.5 we obtain:

$$2s+1 = \pi i \cot \pi i = \pi \coth \pi \quad \text{or} \quad s = \tfrac{1}{2}\pi \coth \pi - \tfrac{1}{2}.$$

4.6.7. If $z \in \text{fr} Q_N$, then $|\sin \pi z|^{-1} \leqslant 1$ for each N, hence

$$\left| \int_{\partial Q_N} (\sin \pi z)^{-1} f(z) dz \right| \leqslant \int_{\partial Q_N} |f(z)| \, |dz| \leqslant 8|\zeta_N f(\zeta_N)| = o(1)$$

as $N \to +\infty$ (ζ_N being the point yielding the maximum of $|f|$ on $\text{fr} Q_N$). If all a_ν are inside Q_N, then

$$\int_{\partial Q_N} (\sin \pi z)^{-1} \pi f(z) dz = 2\pi i \left\{ \sum_{n=-N}^{N} \text{res}\left[n; \frac{\pi f(z)}{\sin \pi z}\right] + \sum_{k=1}^{m} \text{res}\left[a_k; \frac{\pi f(z)}{\sin \pi z}\right] \right\} = o(1)$$

as $N \to +\infty$. Note that

$$\text{res}\left[n; \frac{\pi f(z)}{\sin \pi z}\right] = \frac{f(n)}{\cos n\pi} = (-1)^n f(n).$$

4. SEQUENCES AND SERIES

4.6.8. E.g. (iv). The function $f(z) = z^2/(z^4+a^4)$ satisfies the assumptions of Exercise 4.6.7. It has 4 simple poles $a_k = 2^{-1/2}a(\mp 1 \mp i)$, $k = 1, \ldots, 4$ with residues $(4a_k)^{-1}$. Hence

$$s = -\tfrac{[1]}{[4]}\pi \sum_{k=1}^{4}(a_k \sin \pi a_k)^{-1} = -\tfrac{1!}{2}\pi[(a_1 \sin \pi a_1)^{-1}+(a_2 \sin \pi a_2)^{-1}],$$

where a_1, a_2 are situated in the upper half-plane. Now, $a_1 a_2 = -a^2$ and therefore

$$s = (2a^2)^{-1}\pi(a_2 \sin \pi a_2 + a_1 \sin \pi a_1)/(\sin \pi a_1 \sin \pi a_2)$$

and the result follows.

4.6.9. We have:

$$|\sin az/\sin \pi z|^2 = (\sin^2 ax + \sinh^2 ay)(\sin^2 \pi x + \sinh^2 \pi y)^{-1},$$

$z = x+iy$; therefore $|\sin az/\sin \pi z| \leq 1$ on vertical sides since $\sin^2 \pi x = 1 \geq \sin^2 \pi a$, $\sinh^2 \pi y \geq \sinh^2 \pi a$. Moreover, $|\sin az/\sin \pi z| \to 0$ uniformly as $|y| \to +\infty$. Hence $\int_{\partial Q_N} f(z)dz \to 0$ as $N \to +\infty$ ($f(z) = \pi \sin az(z^3 \sin \pi z)^{-1}$). Now,

$$\int_{\partial Q_N} = 2\pi i \sum_{n=-N}^{N} \text{res}[n;f], \quad \text{res}[0;f] = a(\pi^2-a^2)/6, \quad \text{res}[n;f] = (-1)^n n^{-3}\sin na,$$

$n = \mp 1, \mp 2, \ldots$ and consequently

$$a(\pi^2-a^2)/6 + 2\sum_{n=1}^{N}(-1)^n n^{-3}\sin na \to 0 \quad \text{as} \quad N \to +\infty;$$

(ii) put $a = \pi/2$ in (i).

4.6.10. Similarly as in Exercise 4.6.9. we verify that $|\cos az/\sin \pi z| \leq 1$ on vertical sides of Q_N, whereas $|\cos az/\sin \pi z| \to 0$ uniformly as $|\text{im } z| \to +\infty$. Hence $I_N \to 0$ as $N \to +\infty$. If $f(z) = \pi \cos az \csc \pi z/(x^2-z^2)$, then $\text{res}(n;f) = (-1)^n \times \cos na/(x^2-n^2)$, $\text{res}(x;f) = -\pi \cos ax/(2x \sin \pi x) = \text{res}(-x;f)$, and consequently for all N sufficiently large

$$I_N = 2\pi i \left\{ -\pi \cos ax/(x \sin \pi x) + \sum_{n=-N}^{N}(-1)^n \cos na/(x^2-n^2) \right\} \to 0.$$

4.6.11. If $f(z) = z(\sinh \pi az \sin \pi z)^{-1}$, then for real $a \neq 0$ and z on the vertical side of R_N we have:

$$|zf(z)| \leq (N+\tfrac{1}{2})^2(1+a^{-2})/\sinh \pi|a|(N+\tfrac{1}{2}) \to 0 \quad \text{as} \quad N \to +\infty;$$

for z on the horizontal sides of R_N we have:

$$|zf(z)| \leq (N+\tfrac{1!}{2})^2(1+a^{-2})/\sinh \pi|a|^{-1}(N+\tfrac{1}{2}) \to 0 \quad \text{as} \quad N \to +\infty.$$

This implies $I_N \to 0$.

(i) There are following singularities of f inside R_N: simple poles $-N$, $-N+1, \ldots, -1, 0, 1, \ldots, N$, with residues $(-1)^n n/(\pi \sinh \pi an)$, $n \neq 0$, as well as simple poles $-Ni/a, \ldots, -i/a, i/a, \ldots, Ni/a$ with residues $(-1)^m m \times [\pi a^2 \sinh(\pi m/a)]^{-1}$; $\operatorname{res}(0;f) = 1/\pi^2 a$. Hence

$$I_N = 2\pi i \left[(\pi^2 a)^{-1} + 2\sum_{m=1}^{N} (-1)^m m (\pi \sinh \pi am)^{-1} \right.$$
$$\left. + 2\sum_{m=1}^{N} (-1)^m m \left(\pi a^2 \sinh(\pi m/a)\right)^{-1} \right] \to 0 \quad \text{as} \quad N \to +\infty.$$

Note that each sum has a finite limit.

(ii) Put $a = 1$ in (i).

4.6.12. Suppose that h, θ are arbitrary real numbers satisfying: $h > 1$, $0 < \theta < \pi/2$ and $D(h, \theta)$ is the bounded, closed domain whose boundary consists of:
1° the segment l on the straight line $\operatorname{re} z = h$ such that $-\theta < \arg z < \theta$,
2° the circular arc γ arising from l by inversion,
3° two segments joining the end points of l and γ.

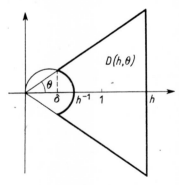

Fig. 4

Obviously $a \in D(h, \theta)$ implies $a^{-1} \in D(h, \theta)$. If $\delta = h^{-1}\cos^2\theta$, then $\operatorname{re} z \geqslant \delta$ for each $z \in D(h, \theta)$ and therefore

$$|m/\sinh \pi mz| \leqslant |m/\sinh \pi m\delta| \leqslant A m^{-2}$$

for all sufficiently large m and all $z \in D(h, \theta)$. This implies that both series of Exercise 4.6.11 (i) are uniformly convergent in $D(h, \theta)$ and represent analytic functions of a in the right half-plane. By Exercise 4.6.11 (i) the difference of both sides vanishes identically for real $a \in [h^{-1}, h]$, hence being analytic in the right half-plane, it vanishes there identically. An analogous identity for the

4. SEQUENCES AND SERIES

left half-plane is obtained by a change of sign. On the imaginary axis both series are divergent.

4.6.13. If $N+\tfrac{1}{2} > |x|$, the sum of integrals over the horizontal sides does not exceed

$$4(N+\tfrac{1}{2})[1+\sinh|a|(N+\tfrac{1}{2})]\sqrt{2(N+\tfrac{1}{2})}[\sinh\pi(N+\tfrac{1}{2})]^{-1}[(N+\tfrac{1}{2})^2-|x|^2]^{-1}$$

which tends to 0 as $N \to +\infty$. The parametrized integral over the vertical right-hand side ($\theta = \arg z \in [-\tfrac{1}{4}\pi, \tfrac{1}{4}\pi]$) has the form $\int_{-\delta}^{\delta} + \int_{-\pi/4}^{-\delta} + \int_{\delta}^{\pi/4}$, where the first integral can be made arbitrarily small since the integrand is bounded independently of N and in the remaining intervals $|\sin az|/|\sin \pi z| \to 0$ uniformly as $N \to +\infty$. This implies that the integral over the vertical right-hand side (and also over the left-hand side) tends to zero. Hence

$$\mathrm{res}(x;f) + \sum_{n=1}^{N} \mathrm{res}(n;f) \to 0.$$

Note that

$$\mathrm{res}(x;f) = -\pi\sin ax/(2\sin\pi x), \quad \mathrm{res}(n;f) = (-1)^n \frac{n\sin an}{x^2-n^2}.$$

4.6.14. (i) Put $a = \pi/2$, $x = 2z$ in the formula of Exercise 4.6.13;
(ii) replace z by iz in (i).

4.6.15. $\cot\pi z$ and $\coth\pi z$ are uniformly bounded on $\mathrm{fr}\,Q_N$, where Q_N is the square of Exercise 4.6.1. Hence

$$\int_{\partial Q_N} f(z)\,dz \to 0 \quad \text{as} \quad N \to +\infty.$$

The integrand has following singularities inside Q_N: $2N$ simple poles on the real axis, $2N$ simple poles on the imaginary axis with $\mathrm{res}(n;f) = \mathrm{res}(ni;f) = n^{-7}\coth\pi n$ and a pole of order 9 at the origin. In order to evaluate $\mathrm{res}(0;f)$, we find the Laurent series expansion of the integrand near $z = 0$ by using the formulas:

$$\pi z\cot\pi z = 1 - B_1(2\pi)^2 z^2/2! - B_2(2\pi)^4 z^4/4! + \cdots,$$

$$\pi z\coth\pi z = 1 + B_1(2\pi)^2 z^2/2! - B_2(2\pi)^4 z^4/4! + \cdots.$$

Hence

$$\mathrm{res}(0;f) = \pi^{-1}(2\pi)^8[(B_2/4!)^2 - B_1 B_3/6! - 2B_4/8!] = 2^8\pi^7(-19)/(6!7!).$$

Observe that

$$4\sum_{n=1}^{N} n^{-7}\coth\pi n + \mathrm{res}(0;f) \to 0 \quad \text{as} \quad N \to +\infty.$$

4.7.1. Suppose F is a compact subset of G and $\sum u_n$ an arbitrary convergent series with decreasing positive terms. Suppose $\{A_n\}$ is an increasing sequence such that
$$\left|\int_{A_n}^{b} W(z,t)dt\right| \leq u_n \quad \text{for any} \quad b > A_n \quad \text{and any} \quad z \in F.$$
The series $\sum_{n=1}^{\infty} \int_{A_n}^{A_{n+1}} W(z,t)dt$ is uniformly convergent on F, hence $\int_{a}^{+\infty} = \int_{a}^{A_1} + \sum_{n=1}^{\infty} \int_{A_n}^{A_{n+1}}$ is analytic in G.

4.7.2. Suppose F is a compact subset of the right half-plane. We have: $\operatorname{re} z \geq \delta > 0$ for all $z \in F$; if $\varepsilon > 0$, arbitrary, and $2\delta^{-1}e^{-\delta A} < \varepsilon$, then $\left|\int_{b}^{B} e^{-zt}dt\right| < \varepsilon$ for any $z \in F$ and any $b, B > A$, hence the integral is a.u. convergent in the right half-plane. Evidently
$$\int_{0}^{A} e^{-zt} dt = -z^{-1}e^{-Az} + z^{-1} \to z^{-1} \quad \text{as} \quad A \to +\infty,$$
hence
$$\int_{0}^{+\infty} e^{-xt}\cos yt\, dt = \frac{x}{x^2+y^2}, \quad \int_{0}^{+\infty} e^{-xt}\sin yt\, dt = \frac{y}{x^2+y^2}.$$

4.7.3. The integral is a.u. convergent in $\mathbf{C}$ and can be differentiated under the sign of integral.

4.7.4. Put $g(z) = \int_{-\infty}^{+\infty} (x-z)^{-2}\varphi(x)dx$ and verify that
$$|g(z) - [f(z+h) - f(z)]/h| = O(h) \quad \text{as} \quad h \to 0.$$

4.7.5. Suppose that z_0 is a pole of Γ and z_0 is not an integer. Since $\Gamma(z_0+n) = (z_0+n-1) \cdots (z_0+1)z_0\Gamma(z_0)$, also z_0+n would be a pole of Γ which is impossible for Γ is analytic in the right half-plane.

4.7.6. Consider $G(z)/\Gamma(z)$.

4.7.7. Put $x^{\alpha} = t$.

4.7.8. If $z = Re^{i\theta}$ and $s = \sigma + i\tau$, we have:
$$|e^{-z}z^{s-1}| \leq R^{\sigma-1}\exp(-R\cos\theta);$$
hence by Exercise 3.7.2:
$$\left|\int_{\gamma(R)}\right| \leq R^{\sigma}\int_{0}^{\pi/2} \exp(-R\cos\theta)d\theta \leq \tfrac{1}{2}\pi R^{\sigma-1} \to 0$$

4. SEQUENCES AND SERIES

as $R \to +\infty$ (for $\sigma < 1$). Similarly

$$\left| \int_{\gamma(\delta)} \right| \leq \delta^\sigma \int_0^{\pi/2} \exp(-\delta\cos\theta)\,d\theta \leq \tfrac{1}{2}\pi\delta^\sigma \to 0$$

as $\delta \to 0$ (for $0 < \sigma$). Here $\gamma(t) = te^{i\theta}$, $0 \leq \theta \leq \dfrac{\pi}{2}$. Now,

$$\int_{\partial D(\delta,R)} = 0 = \int_0^{+\infty} e^{-x}x^{s-1}\,dx - \int_0^{+\infty} e^{-iy}\exp[(s-1)(\log y + i\pi/2)]i\,dy$$

and finally

$$\Gamma(s) = \exp(\pi i s/2) \int_0^{+\infty} e^{-iy} y^{s-1}\,dy.$$

4.7.9. Put $s-1 = \alpha$ in Exercise 4.7.8 and separate real and imaginary parts. We have:

$$\lim_{\alpha \to 1} \Gamma(1-\alpha)\cos(\pi\alpha/2) = \lim_{s \to 0} \Gamma(s)\sin(\pi s/2) = \pi/2 = \int_0^{+\infty} x^{-1}\sin x\,dx.$$

4.7.10. Putting $t = \sin^2\theta$ we obtain the integral

$$\tfrac{1}{2}\int_0^1 t^{p/2-1}(1-t)^{q/2-1}\,dt = \tfrac{1}{2}B(p/2, q/2)$$

which is equal to $\tfrac{1}{2}\Gamma(p/2)\Gamma(q/2)/\Gamma(\tfrac{1}{2}(p+q))$.

4.7.11. (i) Put $p = \alpha+1$, $q = 1-\alpha$ in Exercise 4.7.10;
(ii) put $p = 3/2$, $q = 1$.

4.7.12. (i) $\tfrac{1}{4}\pi^{-1/2}\Gamma^2(\tfrac{1}{4})$;
(ii) $2\pi^{3/2}[\Gamma(\tfrac{1}{4})]^{-2}$.

4.7.13. If $\varepsilon = \exp(2\pi i/n)$, then $(z-1)(z-\varepsilon)(z-\varepsilon^2) \ldots (z-\varepsilon^{n-1}) = z^n - 1$ and hence

$$(1-\varepsilon)(1-\varepsilon^2) \ldots (1-\varepsilon^{n-1}) = \lim_{z \to 1}(z^n-1)/(z-1) = n.$$

Now, $|1-\varepsilon^k| = 2\sin(k\pi/n)$ which gives the identity $\prod_{k=1}^{n-1}\sin(k\pi/n) = n2^{1-n}$. Use now the formula $\Gamma(z)\Gamma(1-z) = \pi/\sin\pi z$ with $z = \pi/n, 2\pi/n, \ldots, (n-1)\pi/n$.

4.7.14. If f is nonnegative and decreasing in $(0, 1]$ then

$$\int_{1/n}^1 f(x)\,dx \leq n^{-1}[f(1/n)+f(2/n)+ \ldots +f(n/n)] \leq \int_0^{1-1/n} f(x)\,dx$$

which implies that the improper integral $\int_0^1 f(x)dx$ exists, iff the finite limit $\lim_{n\to\infty} n^{-1}[f(1/n)+f(2/n)+ \ldots +f(n/n)]$ exists and both are equal. Note that $\log \Gamma(x)$ is nonnegative and decreasing in $(0, 1]$ and use the equality of Exercise 4.7.13.

4.7.15. We have: $I'(a) = \log \Gamma(a+1) - \log \Gamma(a) = \log a$, therefore $I(a) = a\log a - a + b$; $b = I(0+) = \frac{1}{2}\log 2\pi$.

4.8.1. If l is an arc in the z-plane through z_0, $|l|$ its length and l_1 is the spherical image of $f(l)$, then

$$\rho(z_0, f) = \lim |l_1|/|l| \quad \text{as } l \text{ shrinks to } z_0.$$

4.8.2. $\rho(0, f) = 2|a|/(1+|b|^2)$ is unbounded.

4.8.3. Note that any normal family in which the case of a.u. divergence to ∞ is excluded, is necessarily compact.

4.8.4. Suppose that $\overline{K}(z_0; 2r) \subset D$ and M is the common bound of $|f|$ in this disk. Then for $z \in \overline{K}(z_0; r)$ by Cauchy's formula:

$$|f'(z)| = \left|(2\pi i)^{-1} \int_{C(z_0; 2r)} (\zeta-z)^{-2} f(\zeta) d\zeta\right| \leqslant 2Mr^{-1};$$

hence f' are locally uniformly bounded and form a compact family. Marty's condition shows that $n(z^2-n^2)$ form a normal family, whereas their derivatives do not.

4.8.5. If H is a compact subset of D, then $|f(z)| \leqslant R$ for all $z \in H$, $f \in \mathscr{F}$. Hence $|F \circ f(z)| \leqslant \sup_{|w| \leqslant R} |F(w)|$ for all $z \in H$. Hence $\{F \circ f\}$ is a compact family.

4.8.6. $\{az\}$ form in $|z| > 1$ a normal family by Marty's condition. Consider the sequence $f_n(z) = F[(2n+\frac{1}{2})z]$; we have:

$$f_n(\pi) = \exp[(2n+\tfrac{1}{2})\pi] \to +\infty, \quad f_n(2\pi) = 0$$

which shows that $\{F \circ f\}$ is not a normal family.

4.8.7. By Marty's criterion $\{1/f\}$ is a normal family. Hence $1/f_n$ is either a.u. convergent in D and the limit function $1/g$ is analytic which makes $g(z) = 0$ impossible, or $1/f_n$ tends a.u. to ∞ which gives $g = 0$.

4.8.8. By Ex. 4.8.7, $g(z) - g(a)$ is either never 0 in $D \setminus a$ (which means that g is univalent since a is arbitrary), or $g(z) - g(a) \equiv 0$, i.e. $g = \text{const}$.

4.8.9. The sequence $n^{-1}(e^{nz}-1)$ is equal to 0 at $z = 0$ and tends to ∞ at $z = 1/2$, hence the family considered cannot be normal.

4.8.10. Let $g = \log f(z)/z$ with $f \in T_0$ be this branch which is equal to 0 at the origin. Since g does not take the values $\mp 2\pi i$, so $\{g\}$ is normal and even compact (cf. Ex. 4.8.3), since $g(0) = 0$. By Exercise 4.8.5 T_0 is compact.

4.8.11. The functions g_n form a normal family ($g_n \neq \mp 2\pi i$). Since $g_n(0) = 0$, g_n contains a subsequence g_{n_k} a.u. convergent in $K(0; 1)$ whose limit g is analytic and $g'_{n_k}(0) \to g'(0) = \beta$ with finite β. Now, $g'_{n_k}(0) = -1/\alpha_{n_k} \to \infty$ which is a contradiction.

4.8.12. Obviously we can assume $\alpha = 0$. Since a family is normal, iff it is locally normal, we can also assume that $D = K(0; 1)$. Now, $\log[f(z)/f(0)] = g(z)$ are analytic in $K(0; 1)$ and omit the values $\mp 2\pi i$; hence $\{g\}$ is normal, and also compact (Ex. 4.8.3; $g(0) = 0$). By Exercise 4.8.5 $\{f(z)/f(0)\}$ form a compact family, as well as $\{f(0)/f(z)\}$. Hence there exists a finite $m(r)$ such that $|f(z)| \leq |f(0)|m(r)$ and also $|f(0)|(m(r))^{-1} \leq |f(z)|$ in $\bar{K}(0; r)$. From this the normality readily follows.

4.8.13. $(f(z_0))^2 = (2\pi)^{-1} \int_0^{2\pi} f^2(z_0 + re^{i\theta}) d\theta$, by Cauchy's formula: hence

$$|f(z_0)|^2 \leq (2\pi)^{-1} \int_0^{2\pi} |f(z_0 + re^{i\theta})|^2 d\theta.$$

After multiplying by r and integrating w.r.t. r over $[0, R]$ we obtain

$$\pi R^2 |f(z_0)|^2 \leq \iint_{\bar{K}(z_0; R)} |f(z)|^2 dx\, dy \leq \iint_D |f(z)|^2 dx\, dy \leq M$$

which shows that

$$|f(z_0)| \leq \sqrt{M/\pi R^2} \quad \text{where} \quad R = \text{dist}(z_0, \mathbf{C} \setminus D).$$

CHAPTER 5

Meromorphic and Entire Functions

5.1.1. In what follows H denotes an arbitrary entire function.

(i) $F(z) = H(z) + \sum_{n=1}^{\infty} z^2 [n(z-n)]^{-1}$;

(ii) $F(z) = H(z) + \sum_{n=1}^{\infty} z/(z-a^n)$;

(iii) $F(z) = H(z) + z^{-1} + \sum_{n=1}^{\infty} z^2/[n(z-\sqrt{n})]$;

(iv) we have $n(z-n)^{-2} = n^{-1}[1+2(z/n)+3(z/n)^2 + \ldots], |z| < n$, and since $|n(z-n)^{-2} - n^{-1}| \leqslant (n-\sqrt{n})^{-2}$ in $K(0; \sqrt{n})$, therefore

$$F(z) = H(z) + \sum_{n=1}^{\infty} z(2n-z)/[n(z-n)^2];$$

(v) $F(z) = H(z) + \sum_{n=1}^{\infty} [n^2(z-n)^{-2} - 1 - 2z/n + (z-n)^{-1} + 1/n]$;

(vi) $F(z) = H(z) + z^2 \sum_{n=1}^{\infty} [n(z-n)]^{-1} + z^2 \sum_{n=1}^{\infty} [(z+n)]^{-1}$

$= H(z) + 2z^3 \sum_{n=1}^{\infty} [n(z^2-n^2)]^{-1}$;

(vii) $F(z) = H(z) + \pi \cot \pi z$ (cf. Ex. 4.5.16);

(viii) $F(z) = H(z) + z^{-1} - z \sum_{n=1}^{\infty} [n(z+n)]^{-1}$.

5.1.2. $F(z) = H(z) + \sum_{n=1}^{\infty} (-1)^n/(z+n)$.

5.1.3. $(z-w)^{-1}$ is the principal part at a pole w and

$$|(z-w)^{-1} + w^{-1} + zw^{-2}| \leqslant 2|z|^2|w|^{-3} \quad \text{for} \quad |z| < \tfrac{1}{2}|w|.$$

There are $8k$ poles on the boundary of the square with vertices $k(\mp 1 \mp i)$ and

5. MEROMORPHIC AND ENTIRE FUNCTIONS

grouping together all corresponding terms, we obtain after dropping a finite number of initial terms an absolutely and uniformly convergent series. Hence

$$F(z) = H(z) + z^{-1} + \sum_{m,n}' [(z-m-ni)^{-1} + (m+ni)^{-1} + z(m+ni)^{-2}].$$

5.1.4. We have: $e^{-t}t^{z-1} = \sum_{n=0}^{\infty} (-1)^n t^{n+z-1}/n!$ and integrating term by term, we obtain the desired result.

5.1.5. From the equality $\Gamma(z) = \int_0^{+\infty} e^{-t}t^{z-1} dt = \int_0^1 + \int_1^{+\infty}$ and the equality of Exercise 5.1.4 we obtain

$$\Gamma(z) = H(z) + \sum_{n=0}^{\infty} (-1)^n [(z+n)n!]^{-1},$$

where $H(z) = \int_1^{+\infty} e^{-t}t^{z-1} dt$ is an entire function.

5.2.1. From 2° and 3° it follows that for all n sufficiently large the points $0, z$ lie inside C_n. If $\varphi(\zeta) = [\zeta(\zeta-z)]^{-1} f(\zeta)$, then $\int_{C_n} \varphi(\zeta) d\zeta \to 0$ as $n \to +\infty$ which is a consequence of 1°, 4° and 5°. On the other hand, the integral can be evaluated by means of the theorem of residues. Since $\mathrm{res}(0; \varphi) = -f(0)/z$, $\mathrm{res}(z; \varphi) = f(z)/z$, $\mathrm{res}(a_n; \varphi) = A_n[a_n(a_n-z)]^{-1}$, we have

$$2\pi i \{-f(0) + f(z) + z \sum_{n=1}^{\infty} A_n [a_n(a_n-z)]^{-1}\} = 0.$$

5.2.2. The function $g(z) = f(z) - G(1/z)$ is analytic, at $z = 0$, hence we can apply Exercise 5.2.1 for g and then add $G(1/z)$ to both sides.

5.2.3. (i) If Q_n are squares with vertices $(n+\frac{1}{2})(\mp 1 \mp i)$, then $|z^{-1} - \pi/\sin \pi z| \leq \pi + 2$ for all positive integers n and all $z \in \mathrm{fr} Q_n$ (note that $|z| \geq \frac{1}{2}$). Moreover, $z = 0$ is a removable singularity of $z^{-1} - \pi/\sin \pi z$, hence by Exercise 5.2.1:

$$\pi/\sin \pi z - z^{-1} = \lim_{m \to \infty} \sum_{n=-m}^{m} (-1)^n [(z-n)^{-1} + n^{-1}] = 2z \sum_{n=1}^{\infty} (-1)^{n+1}/(n^2-z^2);$$

(ii), (iii) can be solved in an analogous manner.

5.2.4. By Exercise 5.2.3 (iii) we have

$$\pi \tan \pi z = \sum_{n=0}^{\infty} [(n+\tfrac{1}{2}-z)^{-1} - (n+\tfrac{1}{2}+z)^{-1}]$$

and hence
$$\sum_{n=0}^{\infty}[(n\alpha+\beta)^{-1}-(n\alpha+\alpha-\beta)^{-1}] = \frac{\pi}{\alpha}\tan\frac{\pi}{2}\left(1-\frac{2\beta}{\alpha}\right) = \frac{\pi}{\alpha}\cot\frac{\pi\beta}{\alpha};$$
next we put $\alpha = 3$, $\beta = 1$.

5.2.5. Using the results of Exercise 5.2.3, the equality of Exercise 4.5.16 as well as the identities of Exercise 2.7.15 we obtain:
$$\pi/\sinh\pi z = z^{-1} + 2z\sum_{n=1}^{\infty}(-1)^n(n^2+z^2)^{-1};$$
$$\pi/\cosh\pi z = 2\sum_{n=0}^{\infty}(-1)^n(n+\tfrac{1}{2})[(n+\tfrac{1}{2})^2+z^2]^{-1};$$
$$\pi\tanh\pi z = 2z\sum_{n=0}^{\infty}[(n+\tfrac{1}{2})^2+z^2]^{-1};$$
$$\pi\coth\pi z = z^{-1} + 2z\sum_{n=1}^{\infty}(n^2+z^2)^{-1}.$$

5.2.6. The function $f(z) = (\sin z \sinh z)^{-1} - z^{-2}$ has a removable singularity at the origin and we may assume $f(0) = 0$. On the boundaries of the squares Q_n with vertices $\pi(n+\tfrac{1}{2})(\mp 1 \mp i)$ we have $|f(z)| \leq 1+4/\pi^2$ since $|z| \geq \pi/2$; f has $4n$ simple poles $m\pi, m\pi i$ ($m = \mp 1, \mp 2, \ldots, \mp n$) inside Q_n with residues: $\mathrm{res}(m\pi;f) = (-1)^m /\sinh m\pi$, $\mathrm{res}(m\pi i;f) = (-1)^{m-1} i/\sinh m\pi$. Hence
$$f(z) = \sum_{m=1}^{\infty}(-1)^m 4m\pi z^2 [(z^4-m^4\pi^4)\sinh m\pi]^{-1}.$$

5.2.7. We have:
$$\cosh z - \cos z = \cos iz - \cos z = 2\sin((1+i)z/2)\sin((1-i)z/2)$$
$$= -2\sin\zeta\sin i\zeta = -2i\sin\zeta\sinh\zeta,$$
where $\zeta = (1+i)z/2$. Hence
$$(\cosh z - \cos z)^{-1} = \tfrac{1}{2}i(\sin\zeta\sinh\zeta)^{-1};$$
use now Exercise 5.2.6.

5.2.8. $f(z) = (\sin z - z\cos z)^{-1} z\sin z - 3/z = O(z)$ as $z \to 0$ so that we may assume $f(0) = 0$. Furthermore, $f(z) = (\tan z/z - 1)^{-1}\tan z - 3/z$ is uniformly bounded on $\mathrm{fr}Q_n$, where Q_n are squares with vertices $n\pi(\mp 1 \mp i)$ and has simple poles at λ_n with residues 1. Hence
$$f(z) = \sum_{n=-\infty}^{\infty}{'}[(z-\lambda_n)^{-1}+1/\lambda_n] = 2z\sum_{n=1}^{\infty}(z^2-\lambda_n^2)^{-1}.$$

5.2.9. The functions $\tan z$, $(\cos z)^{-1}$ are uniformly bounded on $\mathrm{fr}Q_n$, hence $\int_{\partial Q_n} \to 0$ as $n \to +\infty$. The integrand f has the following singularities inside Q_n: a simple pole z with $\mathrm{res}(z;f) = \sin z/z\cos^2 z$, as well as double poles $(m+\tfrac{1}{2})\pi$ with residues

$$(-1)^{m-1}[(2m+1)\pi-z]\{\pi(m+\tfrac{1}{2})[(m+\tfrac{1}{2})\pi-z]\}^{-2}.$$

Since $\sum_{m=-\infty}^{\infty}(-1)^m (m+\tfrac{1}{2})^{-2} = 0$, the desired result follows from the theorem of residues.

5.2.10. $f(z) = e^{az}(e^z-1)^{-1} - z^{-1} \to a - \tfrac{1}{2}$ as $z \to 0$; moreover f is uniformly bounded on $\mathrm{fr}Q_n$, where Q_n are squares with corners $\pi(2n+1)(\mp 1 \mp i)$ and has at $z = 2m\pi i$ ($m = \mp 1, \mp 2, \ldots$) simple poles with residues $\exp(2am\pi i)$. Hence by Exercise 5.2.1:

$$f(z) = a - \tfrac{1}{2} + \sum_{m=-\infty}^{\infty}{}' \exp(2am\pi i)[(z-2m\pi i)^{-1} + 1/2m\pi i]$$

$$= \sum_{m=1}^{\infty}(z^2+4m^2\pi^2)^{-1}(2z\cos 2am\pi - 4m\pi\sin 2am\pi)$$

since

$$-a + \tfrac{1}{2} = \sum_{m=1}^{\infty}(m\pi)^{-1}\sin 2am\pi \quad \text{for} \quad 0 < a < 1.$$

5.3.1. If

$$\varphi(z) = f(z)\prod_{k=1}^{m}(r^2-\bar{a}_k z)[r(z-a_k)]^{-1}\prod_{k=1}^{n}r(z-b_k)(r^2-\bar{b}_k z)^{-1},$$

then φ is analytic in $K(0;r)$ and has no zeros. Moreover, $|\varphi| = |f|$ on $C(0;r)$. Since $\log|\varphi|$ is harmonic in $K(0;r)$, we have by the Gauss formula (cf. Ex. 3.2.5):

$$(2\pi)^{-1}\int_0^{2\pi}\log|f(re^{i\theta})|\,d\theta = (2\pi)^{-1}\int_0^{2\pi}\log|\varphi(re^{i\theta})|\,d\theta = \log|\varphi(0)|$$

which gives Jensen's formula.

5.3.2. It is sufficient to prove that $\int_0^{2\pi}\log|1-e^{i\theta}|\,d\theta = 0$. This can be done e.g. by integrating $(iz)^{-1}\mathrm{Log}(1-z)$ round the boundary of $K(0;1)$ with an indentation at $z = 1$ so as to leave $z = 1$ outside. If the indentation γ_r has the equation:

$z = 1-re^{i\varphi}$, $\varphi_1 \leqslant \varphi \leqslant \varphi_2$, and $\varepsilon = \varepsilon(r) = \operatorname{Arg} z_1$ for its upper end point z_1, then

$$\int_{\varepsilon}^{2\pi-\varepsilon} \operatorname{Log}(1-e^{i\theta})d\theta + \int_{\varphi_1}^{\varphi_2} (1-re^{i\varphi})^{-1}\operatorname{Log}(re^{i\varphi})re^{i\varphi}d\varphi = 0.$$

Since the latter integral tends to 0 as $r \to 0$, the result follows.

5.3.3. If $f(z) = z^\lambda F(z)$, then $\log|f| = \lambda \log r + \log|F|$ and

$$(2\pi)^{-1}\int_0^{2\pi} \log|f(re^{i\theta})|\,d\theta$$
$$= \lambda \log r + \log|F(0)| + \log[r^m/|a_1 a_2 \ldots a_m|] - \log[r^n/|b_1 b_2 \ldots b_n|].$$

5.3.4. (i) $|a_1| = 1$, $|a_2| = |a_3| = \sqrt{3}$, therefore

$$\Phi(r) = \begin{cases} \log 3 & \text{for } r \in (0, 1], \\ \log 3 + \log r & \text{for } r \in [1, \sqrt{3}], \\ 3\log r & \text{for } r \geqslant \sqrt{3}; \end{cases}$$

(ii) by Exercise 5.3.3 we have:

$$\Phi(r) = (2n+1)\log r - 2n\log \pi - 2\log n!$$

for $n\pi \leqslant r \leqslant (n+1)\pi$;

(iii) we have for $\cos z$:

$$\Phi(r) = 2n\log r - 2n\log \pi - 2\log[\tfrac{1}{2} \cdot \tfrac{3}{2} \ldots (n-\tfrac{1}{2})],$$

$(n-\tfrac{1}{2})\pi \leqslant r \leqslant (n+\tfrac{1}{2})\pi$; now subtract both expressions for Φ for cosine and sine.

5.3.5. Obviously $\log x = \log^+ x - \log^+(1/x)$. Since the zeros of f are poles of $1/f$, we have only to show that

$$\log[r^n/|b_1 b_2 \ldots b_n|] = \int_0^r t^{-1}n(t,f)dt.$$

If $n_k(t) = 0$ for $t < |b_k|$, $n_k(t) = 1$ for $t \geqslant |b_k|$, then

$$\int_0^r t^{-1}n_k(t)dt = \log[r/|b_k|].$$

Observe now that $\sum_{k=1}^\infty n_k(t) = n(t,f)$.

5. MEROMORPHIC AND ENTIRE FUNCTIONS

5.3.6. f is analytic and $\neq 0$ at the origin and moreover $n(r,f) = n(r, 1/f)$.

5.3.7. $n(r,f) = 2[r/\pi + 1/2]$; $n(r, 1/f) = 2[r/\pi]$.

5.3.8. Since $n(r,f) \equiv 0$, we have
$$T(r,f) = m(r,f) = (2\pi)^{-1} \int_{-\pi/2}^{\pi/2} r\cos\theta \, d\theta = r/\pi.$$

5.3.9. $P(z) = |a|r^n \exp i(\alpha + n\theta) + O(r^{n-1})$ where $a = |a|e^{i\alpha}$;
$$\operatorname{re} P(z) = |a|r^n \cos(\alpha + n\theta) + O(r^{n-1}).$$

Thus
$$T(r,f) = m(r,f) = (2\pi)^{-1}|a|r^n \int_0^{2\pi} \{\cos(\alpha + n\theta)\}^+ d\theta + O(r^{n-1}) = |a|r^n/\pi + O(r^{n-1}),$$
where $\{f\}^+ = \tfrac{1}{2}(f + |f|)$. Hence $r^{-n}T(r,f) \to |a|/\pi$.

5.3.10. $\log|f(re^{i\theta})| = \operatorname{re}\operatorname{Log} f(re^{i\theta}) = r\sin\theta/(1 + r^2 - 2r\cos\theta)$ which is ≥ 0 for $0 \leq \theta \leq \pi$. Hence
$$m(r,f) = (2\pi)^{-1} \int_0^{\pi} (1 + r^2 - 2r\cos\theta)^{-1} r\sin\theta \, d\theta$$
$$= (4\pi)^{-1}[\log(1 + r^2 - 2r\cos\theta)]_0^{\pi} = (2\pi)^{-1}\log[(1+r)/(1-r)],$$
$0 < r < 1$.

5.3.11. If $f(z) = c_\lambda z^\lambda + \ldots$, $|c_\lambda| > 0$, then
$$\Phi(r) = \lambda \log r + \log|c_\lambda| + \int_0^r t^{-1} n(t, z^\lambda/f) \, dt$$
which shows that Φ is an increasing function of $\log r$; we have $d\Phi(r)/d\log r = \lambda + n(r, z^\lambda/f)$ at continuity points of n which is an increasing function of r; moreover, the right-hand side derivative is greater or equal to the left-hand side derivative.

5.3.12. The integral $\int_0^1 t^{-1}\mu(t)\,dt$ is finite (cf. Ex. 5.3.1) and this implies that also $\int_{1/2}^1 \mu(t)\,dt$ is finite. Since $\mu(t)$ is increasing, we have $\lim_{t \to 1-}(1-t)\mu(t) = 0$. Observe now that $-\log r \sim 1 - r$ as $r \to 1$.

5.3.13. (i) By Exercises 5.3.11, 5.3.12 finite limits $\lim_{r \to 1-} \Phi(r)$, $\lim_{r \to 1-} \log r^{\mu(t)} = 0$ exist. Using this and Exercise 5.3.1 we see that also a finite limit $\lim_{n \to +\infty} \log|a_1 a_2 \ldots a_n|$ exists.

(ii) $|\log|a_n||/(1-|a_n|) \to 1$ because $|a_n| \to 1$. Now, the series $\sum \log|a_n|$ is absolutely convergent by (i) and this implies the convergence of $\sum (1-|a_n|)$.

5.3.14. The function $F = (f-g)z^{-\lambda}$ is analytic and bounded in $K(0;1)$ and does not vanish at the origin. It vanishes at $z = a_n$ and if we had $f-g \neq 0$, then the series $\sum (1-|a_n|)$ would be convergent by Exercise 5.3.13 (ii).

5.4.1. We have $0 < A \leq |z^{-2}[\text{Log}(1+z)-z]| \leq B$ for $z \in K(0;r)$ and sufficiently small $r > 0$. Therefore
$$A|u_n|^2 + u_n \leq \text{Log}(1+u_n) \leq B|u_n|^2 + u_n \quad \text{for all} \quad n \geq n_0.$$

5.4.2. The series $\sum u_n$ is convergent as an alternating series, whereas $\sum u_n^2 = \sum 1/n$ is divergent (cf. Ex. 5.4.1).

5.4.3. (i) $P_n = p_1 p_2 \cdots p_n = (n+1)/2n$;
(ii) $P_n = \frac{2}{3}[1+1/n(n+1)]$;
(iii) $P_{2k} = 1$, $P_{2k+1} = (2k+2)/(2k+1)$.

5.4.4. $\prod_{n=0}^{m} (1+z^{2^n}) = 1+z+z^2+\cdots+z^{2^{m+1}-1}$.

5.4.5. (i) The finite plane; (ii) $K(0; 1/e)$; (iii) the finite plane; (iv) $E \setminus N_1$, N_1 is the set of all negative integers.

5.4.6. Put $u_n(z) = (z-a_n)/(z-\bar{a}_n^{-1}) - 1$. If $0 < r < 1$, and $z \in \bar{K}(0;r)$, we have $|u_n(z)| \leq (1-|a_n|^2)/(1-r) = A_n$. Since $\sum A_n < +\infty$, the product $\prod (1+u_n(z))$ represents a function F analytic in $K(0;1)$. If $z \neq a_n$, then all the factors are different from 0 and consequently $F(z) \neq 0$ which follows from the definition of convergence for infinite products. Moreover,
$$|(z-a_n)/(z-\bar{a}_n^{-1})| = |a_n||(z-a_n)/(1-\bar{a}_n z)| < |a_n| < 1$$
for $z \in K(0;1)$ and this implies $|F(z)| < 1$ in $K(0;1)$.

5.4.7. We have:
$$|(1-z)e^z - 1| = |z|^2|(1-1/2!) + (1/2!-1/3!)z + (1/3!-1/4!)z^2 + \cdots|$$
$$\leq |z|^2[1-1/2! + (1/2!-1/3!) + \cdots] = |z|^2$$
for $z \in K(0;1)$. Suppose now that $z \in K(0;R)$; if $n > R$, then $|z|/n < R/n < 1$ and this implies
$$|u_n(z)| = |(1+z/n)e^{-z/n} - 1| \leq |z^2|/n^2 < R^2/n^2.$$

5.4.8. We have:
$$h_n(z) = z\exp[z(1+1/2+\cdots+1/n-\log n)]\prod_{k=1}^{n}(1+z/k)e^{-z/k}.$$

5. MEROMORPHIC AND ENTIRE FUNCTIONS

Note that $\gamma_n = 1+1/2+ \ldots +1/n-\log n \to \gamma = 0.5772\ldots$ (γ is the so-called Euler's constant) and use the result of Exercise 5.4.7.

5.4.9. (i) With the notation of Exercise 5.4.8 we have:
$h_n(z+1) = z^{-1}h_n(z)(z+n+1)/n$ and therefore $\Gamma(z+1) = z\Gamma(z)$;
(ii) $h_n(1) = 1+1/n \to 1$, hence $\Gamma(1) = 1$; now apply (i).

5.5.1. (i) $q = 2$; (ii) no finite genus exists; (iii) $q = 2$; (iv) $q = 3$; (v) $q = 4$; (vi) $q = 3$.

5.5.2. (i) $z \exp g(z) \prod_{n=-\infty}^{\infty} (1-z/n)e^{z/n}$;

(ii) $z \exp g(z) \prod_{n=1}^{\infty} E\big(z/\log(1+n), n\big)$, $\quad m_n = n$;

(vi) $z \exp g(z) \prod_{n=1}^{\infty} [E(z/n, 2)]^n$.

5.5.3. If F is an entire function with zeros $a_1, a_2, \ldots, a_n, \ldots$ ($0 < |a_1| \leqslant |a_2| \leqslant \ldots$) written as many times as their order shows, then F'/F is a meromorphic function which has at a_n simple poles with principal parts $(z-a_n)^{-1}$ (for a zero of order k, we have k terms of this kind). Now, according to Mittag-Leffler formula,

$$\frac{F'(z)}{F(z)} = \sum_{n=1}^{\infty} \left\{ \frac{1}{z-a_n} + \frac{1}{a_n}\left[1 + \frac{z}{a_n} + \ldots + \left(\frac{z}{a_n}\right)^{m_n-1}\right]\right\} + h'(z)$$

where h is an entire function. If G is a simply connected domain such that $0, z \in G$ and all a_n belong to $\mathbf{C} \backslash G$ and $\gamma \subset G$ joins 0 to z, then

$$\int_0^z \frac{F'(\zeta)}{F(\zeta)} d\zeta = \log \frac{F(z)}{F(0)}$$

$$= \sum_{n=1}^{\infty} \log\left\{\left(1-\frac{z}{a_n}\right)\exp\left[\frac{z}{a_n}+\ldots+\frac{1}{m_n}\left(\frac{z}{a_n}\right)^{m_n}\right]\right\} + \log\exp\big(h(z)/h(0)\big),$$

where log denotes a single-valued branch of logarithm in G. The term by term integration is admissible by uniform convergence on γ.

5.5.4. If $b_0 = 0, b_1, b_2, \ldots, b_n, \ldots$ are poles of f and h is the Weierstrass product for $\{b_n\}$, then $H = fh$ is an entire function.

5.5.5. Suppose that $0 < |a_1| \leqslant |a_2| \leqslant \ldots$,

$$K_n = \overline{K}(0; \tfrac{1}{2}|a_n|) \quad \text{and} \quad M_n = \sup_{z \in K_n} \left|\frac{\eta_n \omega(z)}{\omega'(a_n)(z-a_n)}\right|.$$

Choose q_n so that $2^{-q_n}M_n \leq 2^{-n}$. If $N > n$, then the Nth term of (A) is dominated by 2^{-N} in K_n and this implies a.u. convergence of (A). If $z \to a_n$, then the only term with a removable singularity at a_n tends to η_n, while all the remaining terms tend to zero because they contain the factor $\omega(z)$.

5.5.6. We can take $\sin z$ instead of ω in Exercise 5.5.5. We have $|\sin \pi z| < \exp(n\pi/2)$ in $\overline{K}(0; n/2)$ hence e.g.

$$F(z) = \sum_{n=1}^{\infty} (-1)^n \frac{(n-1)!\sin \pi z}{\pi(z-n)} \left(\frac{z}{n}\right)^{n^2}.$$

5.6.1. The genus of the sequence of zeros is equal to 2, hence

$$\sin \pi z = \pi z \exp g(z) \prod_{n=-\infty}^{\infty}{}' (1-z/n)e^{z/n} = \pi z \exp g(z) \prod_{n=1}^{\infty} (1-z^2/n^2).$$

Taking the logarithmic derivative of both sides and using Exercise 4.5.16 we see that $g = 0$.

5.6.2. (i) Replace z by iz in Exercise 5.6.1;

(ii) use Exercise 5.6.1 and the identity:

$$\cos iz - \cos z = 2\sin[(1+i)z/2]\sin[(1-i)z/2];$$

(iii) a particular case of (iv);

(iv) replace z by $(a-b)z/2\pi$ in (i).

5.6.3. Cf. Exercise 5.6.1.

5.6.4. (i) $(\cosh z\pi \sqrt{2} - \cos z\pi \sqrt{2})[2\pi^2 z^2(1+z^4)]^{-1}$ (cf. Ex. 5.6.2 (ii));

(ii) $(\sin \pi z \sinh \pi z)[\pi^2 z^2(1-z^4)]^{-1}$; take the limits of both sides as $z \to 1$.

5.6.5. (i) If $\eta = \frac{1}{2}(1+i\sqrt{3})$, then

$$(1-z^2\eta^2/n^2)(1-z^2\bar{\eta}^2/n^2) = 1+(z/n)^2+(z/n)^4$$

and using Exercise 5.6.1. we obtain

$$\prod_{n=1}^{\infty} [1+(z/n)^2+(z/n)^4] = (\pi z)^{-2} \sin \pi z\eta \sin \pi z\bar{\eta};$$

(ii) if $\tau = \frac{1}{2}(\sqrt{3}+i)$, then

$$1+(z/n)^6 = [1+(z/n)^2][1-(\tau z/n)^2][1-(\bar\tau z/n)^2]$$

and using Exercises 5.6.1, 5.6.2 (i), we obtain that the value of the product is equal to

$$(\pi z)^{-3}\sinh \pi z \sin \pi\tau z \sin \pi\bar\tau z;$$

(iii) $(\pi z)^{-3}\sin \pi z \sin \pi\eta z \sin \pi\bar\eta z$.

5. MEROMORPHIC AND ENTIRE FUNCTIONS

5.6.6. (i), (ii). Put $z = 1$ in Exercise 5.6.5 (i), (ii).

5.6.7. Integrate both sides of the formula of Exercise 5.2.8 similarly as in Exercise 5.5.3.

5.6.8. If $z = \zeta + \tfrac{1}{2}$, then by Exercise 5.6.3 we have:

$$\frac{\sin \pi z}{\pi z(1-z)} = \frac{4\cos\pi\zeta}{\pi(1-4\zeta^2)} = \frac{4}{\pi} \prod_{n=1}^{\infty} \frac{n^2+n}{(n+\tfrac{1}{2})^2}\left(1 + \frac{z-z^2}{n+n^2}\right)$$

making $z \to 0$ we obtain $1 = (4/\pi) \prod_{n=1}^{\infty} (n^2+n)(n+\tfrac{1}{2})^{-2}$ which gives the desired result.

5.6.9. Taking the logarithmic derivative of the right-hand side we obtain:

$$\sum_{n=-\infty}^{\infty}{}'\left[\left(\frac{3}{n\pi+3z} - \frac{3}{n\pi}\right) - 2\left(\frac{1}{n\pi+z} - \frac{1}{n\pi}\right) + \left(\frac{1}{z-n\pi} + \frac{1}{n\pi}\right)\right] = 3\cot 3z - \cot z$$

which is the logarithmic derivative of the left-hand side. The two expressions differ from each other only by a constant factor which shows to be equal 1 (put $z = \pi/2$).

5.6.10. Integrate both sides of the equation of Exercise 5.2.1:

$$\frac{F'(z)}{F(z)} = \frac{F'(0)}{F(0)} + \sum_{n=1}^{\infty}\left(\frac{1}{z-a_n} + \frac{1}{a_n}\right) \quad (A_n = 1).$$

5.6.11. The function $F(z) = \sin\pi(z+a)/\sin\pi a$ vanishes at $z = -a-n$, n being an integer, moreover its logarithmic derivative $F'(z)/F(z) = \pi\cot\pi(z+a)$ is uniformly bounded on squares with corners $(n+\tfrac{1}{2})(\mp 1 \mp i) - a$. Hence by Exercise 5.6.10:

$$F(z) = \exp(z\pi\cot\pi a) \prod_{n=-\infty}^{\infty}\left(1 + \frac{z}{a+n}\right)\exp[-z/(a+n)]$$

$$= \left(1 + \frac{z}{a}\right)\exp z(\pi\cot\pi a - a^{-1}) \prod_{n=-\infty}^{\infty}{}'\left(1 + \frac{z}{a+n}\right)\exp\left(-\frac{z}{n} + \frac{z}{n} - \frac{z}{a+n}\right)$$

$$= \left(1 + \frac{z}{a}\right)\exp z\left\{\pi\cot\pi a - a^{-1} - \sum_{n=-\infty}^{\infty}{}'[(a+n)^{-1} - n^{-1}]\right\} \times$$

$$\times \prod_{n=-\infty}^{\infty}{}'\left(1 + \frac{z}{a+n}\right)\exp(-z/n).$$

Note that the expression $\{\ldots\}$ vanishes identically.

5.6.12. $F(z) = \cos(\pi z/4) - \sin(\pi z/4) = \sqrt{2}\sin[\pi(1-z)/4]$;
$F'(z)/F(z) = -\frac{1}{4}\pi\cot[\pi(1-z)/4]$ is uniformly bounded on the boundary of squares with vertices $(4n+2)(\mp 1 \mp i)+1$, moreover $F(z) = 0$ for $z = 4n+1$, where n is an integer. Hence

$$F(z) = \exp(-\pi z/4)(1-z)e^z(1+z/3)e^{-z/3}(1-z/5)e^{z/5}\ldots$$
$$= (1-z)(1+z/3)(1-z/5)\ldots$$

because $1 - 1/3 + 1/5 - \ldots = \pi/4$.

5.6.13. We have

$$[\Gamma(z)\Gamma(1-z)]^{-1} = [-z\Gamma(z)\Gamma(-z)]^{-1}$$
$$= (-z)^{-1}ze^{\gamma z}\prod_{n=1}^{\infty}(1+z/n)e^{-z/n}(-z)e^{-\gamma z}\prod_{n=1}^{\infty}(1-z/n)e^{z/n} = \sin\pi z/\pi.$$

5.6.14. (i) $[\Gamma(-\frac{1}{2})]^{-1} = -\frac{1}{2}e^{-\gamma/2}\prod_{n=1}^{\infty}(1-1/2n)e^{1/2n}$; now, $\Gamma(-\frac{1}{2})(-\frac{1}{2})$
$= \Gamma(\frac{1}{2}) = \sqrt{\pi}$, and therefore

$$\prod_{n=1}^{\infty}(1-1/2n)e^{1/2n} = \sqrt{e^{\gamma}/\pi};$$

(ii) $[\Gamma(-z/2)]^{-1} = -\frac{1}{2}z\exp(-\gamma z/2)\prod_{n=1}^{\infty}(1-z/2n)e^{z/2n}$,

$[\Gamma((z-1)/2)]^{-1}$
$= \frac{1}{2}(z-1)\exp[\gamma(z-1)/2]\prod_{n=1}^{\infty}(1+z/(2n-1))(1-1/2n)\exp[(1-z)/2n]$

and hence

$$\prod_{n=1}^{\infty}\left(1+\frac{z}{2n-1}\right)\left(1-\frac{z}{2n}\right) = \frac{4\sqrt{\pi}}{z(1-z)}[\Gamma(-z/2)\Gamma((z-1)/2)]^{-1};$$

(iii) putting $z = 1/2$ in (ii) and using the equality $-\frac{1}{4}\Gamma(-\frac{1}{4}) = \Gamma(\frac{3}{4})$ we obtain:

$$(1+\tfrac{1}{2})(1-\tfrac{1}{4})(1+\tfrac{1}{6})\ldots = \sqrt{\pi}[\Gamma(\tfrac{3}{4})]^{-2}.$$

5.7.1. $M(r) = \exp r^k$, hence $\log\log M(r)/\log r = k$.

5.7.2. $M(r) = \exp e^r$, hence $\log\log M(r)/\log r = r/\log r \to +\infty$.

5.7.3. (i) $M(r) = \cosh\sqrt{r}$, hence

$\log M(r) = \sqrt{r} + o(1)$, or $\log\log M(r)/\log r = \tfrac{1}{2} + o(1)$;

5. MEROMORPHIC AND ENTIRE FUNCTIONS

(ii) if $P(z) = az^n + \ldots + a_0$, then
$$\log M(r) = \log|a| + n\log r + o(1) = n\log r(1+o(1))$$
and consequently
$$\log\log M(r) = \log n + \log\log r + o(1) = o(\log r).$$

5.7.4. Suppose that there exists a positive integer k and a sequence $\{r_n\}$, $r_n \to +\infty$, such that
$$\log M(r_n, f)/\log r_n \leqslant k, \quad \text{i.e.} \quad M(r_n, f) \leqslant r_n^k.$$
From Cauchy's formula for coefficients it follows that $a_{k+1} = a_{k+2} = \ldots = 0$, where a_m are Taylor's coefficients of f at the origin. This means that f is a polynomial of degree at most k.

5.7.5. $\log M(r, f) = k\log r + \log M(r, g) = \log M(r, g)[1+o(1)]$ by Exercise 5.7.4.

5.7.6. If f is a polynomial, then $\rho(f) = \rho(f') = 0$; if $\rho(f) > 0$, then $\rho(zf') = \rho(f')$ by Exercise 5.7.5. Using the formula (5.7A) we easily verify that $\rho(zf') = \rho(f)$.

5.7.7. We may assume (cf. Ex. 5.7.5) that $f(0) \neq 0$. Then for r sufficiently large
$$\int_0^r t^{-1}m(t)\,dt = -\log|f(0)| + (2\pi)^{-1}\int_0^{2\pi}\log|f(re^{i\theta})|\,d\theta < Kr^{\rho+\varepsilon}$$
($\varepsilon > 0$ is arbitrary, K depends on ε only) since $\log|f(re^{i\theta})| \leqslant \log M(r) < Kr^{\rho+\varepsilon}$ for r sufficiently large. Moreover, $\int_0^{2r} t^{-1}m(t)\,dt < 2^{\rho+\varepsilon}r^{\rho+\varepsilon}K$ and consequently
$$m(r)\log 2 = m(r)\int_r^{2r} t^{-1}\,dt < \int_r^{2r} t^{-1}m(t)\,dt < \int_0^{2r} t^{-1}m(t)\,dt < 2^{\rho+\varepsilon}r^{\rho+\varepsilon}K$$
since $m(t)$ is increasing. Hence $m(r) < (\log 2)^{-1}2^{\rho+\varepsilon}r^{\rho+\varepsilon}K$.

5.7.8. If $\rho < \beta < \alpha$, then $m(r) < Ar^\beta$ and putting $r = |a_n|$ we obtain $m(|a_n|) = n < A|a_n|^\beta$, or $|a_n|^{-\alpha} < (A/n)^{\alpha/\beta}$ for all n sufficiently large.

5.7.9. It is sufficient to apply Weierstrass factorization theorem and take an integer m such that $m \leqslant \rho < m+1$. Then by Exercise 5.7.8 the series $\sum_{n=1}^{\infty} |a_n|^{-m-1}$ is convergent and therefore the product $\prod_{n=1}^{\infty} E(z/a_n, m)$ is convergent.

5.7.10. (i) $-n\log n/\log|c_n| = 1$ for all $n \geqslant 2$, hence $\rho = 1$ (cf. 5.7A); similarly: (ii) $\rho = a$; (iii) $\rho = 0$; (iv) $\rho = 1$.

CHAPTER 6

The Maximum Principle

6.1.1. If $|f|$ has a local maximum at $a \in D$ and $f(z) = \sum_0^\infty A_n(z-a)^n$, then $|f(a)| = |A_0|$ and $|f(a+re^{i\theta})|^2 \leq |A_0|^2$ for all real θ and all r sufficiently small. Hence

$$\int_0^{2\pi} |f(a+re^{i\theta})|^2 d\theta = 2\pi(|A_0|^2 + |A_1|^2 r^2 + \ldots) \leq 2\pi|A_0|^2$$

and consequently $A_1 = A_2 = \ldots = 0$.

6.1.2. Consider $1/f$ and cf. Exercise 6.1.1.

6.1.3. $\bar{D}$ is a compact set on the sphere, hence $|f|$ attains in $\bar{D}$ its lower and upper bounds m, M. If $z_0 \in D$, then the equalities $|f(z_0)| = m$, $|f(z_0)| = M$ are impossible.

6.1.4. If we had $f(z) \neq 0$ for all $z \in D$, then $|f(z)|$ would be a constant by Exercise 6.1.3.

6.1.5. Each component is bounded by a system of curves where $|f| = A$, hence by Exercise 6.1.4 there is at least one zero of f in each component.

6.1.6. A corollary of Exercises 6.1.3, 6.1.1.

6.1.7. $\varepsilon > 0$ being given, choose the integers p, q so that q is positive and $1 < (M_1/M_2)^q (r_1/r_2)^p < 1+\varepsilon$. Then the absolute value of $(f(z))^q z^{-p}$ attains its maximum on $C(0; r_1)$, i.e. $[M(r)]^q r^{-p} < M_1^q r_1^{-p}$ and hence $\log M(r) < \log M_1 + (p/q)\log(r/r_1)$. Now, $0 < q\log(M_1/M_2) + p\log(r_1/r_2) < \log(1+\varepsilon) < \varepsilon$ and consequently $p/q < [(\log(M_2/M_1) + q^{-1}\varepsilon] : \log(r_2/r_1)$. Thus

$$\log M(r) < \log M_1 + \log(r/r_1)[\log(M_2/M_1) + q^{-1}\varepsilon] : \log(r_2/r_1)$$

and by making $\varepsilon \to 0$ we obtain our result.

6.1.8. After a substitution $\zeta = z^{-1}$ we obtain a function F analytic in $K(0; 1) \setminus 0$ with a removable singularity at the origin. Note that F cannot have a maximum at $\zeta = 0$ and use Exercise 6.1.6.

6. THE MAXIMUM PRINCIPLE

6.1.9. By Exercise 6.1.8 the maximum of $|z^{-n}P(z)|$ in $\bar{K}(\infty; 1)$ is attained on $C(0; 1)$, i.e. $|z^{-n}P(z)| \leqslant M$ for $z \in \bar{K}(\infty; 1)$.

6.1.10. The rational function $g(w) = P(\frac{1}{2}(w+w^{-1}))$ is analytic in $\bar{K}(\infty; 1)$ and satisfies $|g(w)| \leqslant M$ on $C(0; 1)$. Hence by Exercise 6.1.9 we have: $|w^{-n}P(\frac{1}{2}(w+w^{-1}))| \leqslant M$ in $\bar{K}(\infty; 1)$. The image domain of the annulus $\{w: 1 < |w| < R\}$ under $\frac{1}{2}(w+w^{-1})$ is an ellipse H with semiaxes $a, b = \frac{1}{2}(R \mp R^{-1})$ slit along $[-1, 1]$, hence for $z \in H$ we have:

$$|P(z)| = |P(\tfrac{1}{2}(w+w^{-1}))| \leqslant MR^n = M(a+b)^n.$$

6.1.11. Take a positive integer n such that $2\pi/n < \beta - \alpha$; if z approaches $C(0; 1)$ within the angle $\alpha + 2k\pi/n \leqslant \arg z \leqslant \beta + 2k\pi/n$, then $f(\omega^k z)$, and also φ tend uniformly to 0. Hence $\lim_{r \to 1} M(r) = 0$, i.e. $\varphi \equiv 0$ by Exercise 6.1.6.

6.1.12.

$$|\varphi(z)| < \exp \lambda|y| |\exp \varepsilon(x^2 - y^2 + 2ixy)| \leqslant \exp(\varepsilon a^2 + \lambda|y| - \varepsilon y^2) < C,$$

if $|y|$ is large enough. Making $\varepsilon \to 0$ for a fixed z, the result follows.

6.2.1. $z^{-1}f(z)$ is analytic in $K(0; 1)$ since it has a removable singularity at the origin, hence the maximum of $|z^{-1}f(z)|$ in $\bar{K}(0; r)$ does not exceed $1/r$ and making $r \to 1$ we obtain $|z^{-1}f(z)| \leqslant 1$ in $K(0; 1)$. If $|z^{-1}f(z)| = 1$ for some $z \in K(0; 1)$, then $z^{-1}f(z) \equiv e^{i\alpha}$.

6.2.2. The function $[f(z) - f(0)][1 - \overline{f(0)}f(z)]^{-1}$ satisfies the assumptions of Exercise 6.2.1.

6.2.3. From Exercise 6.2.2 it follows that $|f'(0)| \leqslant 1 - |f(0)|^2$, hence $|f'(0)| \leqslant 1$. Moreover, if $|f'(0)| = 1$, then $f(0) = 0$, thus f satisfies the assumptions of Exercise 6.2.1 and $|f(z)/z|$ attains a maximum at $z = 0$, hence it is a constant.

6.2.4. By Schwarz's lemma for $f^{-1}(w)$ we have $|f'(0)|^{-1} \leqslant 1$; if $|f'(0)| = 1$, then $f(z) = e^{i\alpha}z$ by Exercise 6.2.3.

6.2.5. $F(z) = M[f(z) - a_0][M^2 - \bar{a}_0 f(z)]^{-1}$ satisfies the assumptions of Exercise 6.2.1, hence $\lim_{z \to 0} |F(z)/z| \leqslant 1$.

6.2.6. Put $\omega = \exp 2\pi i/(n-m)$, $F(z) = z^{n-2m}f(z)$ and $\zeta = z^{n-m}$ and consider the function

$$[F(z) + F(\omega z) + \ldots + F(\omega^{n-m-1}z)]: [(n-m)z^{n-m}] = a_m + a_n \zeta + a_{2n-m}\zeta^2 + \ldots$$

which satisfies the assumptions of Exercise 6.2.5.

6.2.7. The function $[\omega(z)-\alpha][1-\overline{\alpha}\omega(z)]^{-1}$ satisfies the conditions of Exercise 6.2.1, hence $|(\omega(z)-\alpha)/(1-\overline{\alpha}\omega(z))| \leqslant |z|$ or $|(w-\alpha)/(w-\overline{\alpha}^{-1})| \leqslant r\alpha$, where $r = |z|$ and $w = \omega(z)$ (cf. now Ex. 1.1.25).

6.2.8. If $\omega(z)$ satisfies $[\omega(z)-\alpha]/[1-\overline{\alpha}\omega(z)] = ze^{i\beta}$ with real β, then its values taken on $C(0;r)$ cover $C(z_0;\rho)$. Suppose now that $w_1, w_2 \in C(z_0;\rho)$ and $w_1 = \omega_1(z)$, $w_2 = \omega_2(z)$. If $0 < \lambda < 1$, then $\Omega(z) = \lambda\omega_1(z)+(1-\lambda)\omega_2(z)$ takes for suitably chosen w_1, w_2, λ each value from $\overline{K}(z_0;\rho)$, for some $z \in C(0;r)$.

6.2.9. The set Ω_r is a bounded, convex domain whose boundary is of circles $C(z_0;\rho)$ of Exercise 6.2.8 with r fixed and α ranging over $[0, 1]$. The envelope consists of the left half of $C(0;r)$ and two circular arcs emanating from $z = 1$ and symmetric with respect to the real axis which intersect at a right angle the imaginary axis at $\mp ir$.

6.2.10. Put $P(z) = zQ(z)$; $K(0;1) \subset H$, where H is an ellipse with semiaxes $a = \sqrt{2}$, $b = 1$ and foci ∓ 1. It follows from Exercise 6.1.10 that $|P(z)| \leqslant M(1+\sqrt{2})^{n+1}$ in H and also in $K(0;1)$ and by Schwarz's lemma

$$|P(z)| = |zQ(z)| \leqslant |z|M(1+\sqrt{2})^{n+1} \quad \text{in} \quad K(0;1).$$

6.2.11.
$$P(z) = (z-\eta)\,Q(z) = (z-\eta)\,(a_0+a_1z+\ldots+a_nz^n)$$
$$= -a_0\eta+(a_0-\eta a_1)z+(a_1-\eta a_2)z^2+\ldots+(a_{n-1}-\eta a_n)z^n+a_nz^{n+1}.$$

Hence $|a_0| \leqslant M$, $|a_0-\eta a_1| \leqslant M$, i.e.

$$|a_1| \leqslant |a_0|+|\eta a_1-a_0| \leqslant 2M, \quad |a_2| \leqslant 3M, \ldots$$

Similarly $|a_n| \leqslant M$, $|a_{n-1}| \leqslant |a_n|+|\overline{\eta}a_{n-1}-a_n| \leqslant 2M$, $|a_{n-2}| \leqslant 3M$, ...
If $n = 2k+1$, then

$$|Q(z)| \leqslant (|a_0|+|a_n|)+(|a_1|+|a_{n-1}|)+\ldots+(|a_k|+|a_{k+1}|)$$
$$\leqslant 2[M+2M+\ldots+(k+1)M] = \tfrac{1}{4}(n+1)(n+3)M < \tfrac{1}{4}(n+2)^2M;$$

if $n = 2k$, then

$$|Q(z)| \leqslant 2(M+2M+\ldots+kM)+(k+1)M = \tfrac{1}{4}(n+2)^2M.$$

6.3.1. If $f(z) = F(\omega(z))$ and $|\omega(z)| < r_1$ in $K(0;r_1)$, then also $|\omega(z)| < r_2$ in $K(0;r_2)$ by Schwarz's lemma for any $r_2 < r_1$ which means that $f \prec_{r_2} F$.

6.3.2. If ω is analytic in $K(0;r)$ and $|\omega(z)| \leqslant |z|$ in $K(0;r)$, then $|\omega'(0)| \leqslant 1$. Now, $f'(0) = F'(0)\omega'(0)$.

6.3.3. If $M(r,f) = |f(re^{i\theta})|$, then $f(re^{i\theta}) = F(r_1e^{i\varphi})$ with $r_1 \leqslant r$ and consequently $|f(re^{i\theta})| \leqslant M(r_1, F) \leqslant M(r, F)$.

6. THE MAXIMUM PRINCIPLE

6.3.4. Note that $\omega = F^{-1} \circ f$ satisfies the conditions $\omega(0) = 0$, $|\omega(z)| \leqslant |z|$, and is analytic in $K(0; 1)$.

6.3.5. The function $F(z) = (2/\pi)\mathrm{Log}[(1+z)/(1-z)]$ maps conformally $K(0; 1)$ onto $\{w: |\mathrm{re}\,w| < 1\}$ and $F(0) = 0$, hence $f \prec F$;
  (i) follows from Exercise 6.3.2;
  (ii) follows from Exercise 6.3.3.

6.3.6. Suppose that $F(z) = a - ar(\bar{a}-rz)/\bar{a}(r-az)$, $r < |a|$ (F maps $K(0; 1)$ onto the outside of $K(a; r)$). Obviously $\omega = F^{-1} \circ f$ is analytic in $K(0; 1)$ since possible poles of f are removable singularities of ω. Moreover ω, satisfies the conditions of Schwarz lemma, hence $|\omega'(0)| \leqslant 1$ which implies the desired result.

6.3.7. If $|\eta| = 1$, then $p \prec F$, where $F(z) = F_\eta(z) = (1+\eta z)/(1-\eta z)$ and also $p \prec_r F$ which means that the value $p(z)$ is situated in the closed disk $\bar{K}_r$ whose boundary is given by the equation: $w = (1+\eta r)/(1-\eta r)$, where η is a variable parameter with $|\eta| = 1$. If $w \in \bar{K}_r$, then $w = \lambda w_1 + (1-\lambda) w_2$, where $0 \leqslant \lambda \leqslant 1$ and $w_1, w_2 \in \mathrm{fr}\, K_r$. For any z with $|z| = r$ we can choose η_1, η_2 with $|\eta_1| = |\eta_2| = 1$ such that $F_{\eta_k}(z) = w_k$ ($k = 1, 2$). Note that $\lambda F_{\eta_1} + (1-\lambda) F_{\eta_2} \in \mathscr{P}$.

6.3.8. We have
$$(1-|z|)/(1+|z|) \leqslant |p(z)| \leqslant (1+|z|)/(1-|z|),$$
$$-2\arctan|z| \leqslant \mathrm{Arg}\, p(z) \leqslant 2\arctan|z|,$$
which are precise estimates of w and $\mathrm{Arg}\, w$ in $\bar{K}_r$, $r = |z|$ (cf. Ex. 6.3.7).

6.3.9. $\dfrac{1-|z|}{1+|z|} \leqslant \mathrm{re}\, p(z) \leqslant \dfrac{1+|z|}{1-|z|}$, $|\mathrm{im}\, p(z)| \leqslant \dfrac{2|z|}{1-|z|^2}$ which are precise estimates of $\mathrm{re}\, w$, $\mathrm{im}\, w$ in $\bar{K}_r$.

6.6.10. We may assume that $f(0) > 0$. The function $M/f(z)$ is analytic and does not vanish in $K(0; 1)$, hence we may consider a single-valued branch of $\log M/f(z)$; obviously $[\log M/f(z)]:[\log M/f(0)] \in \mathscr{P}$ and hence
$$(1-|z|)/(1+|z|) \leqslant (\log M/|f(z)|):(\log M/f(0))$$
by Exercise 6.3.9 which gives the desired estimate of $|f(z)|$.

6.3.11. From Exercise 6.3.10 it follows that $|f_n(z)| \leqslant |f_n(0)|^{1/2} M^{1/2}$ i.e.
$$|f_n(z)|^2 \leqslant M|f_n(0)| \quad \text{for} \quad z \in \bar{K}(0; \tfrac{1}{3}).$$

6.3.12. There exists a single-valued branch of $\log[\alpha - f(z)][1-\alpha f(z)]$ in $K(0; 1)$ and obviously
$$\varphi(z) = -\log[\alpha - f(z)][1-\alpha f(z)]^{-1} : (-\log \alpha) \in \mathscr{P},$$
hence $\varphi \prec (1+z)(1-z)^{-1}$ and consequently $|\varphi'(0)| \leqslant 2$ (Ex. 6.3.2).

6.3.13. $\varphi \prec F$, hence $|\varphi'(0)| \leqslant |F'(0)| = 1$ and using the equality $f'(\alpha) = f'(-\alpha)$ we obtain $(1-|\alpha|^2)|f'(\alpha)| \leqslant 1$.

6.3.14. Put $\omega = c_1 z + c_2 z^2 + \ldots$; from Exercise 6.2.5 as applied to $\omega(z)/z$ we obtain $|c_2| \leqslant 1-|c_1|^2$. From the identity $f = F \circ \omega$ we obtain: $a_1 = A_1 c_1$, $a_2 = A_2 c_1^2 + A_1 c_2$, which implies

$$|a_2| \leqslant |A_2|(1-|c_2|) + |A_1||c_2| \leqslant \max(|A_1|, |A_2|).$$

6.4.1. If $K(z_0; r) \subset D$, there exists a function $f = u + iv$ analytic in this disk; obviously the absolute value of $\exp \circ f$ i.e. $\exp \circ u$ has a local extremum $\neq 0$ at z_0, hence $u = $ const (cf. Ex. 6.1.1).

6.4.2. u being continuous on the Riemann sphere attains a lower bound at z_0 and an upper bound at z_1. If $u \neq $ const then $z_0 \neq z_1$ which means that u has a local extremum at a point $\neq \infty$ which is a contradiction (Ex. 6.4.1).

6.4.4. $\bigcup_{\zeta} K(\zeta; \rho(1/n)) = B_n$ is an open set and $H_n = (\mathbf{C} \setminus B_n) \cap D$ a closed set whose boundary points are interior points of D; moreover $u(z) \leqslant M + 1/n$ in H_n. We may assume that $\rho(1/n) \leqslant 1/n$ for all ζ and hence $D = \bigcup H_n$ and $u(z) \leqslant M$ for all $z \in D$. Note that actually $u(z) < M$ by Exercise 6.4.1.

6.4.5. If $\zeta \neq 0$ is a pole of F, then $h \to +\infty$ as $z \to \zeta$; if $|z| \to 1$, then $\overline{\lim} h(z) \geqslant 0$ and if $z \to 0$, then $h(z) \to \log|A|$. Now, $z = 0$ is a removable singularity of $f^{-1} \circ F$ hence either $\log|A| > 0$, i.e. $|A| > 1$, or else $\log|z| = \log|f^{-1} \circ F(z)|$, i.e. $F(z) \equiv e^{i\alpha} f(z)$.

6.4.6. The function

$$h(z) = A(r_1)\left(\log\frac{|z|}{r_2}\right):\left(\log\frac{r_1}{r_2}\right) + A(r_2)\left(\log\frac{|z|}{r_1}\right):\left(\log\frac{r_2}{r_1}\right) - u(z)$$

is harmonic in the annulus $r_1 < |z| < r_2$ ($R_1 < r_1 < r_2 < R_2$) and nonnegative on its boundary hence it is nonnegative on $C(0; r)$ and

$$A(r_1)\left(\log\frac{r}{r_2}\right):\left(\log\frac{r_1}{r_2}\right) + A(r_2)\left(\log\frac{r}{r_1}\right):\left(\log\frac{r_2}{r_1}\right) - A(r) \geqslant 0.$$

6.4.7. We have: $\dfrac{\partial}{\partial \theta} \arg f(re^{i\theta}) = \mathrm{re}[zf'(z)/f(z)] \geqslant 0$ for $z \in C(0; r)$. Since f is univalent, $f(z) \neq 0$ for $z \neq 0$ and therefore $\mathrm{re}[zf'(z)/f(z)]$ is harmonic in $K(0; R)$, nonnegative on $C(0; r)$ and positive in $K(0; r)$. This implies $\dfrac{\partial}{\partial \theta} \arg f(\rho e^{i\theta}) > 0$ for any $\rho < r$.

CHAPTER 7

Analytic Continuation. Elliptic Functions

7.1.1. The former series represents $\log 2 + \text{Log}[1-\frac{1}{2}(1-z)] = \text{Log}(1+z)$ in $K(1; 2)$, whereas the latter series represents $\text{Log}(1+z)$ in $K(0; 1) \subset K(1; 2)$.

7.1.2. Both series represent $(1-z)^{-1}$ in $K(0; 1)$ and $K(i; \sqrt{2})$, resp.;
$$K(0; 1) \cap K(i; \sqrt{2}) \neq \emptyset.$$

7.1.3. The former series represents $-\text{Log}(1-z)$ in the disk $K(0; 1)$, while the latter one represents $\pi i - \text{Log}(z-1)$ in $K(2; 1)$, both disks being disjoint. The function element $(-\text{Log}(1-z), \{z: \text{im } z > 0\})$ is a direct analytic continuation of both series which is easily verified for z approaching 0 through the upper half-plane.

7.1.4. Suppose that $D_1 = \mathbf{C} \setminus \overline{(-;+)}$, $D_2 = \mathbf{C} \setminus \overline{(+;-)}$, and
$$f_1(z) = \log|z| + i \arg z, \quad -\pi < \arg z < \tfrac{1}{2}\pi,$$
while
$$f_2(z) = \log|z| + i \arg z, \quad 0 < \arg z < \tfrac{3}{2}\pi.$$
Obviously $f_1 = f_2$ in $(+, +)$, while $f_1 \neq f_2$ in $(-, -)$.

7.1.5. $f(z) = -z^{-1}\text{Log}(1-z)$, $z \in K(0; 1)$; if this element is continued along $C(1; 1)$, then after encircling the point 1 we obtain the element
$$f_1(z) = -[\mp 2\pi i + \text{Log}(1-z)]z^{-1}, \quad z \in K(0; 1) \setminus 0$$
(the sign depends on the sense of encircling).

7.1.6. No radial limits at points $z = \exp(2\pi i m 2^{-n})$ which form a dense subset of $C(0; 1)$ do exist.

7.1.7. Suppose that the point $z = 1$ is a regular point of $f(z) = \sum_{n=0}^{\infty} a_n z^n$. Then the Taylor series of f with center h: $\sum_{n=0}^{\infty} (z-h)^n f^{(n)}(h)/n!$

must be convergent in $K(h;r)$, where $0<h<1$ and $h+r>1$. Obviously $|f^{(n)}(he^{i\theta})| \leqslant f^{(n)}(h)$ and this implies that also the series

$$\sum_{n=0}^{\infty} (z-he^{i\theta})^n f^{(n)}(he^{i\theta})/n!$$

is convergent in $K(he^{i\theta};r)$ for any real θ. This implies that f is analytic in $K(0;h+r)$ which is a contradiction.

7.1.8. The function f is analytic in H_+ and in H_- which is easily proved by using the M-test. Suppose that there exists a disk $K(x_0;r)$, x_0 real and a function F analytic in $K(z_0;r)$ and such that $f = F$ in $H_+ \cap K(z_0;r)$. There exists a rational number $w_k \in K(z_0;r)$ and by our assumptions $\lim_{y \to 0+} f(w_k+iy) = F(w_k)$. However,

$$|\mathrm{im} f(w_k+iy)| = \sum_{n=1}^{\infty} \frac{c_n y}{(w_k-w_n)^2+y^2} > \frac{c_k}{y} \to +\infty \quad \text{as} \quad y \to 0+$$

which is a contradiction.

7.1.9. E.g. $\sum_{n=1}^{\infty} 2^{-n} z^{2^n}$; the derivative cannot be continued beyond the unit disk and the same necessarily holds for the function itself.

7.2.1. After reflections w.r.t. the real axis we obtain $\overline{f(z)} = f(\bar{z})$ and after reflections w.r.t. the imaginary axis we obtain $f(\bar{z}) = \overline{-f(-z)}$ which gives $f(z)+f(-z) = 0$.

7.2.2. After reflections we obtain a function meromorphic in the extended plane, i.e. a rational function.

7.2.3. If $f(z) = \sum_{n=-\infty}^{\infty} A_n (z-a)^n$, then

$$\overline{f(\bar{z})} = \sum_{n=-\infty}^{\infty} \bar{A}_n (z-\bar{a})^n.$$

7.2.4. We have: $z = a + R^2/(\bar{z}^*-\bar{a})$, $b = a + R^2/(\bar{b}^*-\bar{a})$;

$$z-b = R^2 (\bar{b}^*-\bar{z}^*) [(\bar{z}^*-\bar{a})(\bar{b}^*-\bar{a})]^{-1}$$

and consequently

$$f(z^*) = \overline{f(z)} = \sum_{k=1}^{n} \frac{\bar{c}_k (z^*-a)^k (a-b^*)^k}{(z^*-b^*)^k R^{2k}} + g(z^*),$$

7. ANALYTIC CONTINUATION. ELLIPTIC FUNCTIONS

where g is analytic in some neighborhood of b^*. Observe that
$$\text{res}[b^*; \bar{c}_k(z^*-a)^k(a-b^*)^k R^{-2k}(z^*-b^*)^{-k}] = (-1)^k k \bar{c}_k(b^*-a)^{k+1} R^{-2k}.$$

7.2.5. Take $f_1 = \dfrac{1}{2}[f(z)+\overline{f(\bar{z})}]$, $f_2 = \dfrac{1}{2i}[f(z)-\overline{f(\bar{z})}]$.

7.2.6. f is a rational function with poles $z_1, 1/\bar{z}_1$, hence
$$f(z) = (Az^2+Bz+C)[(z-z_1)(1-\bar{z}_1 z)]^{-1}$$
for $z_1 \neq 0$ (note that f is analytic at ∞), or $f(z) = Az+B+C/z$ for $z_1 = 0$; moreover, $\overline{f(z)} = f(1/\bar{z})$ which implies: $C = A$, B is real.

7.2.7. After reflections we obtain a function meromorphic in the extended plane which has two poles $0, \infty$ and maps $\mathbf{C}$ onto a two-sheeted w-plane; hence $f(z) = az+b/z+c$; $f'(\mp 1) = 0$ because the angles with vertices at these points are doubled. Moreover $f(\mp 1) = \mp 2$ which gives $a = b = 1$, $c = 0$.

7.2.8. f is a rational function which has a double pole $z = 1$ (the angles with vertices at $z = 1$ are doubled), moreover, $f(z) \equiv f(1/z)$ which implies $f(0) = f(\infty) = 0$ and finally $f(z) = Az(1-z)^{-2}$. $f(-1) = -\tfrac{1}{4}$ gives $A = 1$.

7.2.9. $Az^k(z-z_1)(z-z_2)\ldots(z-z_l)[(1-\bar{z}_1 z)(1-\bar{z}_2 z)\ldots(1-\bar{z}_l z)]^{-1}$, $k+l = n$, $|A| = 1$.

7.2.10, 11. After reflections we obtain a function f analytic and univalent in $\mathbf{C}$ such that $\infty \leftrightarrow \infty$, which means that $f(z) = az+b$.

7.2.12. After suitable rotations of D and $K(0; 1)$ round the origin we can achieve that the real axis in the w-plane is the axis of symmetry of D and $f'(0) > 0$. Suppose now that $\varphi = f^{-1}$ and consider $\varphi(\overline{f(\bar{z})}) = \psi(z)$; thus ψ is analytic and univalent in $K(0; 1)$. Moreover $\psi[K(0; 1)] = K(0; 1)$, $\psi(0) = 0$, $\psi'(0) > 0$, hence ψ must be identity which implies $f(z) \equiv \overline{f(\bar{z})}$, i.e. f is real on $(-1, 1)$.

7.2.13. If the branch f_2 arises from f_1 by an analytic continuation along an arc l, then the arc l (with both end-points in the upper half-plane H_+) intersects the real axis an even number of times. To each intersecting of the real axis there corresponds a reflection with respect to some boundary arc of D, hence f_2 arises from f_1 by an even number of reflections. Observe that two reflections can be replaced by one linear transformation.

7.2.14. We have:
$$F' = (ad-bc)f'(cf+d)^{-2}, \quad \text{and also} \quad F''/F' = f''/f' - 2cf'(cf+d)^{-1}.$$
Hence
$$(F''/F')' = (f''/f')' + 2c^2 f'^2 (cf+d)^{-2} - 2cf''(cf+d)^{-1},$$

and moreover,
$$(F''/F')^2 = (f''/f')^2 + 4c^2f'^2(cf+d)^{-2} - 4cf''(cf+d)^{-1}$$
which implies $\{F, z\} \equiv \{f, z\}$.

7.3.1. Suppose (f, D) is a function element of the given global analytic function and K is a disk contained in $\{w: \text{re}\, w < 0\}$ such that $z = \exp w$ is univalent in K and $\exp(K) \subset D$. The function element $(f \circ \exp, K)$ can be continued along any arc situated in the left half-plane and hence it determines a single-valued function $F(w)$. Note that $F(w) = f \circ \exp w$, $w \in K$, or $f(z) = F(\log w)$ with suitably chosen branch of log.

7.3.2. Suppose $g(z_0) = w_0$, z_0 being arbitrary. Since $g'(z_0) \neq 0$, there exists a branch of g^{-1} in a disk $K(w_0; r) = K$, say (f, K), such that $f(w_0) = z_0$. By our assumptions the element (f, K) can be continued arbitrarily in $\mathbf{C}$, thus it defines a single-valued inverse function g^{-1} in $\mathbf{C}$. Hence g as an analytic, univalent function mapping $\mathbf{C}$ onto itself must be a similarity transformation.

7.3.3. Since $f(z) \neq 0$, we have $f = \exp h$, where h is an entire function. Hence $f' = h' \exp h$ and $f'(z) \neq 0$ implies $h'(z) \neq 0$ for all $z \in \mathbf{C}$. Suppose h does not take a finite value a. This means that $f(z) \neq e^a$, 0 which is impossible for non-constant f by Picard's theorem. Thus h assumes all finite values, whereas $h'(z) \neq 0$ for all z, and consequently, $h(z) = az + b$ (cf. Ex. 7.3.2).

7.3.4. Let $z = z(t)$, $\alpha \leqslant t \leqslant \beta$, be the equation of γ and $\delta = \text{dist}(\gamma, \mathbf{C} \setminus D)$. Let $\alpha = t_0 < t_1 < \ldots < t_n = \beta$ be a partition of $[\alpha, \beta]$ such that for $m = 1, 2, \ldots, n-1$ both arcs γ_{m-1}, γ_m of γ corresponding to $[t_{m-1}, t_m]$, $[t_m, t_{m+1}]$ are situated inside $K_m = K(z_m; \delta)$, where $z_m = z(t_m)$. Put $F_m(z) = \int_{[z_m, z]} f(\zeta)d\zeta + C_m$, where C_m are such that $F_{m-1} = F_m$ in $K_{m-1} \cap K_m$ and $C_0 = 0$. Evidently F_0 admits analytic continuation along γ, F_m being functions of the chain. Moreover,
$$\int_{\gamma_m} = F_m(z_m) - F_{m-1}(z_{m-1}).$$
Hence
$$\int_\gamma = \sum_{m=1}^n \int_{\gamma_m} = F_n(z_n) = F(Z).$$

7.3.5. Take $K(z_0; r) = K \subset D$ and continue the function element $\left(\int_{[z_0, z]} (\zeta)d\zeta, K\right)$ along γ_j which gives $(F_j, K(Z; \delta))$, $j = 1, 2$. Now $F_1 = F_2 = F$ by theorem of monodromy since γ_1, γ_2 are homotopic w.r.t. D. However,

7. ANALYTIC CONTINUATION. ELLIPTIC FUNCTIONS

$$\int_{\gamma_1} f(\zeta)d\zeta = F_1(Z) = F_2(Z) = \int_{\gamma_2} f(\zeta)d\zeta,$$

thus both integrals have the same value.

7.3.6. Let γ be an arc situated in D and $\Gamma = f(\gamma)$. If $\arg w$ is a continuous branch of argument of Γ, then $\log|f(z)| + i \arg f(z)$ is analytic and single-valued in any sufficiently small disk $K(z; r_z)$, $z \in \gamma$. This implies that any initial element of $\log f(z_0)$, $z_0 \in D$ can be continued along any arc $\gamma \subset D$ starting at z_0 and defines a single-valued branch of $\log f$ in D.

7.3.7. Take any finite z_0 and any arc γ starting at z_0 and let $\Gamma = g(\gamma)$. Since Γ omits 0, 1, with each $\zeta \in \gamma$, we can associate a disk K_ζ such that $\tau \circ g$ is analytic in K_ζ, τ being a branch of λ^{-1}, and moreover the corresponding elements $(\tau \circ g, K_\zeta)$ are identical for ζ sufficiently close to each other. This procedure defines a single-valued entire function $G(z)$ with $\operatorname{re} G(z) > 0$, which evidently implies $G = \text{const}$.

7.3.8. If $f(z) = a, b$, then $g = (f-a)/(f-b)$ is an entire function which omits the values 0, 1 and therefore $g = \text{const}$.

7.3.9, 10. It follows from the definition of λ and from the reflection principle that $\lambda(\tau+2) = \lambda(\tau)$; moreover, $\operatorname{im}(\log w/\pi i) > 0$ in $|w| < 1$. Hence Q is single-valued and analytic in $K(0; 1)$. An open segment $(-\delta, \delta)$, $0 < \delta < 1$, is mapped under Q onto another segment $(-\eta, \eta)$ of the real axis, hence $Q(w) = A_1 w + A_2 w^2 + \dots$ and all A_k must be real. $A_1 > 0$ since the angle of local rotation at $w = 0$ is equal to 0. An element of the inverse function $w = Q^{-1}(W)$, $0 = Q^{-1}(0)$, can be continued on $|W| < 1$, and its values cover a part of $K(0; 1)$, hence $|dw/dW|_{W=0} < 1$, which implies $A_1 > 1$.

7.3.11. The function element $(Q^{-1} \circ f, K(0; \delta))$, where δ is sufficiently small, can be continued along any arc situated inside $K(0; 1)$ and determines a single-valued analytic function ω with $|\omega(z)| < 1$. Hence $f = Q \circ \omega$, i.e. $f \prec Q$.

7.3.12. A consequence of Exercises 7.3.11, 6.3.2 and the remark given in Exercise 7.3.10.

7.4.1. If $f(t) = A \int_0^t \prod_{k=1}^n (\tau - b_k)^{\alpha_k - 1} d\tau + B$ maps $\{t: \operatorname{im} t > 0\}$ onto the inside of D and $z = (b_n - t)^{-1}$, then

$$t - b_k = (b_n - b_k)z^{-1}(z - x_k), \quad k = 1, 2, \dots, n-1, \qquad t - b_n = -z^{-1},$$

where $x_k = (b_n - b_k)^{-1}$. Hence

$$dw/dz = dw/dt \cdot dt/dz = C \prod_{k=1}^{n-1} (z - x_k)^{\alpha_k - 1}.$$

7.4.2. $w = \int_0^z \zeta^{-2/3}(1-\zeta)^{-2/3}d\zeta$; $a = (2\pi)^{-1}3^{1/2}[\Gamma(\tfrac{1}{3})]^3$; $x_1 = 0$, $x_2 = 1$, $x_3 = \infty$; $\alpha_1 = \alpha_2 = \alpha_3 = \tfrac{1}{3}$.

7.4.3. $w = \int_0^z \zeta^{\alpha-1}(1-\zeta^2)^{-\alpha}d\zeta$; $a = \tfrac{1}{4}[\Gamma(\alpha)\cos(\alpha\pi/2)]^{-1}\Gamma^2(\tfrac{1}{2}\alpha)$.

7.4.4. We have $dw = z^{-1}[z/(1-z^2)]^\alpha dz$, hence $\arg dw = \text{const}$ on the circle $z = e^{i\theta}$, as well as on the line $z = iy$, $y > 0$; $z = i$ corresponds to the center of the rhombus.

7.4.5. The image domain is bounded by the polygonal line with interior angles $\tfrac{1}{2}\pi, \tfrac{3}{2}\pi, \tfrac{3}{2}\pi, \tfrac{1}{2}\pi$, hence

$$w = \int_a^z [(\zeta^2-\delta^2)/(\zeta^2-a^2)]^{1/2}d\zeta,$$

$$h = \int_\delta^a [(t^2-\delta^2)/(a^2-t^2)]^{1/2}dt,$$

$$k = 2\int_0^\delta [(\delta^2-t^2)/(a^2-t^2)]^{1/2}dt.$$

If $\delta \to 0$, then $k \to 0$, $h \to a$, $w \to \sqrt{z^2-a^2}$; in the limiting case the image domain is $H_+ \setminus (0; ia]$.

7.4.6. Put $x_1 = -1$, $x_2 = 1$, $\alpha_1 = \tfrac{3}{2}$, $\alpha_2 = \tfrac{1}{2}$ which gives $w = 2i\pi^{-1} \times (\sqrt{1-z^2} - \text{Arc}\sin z)$, the branch being the principal branch of arc sin taking on $(-1, 1)$ the values belonging to $(-\pi/2, \pi/2)$.

7.4.7. $w = A\int_1^z (1+\zeta^{-1})^{1-\delta}d\zeta + B$, where A, B are real constants which can be determined from the equality:

$$\pi i = Ai\int_0^\pi (1+e^{-i\theta})^{1-\delta}e^{i\theta}d\theta + B$$

by separating real and imaginary parts. If $\delta \to 0$, $A(\delta) \to 1$, $B(\delta) \to 2$ and the limiting function $w = z+1+\text{Log}\,z$ maps conformally H_+ onto H_+ slit along the ray: $\text{im}\,w = \pi$, $\text{re}\,w \leqslant 0$.

7.4.8. Applying the reflection principle to the mapping $w = z+1+\text{Log}\,z$ we can continue the mapping through the positive real axis, the image domain being the given domain; then we put $\sqrt{z} = \zeta$ which gives the desired mapping: $w = \zeta^2 + 1 + 2\text{Log}\,\zeta$.

7. ANALYTIC CONTINUATION. ELLIPTIC FUNCTIONS

7.4.9. The mapping of H_+ onto a polygonal domain with interior angles π/n, $2\pi(1-1/n)$, π/n has the following form:

$$f_n(z) = A \int_0^z \frac{(\zeta-a)^{1-2/n}}{(\zeta^2-1)^{1-1/n}} d\zeta + B \Rightarrow A \int_0^z \frac{\zeta-a}{\zeta^2-1} d\zeta + B$$

$$= \tfrac{1}{2} A[(1-a)\operatorname{Log}(z-1) + (1+a)\operatorname{Log}(z+1)] + B.$$

We may expect that after a suitable choice of parameters we obtain the desired mapping function. In fact, the function

$$\frac{v_1}{\pi} \operatorname{Log}(z-1) + \left(1 - \frac{v_1}{\pi}\right) \operatorname{Log}(z+1) + B,$$

where

$$-B = \frac{v_1}{\pi} \log \frac{v_1}{\pi-v_1} + \log \frac{\pi-v_1}{\pi} + \log 2,$$

satisfies our conditions. If $v_1 = \pi/2$, then $B = 0$ and $f(z) = \tfrac{1}{2}\log(z^2-1)$; the branch of logarithm corresponds to $\arg z \in (0, 2\pi)$.

7.4.10. The mapping can be obtained as a limiting case of a mapping of H_+ onto a polygonal domain with interior angles $\pi/2$, π/n, $(2-1/n)\pi$, as $n \to +\infty$.

$$w = 2\zeta/(a^2-1) + \operatorname{Log}[(1+\zeta)/(1-\zeta)], \quad \zeta = \sqrt{z};$$

$$h = 2a/(a^2-1) + \log[(a+1)/(a-1)].$$

7.4.11. $K(0; \delta) \setminus (i\delta, 0]$ being simply connected, the integral $\int_{z_0}^z \zeta^{-1} F(\zeta) d\zeta$ does not depend on the curve of integration and this implies that $W(z)$ is single-valued. If $z = te^{i\theta}$ ($\theta = \text{const}$) we see that $\arg(dW/dt) = \arg F(te^{i\theta}) = \text{const}$ for $\theta = 0, \pi$ and this means that the images of $(-\delta, 0)$, $(0, \delta)$ are situated on straight lines. Since $W \to \infty$ as $z \to 0$, these are necessarily infinite rays. Assuming $\arg F(z) = 0$ for $(-\delta, \delta)$ we obtain that the distance d between the rays is equal to

$$\operatorname{im}[W(-r)-W(r)] = \lim_{r \to 0} \operatorname{im}[W(-r)-W(r)]$$

$$= \operatorname{im} \lim_{r \to 0} \int_0^\pi i[F(0)+o(1)] d\theta = \pi F(0).$$

In the general case $d = \pi |F(0)|$.

7.4.12. Suppose that $F(z) = (z-b)^{-1}(z-a)^{-1/2}$. Obviously $\arg F(z) = -\tfrac{3}{2}\pi$ for all $z < a$. We have $\arg H'(z) = -\pi/2$ on $(-\infty, 0)$, $(0, a)$ hence $H(z)$ carries these intervals into two parallel rays inclined to the positive real axis at the angle

$-\pi/2$. The distance between the rays is equal to $\pi/b\sqrt{a}$. Putting $Z = z-b$, we obtain:

$$w = \int_{z_0}^{z} [\eta(\eta+b)\sqrt{\eta+b-a}]^{-1}d\eta$$

and a similar reasoning shows that the images of (a, b), $(b, +\infty)$ are infinite rays parallel to the negative real axis, the distance between them being $\pi/b\sqrt{b-a}$.

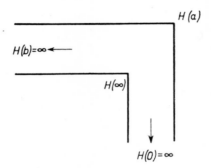

Fig. 5

If $2a = b$, both distances are the same. If $z_0 = a$, two perpendicular rays of the boundary and the negative coordinate axes in the w-plane coincide.

7.4.13. $d = \pi a^{-1/3}$.

7.4.14. After the transformation $W_1 = w^{-1}$ we obtain the domain bounded by the rays $[0, +\infty)$, $(-\tfrac{1}{2}i-\infty, -\tfrac{1}{2}i]$ and the segment $[0, -\tfrac{1}{2}i]$ which contains Q_1. This domain is carried under $W_2 = 4W_1+i$ into a domain of similar type as considered in Exercise 7.4.6 and being the image of H_+ under $W_2 = 2i\pi^{-1} \times (\sqrt{1-z^2}+\text{Arc}\sin z)$.

7.4.15. $w = \pi^{-1} \int_{-1}^{z} (1+\zeta)^\mu \zeta^{-1} d\zeta + i$ (cf. Ex. 7.4.11).

7.4.16. The mapping obtained in Exercise 7.4.15 after a reflection w.r.t. the positive real axis in the z-plane whose image is the whole real axis in the w-plane gives the desired image domain. Put now $z = Z^2$.

7.4.17. The mapping $z = i(1-t)/(1+t)$ carries H_+ into the unit disk in the t-plane; if $(t = c_k) \leftrightarrow (z = x_k)$, then

$$z - x_k = 2i(t-c_k)[(1-c_k)(1-t)]^{-1};$$

7. ANALYTIC CONTINUATION. ELLIPTIC FUNCTIONS

moreover,

$$dw/dt = dw/dz \cdot dz/dt = 2i(1-t)^{-2} A \prod_{k=1}^{n} (z-x_k)^{\alpha_k - 1}$$

$$= A_1(1-t)^{-2}(1-t)^{\alpha_1+\alpha_2+\cdots+\alpha_n-n} \prod_{k=1}^{n} (t-c_k)^{\alpha_k-1}$$

$$= A_1 \prod_{k=1}^{n} (t-c_k)^{\alpha_k-1}$$

since $\alpha_1+\alpha_2+\cdots+\alpha_n = n+2$.

7.4.18. If $c_k = \exp(2\pi i k/n)$ and F carries the sector $0, c_0, c_1$ into the triangle $0, a_0, a_1$ (a_k being the vertices of a regular n-angle) then after reflections F can be continued on the whole disk, which implies that the preimages of vertices a_k also form a regular n-angle (if $F(0) = 0$). We have:

$$F(t) = A \int_0^t \prod_{k=1}^{n} (\tau-c_k)^{\alpha_k-1} d\tau + B$$

with $\alpha_k = 1 - 2/n$. If $A = 1$, $B = 0$ then

$$F(t) = \int_0^t \prod_{k=1}^{n} (\tau-c_k)^{-2/n} d\tau = \int_0^t (1-\tau^n)^{-2/n} d\tau.$$

The radius R_n of the circumcircle is equal to

$$\int_0^1 (1-\tau^n)^{-2/n} d\tau = \frac{1}{n} B\left(\frac{1}{n}, 1-\frac{2}{n}\right),$$

hence the perimeter

$$l_n = 2nR_n \sin(\pi/n) = 2\Gamma(1/n)\Gamma(1-2/n) \sin(\pi/n) [\Gamma(1-1/n)]^{-1}$$
$$= 2\pi \Gamma(1-2/n) [\Gamma(1-1/n)]^{-2}.$$

7.4.19. The alternate interior angles of the n-pointed star are equal: $\pi(1+\lambda)$, $\pi(1-\lambda-2/n)$ and a symmetry reasoning shows that the preimages on $|t|=1$ form a regular $2n$-angle. Hence

$$w = F(t) = \int_0^t (1+\tau^n)^{\lambda}(1-\tau^n)^{-\lambda-2/n} d\tau, \quad 0 < \lambda < 1-2/n.$$

7.4.20. Particular cases of Exercise 7.4.19: (i) take $\lambda = 2/5$, $n = 5$; (ii) take $\lambda = 1/3$, $n = 6$.

7.4.21. 1° If $\theta = \arg z$ then $dw/d\theta = \text{const}$ on both arcs (φ_k, ψ_k), (ψ_k, φ_{k+1}) of $C(0; 1)$; in fact,
$$1-\bar{z}_k z = -2i\exp[\tfrac{1}{2}i(\theta-\varphi_k)]\sin\tfrac{1}{2}(\theta-\varphi_k),$$
$$1-\bar{\zeta}_k z = -2i\exp[\tfrac{1}{2}i(\theta-\psi_k)]\sin\tfrac{1}{2}(\theta-\psi_k),$$
hence
$$\arg(dw/d\theta) = \arg(ie^{i\theta}dw/dz) = \text{const}+\theta+\tfrac{1}{2}n\theta-\tfrac{1}{2}\theta(n+\sum_{k=1}^{n}\beta_k) = \text{const},$$

if there are no z_k, ζ_k on the arc considered; hence the image of any arc (φ_k, ψ_k), (ψ_k, φ_{k+1}) is a half-line (note that $z \to \zeta_k$ implies $w \to \infty$);

2° $\Delta\arg(dw/dz) = -\pi$ when z describes a small semicircle with center at z_k which means that the rays l_k corresponding to circular arcs with a common end point z_k coincide (note the continuity of $w(z)$ at $z = z_k$);

3° $\Delta\arg(dw/dz) = (1+\beta_k)\pi$ when z describes a small semicircle with center at ζ_k which means that the angle between l_k and l_{k+1} is equal to β_k.

If the rays l_k do not intersect each other, the mapping is 1:1 which may be verified by the argument principle. The univalence may be proved either by argument principle, or by using the remark given in Exercise 7.4.24.

7.4.22. The mapping $\Phi(z) = \int_0^z (1+t^4)^{-1/2}dt$ carries $K(0; 1)$ into a square which is a convex domain; we have: $\Phi'(z)/w'(z) = 1-z^4$ which has a positive real part in $K(0; 1)$. Hence the mapping is univalent (cf. Ex. 7.4.24). The images of arcs $(k\pi/4, (k+1)\pi/4)$ of $C(0; 1)$ can be found similarly as in Exercise 7.4.23. The points i^n correspond to $w = \infty$, whereas the points $(1+i)2^{-1/2}i^n$ correspond to the vertices of right angles. Hence
$$a = \sqrt{2}\int_0^1 (1+x^4)^{-1}(1-x^4)^{-1/2}dx.$$

7.4.23. Cf. Exercises 7.4.1, 7.4.17. The domain being convex, we have $0 < \alpha_k < 1$ for the finite vertices; if $1 \leq \alpha_k \leq 2$, then the image of $z = e^{-i\theta_k}$ is the point $w = \infty$;
$$|f(z)| \leq |A|\int_0^r (1-\rho)^{-2}d\rho = |f'(0)|r(1-r)^{-1},$$
$$|f'(z)| \leq |A|\prod_{k=1}^{n}(1-r)^{-\alpha_k} = |f'(0)|(1-r)^{-2}.$$

7.4.24. (i) If $G = \Phi(K)$ and $g = f\circ\Phi^{-1}$, then
$$g' = (f'/\Phi')\circ\Phi^{-1} = p\circ\Phi^{-1},$$

where p is a function of positive real part in K. By Exercise 3.1.14 g is univalent in G and also $f = g \circ \Phi$ must be a univalent mapping.

(ii) In view of Exercise 7.4.23 it is sufficient to prove that

$$\operatorname{re} \prod_{k=1}^{n} (1-\bar{z}_k z)/(1-\bar{\zeta}_k z) \geqslant 0 \quad \text{on } C(0;1) \quad \text{for} \quad z \neq z_k, \zeta_k.$$

We have on $C(0;1)$:

$$\operatorname{re} \prod_{k=1}^{n} (1-\bar{z}_k z)/(1-\bar{\zeta}_k z) = \operatorname{re} \prod_{k=1}^{n} (z-z_k)/(z-\zeta_k).$$

Now, $\arg[(z-z_k)/(z-\zeta_k)] = -\tfrac{1}{2}(\psi_k - \varphi_k)$ if z is situated outside the arc $(z_k; \zeta_k)$, hence for z on an arc $(\zeta_k; z_{k+1})$ we have

$$\sum_{k=1}^{n} \arg[(z-z_k)/(z-\zeta_k)] = -\tfrac{1}{2} \sum_{k=1}^{n} (\psi_k - \varphi_k) = -\pi/2.$$

Moreover, for z situated on the arc $(z_k; \zeta_k)$ we have

$$\arg[(z-z_k)/(z-\zeta_k)] = \pi - \tfrac{1}{2}(\psi_k - \varphi_k),$$

whereas for the remaining arcs $\arg[(z-z_l)/(z-\zeta_l)] = -\tfrac{1}{2}(\psi_l - \varphi_l)$, hence

$$\sum_{k=1}^{n} \arg[(z-z_k)/(z-\zeta_k)] = \pi - \tfrac{1}{2} \sum_{k=1}^{n} (\psi_k - \varphi_k) = \pi/2.$$

Consequently, the images under $p(z) = \prod_{k=1}^{n} (1-\bar{z}_k z)/(1-\bar{\zeta}_k z)$ of arcs $(\zeta_k; z_{k+1})$ are situated on the negative imaginary axis, whereas the images of arcs $(z_k; \zeta_k)$ are situated on the positive real axis, moreover, $p(z_k) = 0$, $p(\zeta_k) = \infty$. In view of the maximum principle and the relation $p(0) = 1$ we have $\operatorname{re} p(z) > 0$ in $K(0;1)$.

7.4.25. The integrand has in $K(0;1) \setminus 0$ the Laurent expansion: $z^{-2} - z^{-1} \sum_{k=1}^{n} \alpha_k \bar{\zeta}_k + \ldots$, hence it does not depend on the line of integration joining in $K(0;1) \setminus 0$ the point 1 to z, iff $\sum_{k=1}^{n} \alpha_k \bar{\zeta}_k = 0$, i.e. iff $\sum_{k=1}^{n} \alpha_k \zeta_k = 0$. Then $W(z)$ is meromorphic in $K(0;1)$ and has a simple pole at the origin. Moreover, $W(z)$ is continuous in $\bar{K}(0;1) \setminus 0$ since all $\alpha_k > -1$. Hence $W(e^{i\theta})$ is a closed curve. If $\theta_k = \arg \zeta_k$, $0 \leqslant \theta_1 < \theta_2 < \ldots < \theta_k < 2\pi$ and $z = e^{i\theta}$, then

$$dW/d\theta = iz^{-1} \prod_{k=1}^{n} (1-ze^{-i\theta_k})^{\alpha_k};$$

since $1-ze^{-i\theta_k} = -2i\exp[\frac{1}{2}(\theta-\theta_k)i]\sin\frac{1}{2}(\theta-\theta_k)$, we have

$$\arg(dW/d\theta) = \text{const} - \theta + \sum_{k=1}^{n} \alpha_k \tfrac{1}{2}(\theta-\theta_k) = \text{const}$$

and consequently L is a polygonal line. If L has no self-intersections and $w_0 \notin L$, then the index $n(L, w_0) = 0, -1$ and since there is one simple pole inside $K(0; 1)$, we see by using the argument principle that the equation $W(z) - w_0 = 0$ has at most one root, i.e. $W(z)$ is univalent in $K(0; 1)$. Moreover, if w_0 lies in the exterior of L, then $n(L, w_0) = 0$ which means that $W(z)$ necessarily takes in $K(0; 1)$ the value w_0.

7.4.26. $\int_{1}^{z} t^{-2} \prod_{k=1}^{3} (t-\eta_k)^{\alpha_k-1} dt$; $\sum_{k=1}^{3} \alpha_k = 5$, $\sum_{k=1}^{3} (\alpha_k-1)\eta_k = 0$.

7.4.27. Cf. Exercises 7.4.26, 1.1.10.

7.4.28. $w = A \int_{1}^{z} t^{-2}(t^n-1)^{2/l} dt$, where

$$A = 2^{-2n} n \sin(\pi/n) \left[\int_{0}^{\pi} (\sin\varphi)^{2/n} d\varphi \right]^{-1}.$$

7.5.1. Put $x = t\sqrt{2}$.

7.5.2. Put $t = \sqrt{1-\zeta^2}$ and verify that $\sqrt{1-z^2} = \operatorname{sn}\sqrt{2}(u-K)$.

7.5.3. Put $t = \sqrt{1-k^2u^2}/k'$ in the integral

$$K'(k') = \int_{1}^{1/k'} [(t^2-1)(1-k'^2t^2)]^{-1/2} dt$$

which gives $K(k) = K'(k')$.

7.5.4. (i) If $w = \operatorname{sn} u$, then

$$u = u(w) = w + \tfrac{1}{6}(1+k^2)w^3 + \tfrac{1}{40}(3+2k^2+3k^4)w^5 + \ldots$$

which can be found from the definition of $u(w)$ by expanding the integrand into a power series and term by term integration; hence (i), (ii), (iii) can be obtained by the method of undetermined coefficients; $R = K'$ since iK' is a singularity nearest the origin.

7.5.5. From the equality $K(k) = \int_{0}^{\pi/2} (1-k^2\sin^2\theta)^{-1/2} d\theta$ we obtain after expanding the integrand according to the binomial formula and term by term integration:

$$K(k) = \frac{\pi}{2}\left[1 + \left(\frac{1}{2}\right)^2 k^2 + \left(\frac{1 \cdot 3}{2 \cdot 4}\right)^2 k^4 + \ldots\right]$$

7. ANALYTIC CONTINUATION. ELLIPTIC FUNCTIONS

and hence

$$K(k) \nearrow +\infty, \quad K(k') = K(\sqrt{1-k^2}) \searrow \pi/2 \quad \text{as} \quad k \to 1.$$

7.5.6. The product $\operatorname{sn} u \operatorname{sn}(u+iK')$ is an analytic, doubly periodic function, hence it is a constant k^{-1} (make $u \to K$). Consequently

$$\operatorname{sn}(u+iK') = k^{-1}[u^{-1} + \tfrac{1}{6}(1+k^2)u + \ldots].$$

7.5.7. Periods: $2K(k)$, $4K'(k)i$; zeros $2mK(k)+2niK'(k)$; poles: $(2m+1)K(k)+2niK'(k)$ (m, n are integers).

7.5.8. The l.h.s. has periods $2K(k)$, $2iK'(k)$ and possibly poles $2mK+2inK'$; in view of Exercise 7.5.4:

$$\operatorname{sn}^{-2}(u, k) = u^{-2}[1 + \tfrac{1}{3}(1+k^2)u^2 + \ldots]$$

and hence the l.h.s. is equal to $1+$(higher powers of u), near the origin; hence it is identically 1.

7.5.9. Put $k = 1/\sqrt{2}$ so that $K(k) = K'(k) = K$;

$$f(u) = \operatorname{sn} Ku \operatorname{sn} K(u+1+i)$$

is a solution which is unique up to a constant factor.

7.5.10. If z describes $COAB$, then v describes the boundary of $(+; +)$ and $(1-v)/(1+v)$ describes the boundary of the lower half of $K(0; 1)$; note that $w = [(1-v)/(1+v)]^{1/2}$.

7.6.1. Grouping the factors corresponding to the pairs $(m; n)$, $(-m; -n)$, we obtain

$$\sigma(z) = z \prod{}' (1-z^2/\Omega_k^2) \exp(z^2/\Omega_k^2);$$

observe that every factor has the form $1-z^4/\Omega_k^4 + \ldots$

7.6.2. If $z = x$, group the factors $m\omega_1 \mp n\omega_2$; if $z = iy$, group the factors $\mp m\omega_1 + n\omega_2$.

7.6.3. A consequence of Exercise 7.6.3 and the reflection principle.

7.6.4. $\zeta'(u+\omega_k) = \zeta'(u)$ since $\wp$ is periodic; hence $\zeta(u+\omega_k) = \zeta(u)+2\zeta(\tfrac{1}{2}\omega_k)$ (put $u = -\tfrac{1}{2}\omega_k$ and use the fact that ζ is odd). Thus

$$\sigma'(u+\omega_k)/\sigma(u+\omega_k) = \sigma'(u)/\sigma(u)+2\zeta(\tfrac{1}{2}\omega_k)$$

and the result follows by integration since σ is odd.

7.6.5. (i) By Exercise 7.6.4 $\sigma(2u)/\sigma^4(u)$ is doubly-periodic and has poles of order three at $m\omega_1+n\omega_2$. Moreover it is equal to $2u^{-3}+Au+\ldots$ near the origin, hence $\wp'(u)+\sigma(2u)/\sigma^4(u) = 0$ since the l.h.s. has only removable singularities; (ii) take logarithmic derivatives of both sides.

7.6.6, 7. Use the definitions of ζ and $\wp$, as well as the identities

$$\sum_{m=-\infty}^{\infty} (a^2+m^2)^{-1} = a^{-1}\pi\coth\pi a, \quad \sum_{n=-\infty}^{\infty} (a+n)^{-2} = \pi^2\operatorname{cosec}^2\pi a.$$

7.6.8. (i) Follows from Exercise 7.6.2 by differentiation;

(ii) $\overline{\wp(\bar z)} = \wp(z)$, hence $\overline{\wp(x+ib)} = \overline{\wp(x-ib)} = \overline{\wp(x+ib)} = \wp(x+ib)$, i.e. $\wp(x+ib)$, as well as $\wp(x+(2n+1)ib)$ is real; by (i) also $\wp(x+2nib)$ is real. A similar reasoning shows that $\wp(na+iy)$ is real;

(iii) $\wp$ is real on the sides of R, moreover $\wp' \neq 0$, hence $\arg\wp' = \text{const}$ on each side; $\wp' = 0$ has simple zeros at three corners of R hence the image of $\partial R \setminus \{p\}$, p being the pole, is the open real axis. The univalence easily follows by the argument principle (p is omitted along a circular arc).

7.6.9. $\wp$ strictly decreases on $(0, \omega)$ from $+\infty$ to e_1 because $\wp' < 0$; if $iv \in (0, \omega')$, then $\wp'(iv)$ strictly increases from $-\infty$ to e_2. Now, the order of boundary points is preserved, hence $e_2 < e_3 < e_1$. We have: $e_1+e_2+e_3 = 0$, hence $e_2 < 0 < e_1$. Conversely, if $e_2 < e_3 < e_1$ and $e_1+e_2+e_3 = 0$, then the function $u = \frac{1}{2}\int_\infty^w [(t-e_1)(t-e_2)(t-e_3)]^{-1/2} dt$ maps $\operatorname{im} w > 0$ onto a rectangle with sides a, b.

After reflections we obtain a function $w = w(u)$ meromorphic and doubly periodic with periods $2a, 2ib$ which can be identified as $\wp$ because it has double poles at $2ma, 2nib$ and the principal part at the origin is u^{-2}.

7.6.10. If one factor has a pole (which is double) then another one has a double zero, hence the l.h.s. is analytic in $\mathbf{C}$ and must be constant. Now put $u = \omega'$.

7.6.11.

$$\wp[u(e_1-e_2)^{-1/2}; \quad 2K(e_1-e_2)^{-1/2}, \quad 2iK'(e_1-e_2)^{-1/2}] = (e_1-e_2)\wp(u; 2K, 2iK')$$

since $\wp$ is homogeneous of degree -2. Hence both sides have periods $2K$, $2iK'$. Moreover, principal parts of both sides at $u = 0$ are equal to $(e_1-e_2)u^{-2}$, hence their difference reduces to a constant which can be found by putting $u = iK'$ which gives e_2.

7.6.12. If z is replaced by $z+\omega_k$, $\zeta(z)$ being an odd function increases by $2\zeta(\frac{1}{2}\omega_k)$; after integration we obtain $2\eta_1\omega_2 - 2\eta_2\omega_1 = \mp 2\pi i$ because $\operatorname{res}(0; \zeta) = 1$; the sign depends on the orientation of parallelogram of integration.

7. ANALYTIC CONTINUATION. ELLIPTIC FUNCTIONS

7.6.13. (i) this is obvious;
(ii) if $\eta_1 = \zeta(\frac{1}{2})$ then by Exercise 7.6.4:

$$\vartheta(z+1) = \exp(-\eta_1 z^2 - 2\eta_1 z - \eta_1)\sigma(z+1)$$
$$= -\exp(-\eta_1 z^2 - 2\eta_1 z - \eta_1)\sigma(z)\exp(2z+1)\eta_1 = -\vartheta(z);$$

(iii) if $\eta = \zeta(\tau/2)$, then by Exercise 7.6.4:

$$\vartheta(z+\tau) = \exp(-\eta_1 z^2 - 2\eta_1 \tau z - \eta_1 \tau^2)\sigma(z+\tau)$$
$$= -\exp(-\eta_1 z^2 - 2\eta_1 \tau z - \eta_1 \tau^2)\exp(2z+\tau)\eta\sigma(z)$$
$$= -\vartheta(z)\exp[-\pi i(2z+\tau)]$$

since $\tau\eta_1 - \eta = \pi i$ (cf. Ex. 7.6.12).

7.7.1. The angles (on the sphere) at $u = 0$ are doubled, hence $u = 0$ is a double pole, the only singularity in: $|\mathrm{re}\,u| < 2a$, $|\mathrm{im}\,u| < b$; f is an even function, hence after a suitable choice of A, $f - A\wp$ is analytic and hence it reduces to a constant B. Putting $u = 2a, ib, 2a+ib$ we obtain for $f(u)$ the values $0, -h, -h^{-1}$ and this enables us to find A, B, h.

7.7.2. Put $z = \log\zeta/Q$, $a = \log Q$, $b = \pi$ in Exercise 7.7.1.

7.7.3. Put $w = e^{i\theta}$, $0 < \theta < \pi$, and verify that $du/d\theta = ie^{i\theta}du/dw$ is real and $\neq 0$.

7.7.4. Use Exercise 7.7.3; note that the image of $[K, K+\frac{1}{2}iK']$ is the segment $[\sqrt{k}, 1]$; horizontal sides of the rectangle are mapped onto the upper and lower half of $C(0; 1)$, whereas vertical sides correspond to the slits $[-1, -\sqrt{k}], [\sqrt{k}, 1]$.

7.7.5. Under the mapping $w = \sqrt{k}\mathrm{sn}(u, k)$ the rectangle R: $|\mathrm{re}\,u| < K$, $|\mathrm{im}\,u| < \frac{1}{2}K'$ is mapped onto $K(0; 1)\setminus[(-1, -\sqrt{k}] \cup [\sqrt{k}, 1)]$. On the other hand, $z = c\,\mathrm{sn}\,\pi u/2K$ maps R conformally onto an ellipse with foci $\mp c$ slit along $[-a, -c]$ and $[c, a]$; if $(u = \frac{1}{2}iK') \leftrightarrow (z = ib)$, i.e. $b = \sqrt{a^2-b^2}\sinh \pi K'/4K$, or $\pi K'/2K = \log\dfrac{a+b}{a-b}$, we obtain the given ellipse H. The slits can be removed by reflection principle.

7.7.6. $w = \sqrt{k}\mathrm{sn}(u, k)$ maps 1:1 conformally the rectangle Q: $|\mathrm{re}\,u| < K$, $0 < \mathrm{im}\,u < \frac{1}{2}K'$, onto $K(0; 1)\setminus\{[-\sqrt{k}, \sqrt{k}] \cup [0, i)\}$; $u = 2Ki\pi^{-1}\mathrm{Log}\,z$ maps $A\setminus(-R, -1]$ onto Q; the slits can be removed by reflection principle.

7.7.7. Consider $W = [w(\sqrt{z})]^{-2}$, where $w(z)$ is the mapping obtained in Exercise 7.7.6.

7.7.8. If $K'(k)/K(k) = \pi/\log Q$, then $w = \text{sn}(K\log z/\log Q, k)$.

7.7.9. Find k from the equality of cross-ratios: $(a_1, a_2, b_1, b_2) = (-k^{-1}, -1, 1, k^{-1})$ and put $(W, a_2, b_1, b_2) = (w, -1, 1, k^{-1})$, where w is the mapping of Exercise 7.7.8 continued by reflections.

7.7.10. Use the reflection principle and show that $f(z)$ has the same periods and singularities as $\text{cn}(z, 1/\sqrt{2})$.

7.7.11. After reflections the mapping function can be continued over the whole plane, has simple zeros at $\xi+i\eta+2ma+2inb$, $-\xi-i\eta+2ma+2inb$ and simple poles at $\xi-i\eta+2ma+2inb$, $-\xi+i\eta+2ma+2inb$; it is moreover, doubly-periodic with periods $2a$, $2ib$. The function

$$\sigma(z-\xi-i\eta)\sigma(z+\xi+i\eta)[\sigma(z-\xi+i\eta)\sigma(z+\xi-i\eta)]^{-1}$$

has analogous properties and is an elliptic function of order 2 (cf. Ex. 7.6.4). Both functions are identical up to a constant factor whose absolute value is equal to 1 (put $z = 0$ and use Exericise 7.6.2).

CHAPTER 8

The Dirichlet Problem

8.1.1. The inverse mapping would be a bounded, entire function, thus a constant by Liouville's theorem.

8.1.2. The sets of values taken by g on $(-\infty, -1), (-1, 0)$ coincide. Moreover, if $u = -v\cot v$ and $v\csc v = \exp(v\cot v)$ which holds for some $v \in (2\pi n + \pi/4, 2\pi n + \pi/2)$ and all n large enough, then $g(u+iv) = g(-1)$. Thus we need to verify that g takes every finite value in $\mathbf{C}$. Suppose that $we^w - a \neq 0$ for some a and all w. Then also $(w+2\pi i)e^w - a \neq 0$ and the quotient $1 + 2\pi i e^w/(we^w - a)$ which is an entire function $\neq 0, 1$ must be a constant. This is an obvious contradiction. If $h(z) = 4z(1-z)^{-2}$, then $h[K(0;1)] = G$ and $f = \frac{1}{4} g \circ h$ has the desired properties.

8.1.3. Suppose h is a (many-valued) conjugate of g and the increment of h over a small circle with center at z_0 and positive orientation is equal to A; then $f = \exp \circ 2\pi A^{-1}(g+ih)$ has all the desired properties. Univalence follows from the argument principle; since A is real and negative, $f(z_0) = 0$.

8.1.4. The mapping $f_1 \circ f_2^{-1}$ carries the unit disk onto itself; the mapping is a homeomorphism in $\overline{K}(0; 1)$ and after reflections it becomes a 1:1 conformal mapping of the extended plane onto itself and consequently it is a linear mapping with 3 fixed points, i.e. an identity.

8.1.5. Both mappings $f(z)$, $(f(z^*))^*$, where the stars indicate corresponding reflections carry z_k into w_k ($k = 1, 2, 3$), hence they are identical. This means that $f(\gamma_1)$ remains unchanged after a reflection w.r.t. γ_2, i.e. $\gamma_2 = f(\gamma_1)$.

8.1.6. Arcs of: $C(1; 1)$, $\operatorname{re} \lambda = \frac{1}{2}$, $C(0; 1)$ situated in the upper half-plane; $\lambda((1+i\sqrt{3})/2) = (1+i\sqrt{3})/2$, $\lambda(i) = 1/2$, $\lambda(1+i) = 2$.

8.1.7. Suppose h is a positive, continuous and nowhere differentiable function with the period 2π and let D be the domain whose boundary D has the equation $r = h(\theta)$, $0 \leq \theta \leq 2\pi$, in polar coordinates. If f maps 1:1 conformally $K(0; 1)$ onto D, then $C(0; 1)$ is obviously a natural boundary of f.

8.1.8. If the function Φ mapping D onto $K(0;1)$ is such that $0 = \Phi(a)$, then all the functions with this property have the form $e^{i\alpha}\Phi(w)$; hence

$$\varphi(w) = e^{i\alpha}\Phi(w)/\Phi'(a) \quad \text{and} \quad r(a;D) = 1/|\Phi'(a)|.$$

8.1.9. In view of Exercise 8.1.8.

$$r(a;G) = |dw/d\zeta|_{\zeta=0}, \quad \text{where} \quad w = f[(\zeta+z_0)/(1+\bar{z}_0\zeta)].$$

8.1.10. (i) $R^{-1}(R^2-|a|^2)$; (ii) $2h$; (iii) $4d$.

8.1.11. If $D_0 = f[K(0;1)]$, $a_0 = f(z_0)$, then

$$r(a_0;D_0) = (1-|z_0|^2)|f'(z_0)| \quad \text{and} \quad r(a;D) = (1-|z_0|^2)|f'(z_0)||\varphi'(a_0)|.$$

8.1.12. Suppose that $z = \Phi(w)$ maps D_0 onto $K(0;r)$. Then $\varphi \circ \Phi^{-1}$ maps conformally $K(0;r)$ onto D and

$$|D| = \iint\limits_{K(0;r)} |(\varphi \circ \Phi^{-1})'|^2 dx\,dy$$

and since $\varphi \circ \Phi^{-1}(z) = z + A_2 z^2 + \ldots$, we have

$$|D| = \pi r^2(1+2|A_2|^2 r^2 + \ldots),$$

(cf. Ex. 4.2.12). Hence $|D|$ is a minimum if $\varphi \circ \Phi^{-1}$ is an identity mapping, i.e. $\Phi = \varphi$.

8.1.13. We may assume that $\infty \in \Gamma_n$. We apply Riemann mapping theorem to the domain $\hat{\mathbf{C}} \setminus \Gamma_n$ and the mapping function carries G into the unit disk minus $n-1$ continua. After a suitable linear transformation one of the boundary continua, say γ_{n-1} will contain ∞. We again apply the Riemann mapping theorem to $\mathbf{C} \setminus \gamma_{n-1}$ and obtain an image domain whose boundary already contains two analytic Jordan curves, $C(0;1)$ and the image curve of a circle under the latter mapping function. After n analogous steps we finally obtain as an image domain of G the unit disk with removed $n-1$ interior domains of closed analytic Jordan curves.

8.2.1. $\mathrm{re}[(z+\zeta)/(z-\zeta)] = \mathrm{re}[(z+\zeta)(\bar{z}-\bar{\zeta})/|z-\zeta|^2] = (|z|^2-|\zeta|^2)/|z-\zeta|^2 = J$.

8.2.2. $f(z) = (2\pi)^{-1} \int\limits_0^{2\pi} U(\theta)(Re^{i\theta}+z)/(Re^{i\theta}-z)d\theta + iv(0)$, where $v(0)$ is an arbitrary real number.

8.2.3. This is a consequence of (8.2A), the mean-value property:

$$u(0) = (2\pi)^{-1} \int\limits_0^{2\pi} u(Re^{i\theta})d\theta$$

8. THE DIRICHLET PROBLEM

and the inequality
$$\frac{R-|z|}{R+|z|} \leqslant \frac{R^2-|z|^2}{|Re^{i\theta}-z|^2} \leqslant \frac{R+|z|}{R-|z|}.$$

8.2.4. $u(z) = \frac{1}{2}\operatorname{re}(1+z^2)$ (cf. Ex. 3.6.3 (i)).

8.2.5. $2\pi u(z) = \int_\alpha (1-|z|^2)|\zeta-z|^{-2}d\theta = \int_\alpha |(\zeta_1-z)/(\zeta-z)|\,d\theta$ because $|\zeta_1-z||\zeta-z| = 1-|z|^2$; moreover, by considering the triangles $z, \zeta, \zeta+\Delta\zeta$ and $z, \zeta_1, \zeta_1+\Delta\zeta_1$

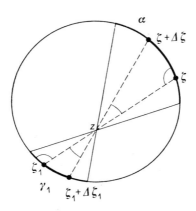

Fig. 6

(cf. Fig. 6) we have $d\theta/|\zeta-z| = d\theta_1/|\zeta_1-z|$, hence $2\pi u(z) = \int_\gamma d\theta =$ length of γ.

8.2.6. If $\bar{K} \subset D$, where D is the domain where the functions u_n are defined, then the formula (8.2A) may be applied, hence the limit function is harmonic, in K and also in D.

8.2.7. In view of Exercise 8.2.6 the function
$$u(z) = u(re^{i\varphi}) = \tfrac{1}{2}a_0 + \sum_{n=1}^\infty (r/R)^n(a_n\cos n\varphi + b_n\sin n\varphi)$$
is harmonic and tends to $U(\varphi)$ as $r \to R-$ (cf. Ex. 4.4.4). By the Euler–Fourier formulas we have:
$$a_n = (2\pi)^{-1}\int_0^{2\pi} U(\theta)\cos n\theta\,d\theta, \quad b_n = (2\pi)^{-1}\int_0^{2\pi} U(\theta)\sin n\theta\,d\theta,$$

and hence

$$u(re^{i\varphi}) = \pi^{-1} \int_0^{2\pi} \left[\tfrac{1}{2} + \sum_{n=1}^{\infty} (r/R)^n \cos n(\varphi-\theta)\right] U(\theta) d\theta$$

$$= (2\pi)^{-1} \int_0^{2\pi} \operatorname{re}\left[1 + 2\sum_{n=1}^{\infty} (z/\zeta)^n\right] U(\theta) d\theta$$

$$= (2\pi)^{-1} \int_0^{2\pi} \operatorname{re}[(\zeta+z)/(\zeta-z)] U(\theta) d\theta,$$

$z = re^{i\varphi}$, $\zeta = Re^{i\theta}$.

8.2.8. $u(z) = u(re^{i\varphi}) = (2\pi)^{-1} \int_0^{2\pi} \operatorname{re}[(z+\zeta)/(z-\zeta)] U(\theta) d\theta$, where $\zeta = Re^{i\theta}$, $z = re^{i\varphi}$, $r > R$.

8.2.9. $u(re^{i\varphi}) = \alpha\pi^{-1} + 2\pi^{-1} \sum_{n=1}^{\infty} n^{-1}(r/R)^n \sin n\alpha \cos n\varphi$.

8.3.1. A function harmonic in $0 < |z| < 1$ and continuous in $\overline{K}(0;1)$ is harmonic in extension to $K(0;1)$ and hence it is determined by its values on $C(0;1)$. If $U(0) \neq (2\pi)^{-1} \int_0^{2\pi} U(e^{i\theta}) d\theta$, the Dirichlet problem has no solution.

8.3.2. (i) $u(z) = A\bigl(\log(|z|/R_2)\bigr) : \bigl(\log(R_1/R_2)\bigr) + B\bigl(\log(|z|/R_1)\bigr) : \bigl(\log(R_2/R_1)\bigr)$;

(ii) the coefficients a_n, b_n, c_n, d_n may be evaluated from Euler–Fourier formulas:

$$a_0 + b_0 \log R_k = \pi^{-1} \int_0^{2\pi} U_k(\theta) d\theta,$$

$$a_n R_k^n + b_n R_k^{-n} = \pi^{-1} \int_0^{2\pi} U_k(\theta) \cos n\theta \, d\theta,$$

$$c_n R_k^n + d_n R_k^{-n} = \pi^{-1} \int_0^{2\pi} U_k(\theta) \sin n\theta \, d\theta,$$

$n = 1, 2, \ldots,\ k = 1, 2$.

8.3.3. The mapping function $\Phi: G \to K(0;1)$ has a homeomorphic extension to $\overline{G}$; if $e^{i\theta} = \Phi(\omega)$, $\omega \in \operatorname{fr} G$, put $U_1(\theta) = U(\omega)$, where U are the given boundary values. If u is the Poisson integral for boundary values U_1, then $u \circ \Phi$ is the solution of the Dirichlet problem for G.

8.3.4. Suppose that $U(\theta) = H\bigl(e^{-i\theta}\sqrt{1+e^{2i\theta}}-1\bigr)$ and u is the Poisson integral for $C(0;1)$ with boundary values U, then $h(w) = u\bigl(2w/(1-w^2)\bigr)$.

8.3.5. $\omega(z) = \operatorname{im} \operatorname{Log}[(z-b)/(z-a)]$;

$$\lim_{z \to \zeta} \omega(z) = \begin{cases} 0 & \text{if } \zeta \in \{(-\infty, a) \cup (b, +\infty)\}, \\ \pi & \text{if } \zeta \in (a, b). \end{cases}$$

8.3.6. $u(z) = \pi^{-1} \sum_{k=1}^{n} u_k \operatorname{im} \operatorname{Log}[1 - \Delta x_k/(z - x_{k-1})]$, $\Delta x_k = x_k - x_{k-1}$.

8.3.7. The continuous function U can be uniformly approximated by a sequence of step functions; in the limiting case we obtain from Exercise 8.3.6:

$$u(z) = \pi^{-1} y \int_{-\infty}^{+\infty} [(x-t)^2 + y^2]^{-1} U(t) \, dt = \operatorname{im} \pi^{-1} \int_{-\infty}^{+\infty} (z-t)^{-1} U(t) \, dt.$$

8.3.8. The integral $f(z) = \pi^{-1} \int_{-\infty}^{+\infty} (z-t)^{-1} U(t) \, dt$ is a.u. convergent for $\operatorname{im} z > 0$, hence f is analytic and $u(z) = \operatorname{im} f(z)$ is harmonic in the upper half-plane. Suppose that ε is an arbitrarily chosen positive number and δ is such that

$$|U(t) - U(t_0)| < \varepsilon \quad \text{for} \quad |t - t_0| < \delta.$$

Put

$$u_1(z) = \pi^{-1} y \int_{-\infty}^{t_0 - \delta} |z-t|^{-2} U(t) \, dt + \pi^{-1} y \int_{t_0 + \delta}^{+\infty} |z-t|^{-2} U(t) \, dt +$$

$$+ \pi^{-1} y \int_{t_0 - \delta}^{t_0 + \delta} |z-t|^{-2} U(t_0) \, dt.$$

If $z \to t_0$, the two initial terms tend to 0 and the third tends to $U(t_0)$ by Exercise 8.3.5. Moreover,

$$|u_1(z) - u(z)| \leq \varepsilon \pi^{-1} y \int_{t_0 - \delta}^{t_0 + \delta} |z-t|^{-2} \, dt < \varepsilon$$

and the existence of a finite limit $\lim_{z \to t_0} u_1(z) = U(t_0)$ implies $\overline{\lim}_{z \to t_0} u(z) - \underline{\lim}_{z \to t_0} u(z) \leq 2\varepsilon$ which shows that $\lim_{z \to t_0} u(z)$ exists and is equal to $U(t_0)$.

8.3.9. The mapping $z = \frac{1}{2}(w + w^{-1})$ gives the annulus $1 < |w| < 2 + \sqrt{3}$, hence $u(z) = \log|z + \sqrt{z^2 - 1}|/\log(2 + \sqrt{3})$ by Exercise 8.3.2 (i).

8.3.10. The mapping $z = \frac{1}{2}(w + w^{-1})$ gives the upper half-plane $\operatorname{im} w > 0$ with the negative and positive real half-lines being the image lines of the slits. Hence

$$u(z) = 2h[\tfrac{1}{2} - \pi^{-1} \operatorname{Arg}(z + \sqrt{z^2 - 1})], \quad \operatorname{im}(z + \sqrt{z^2 - 1}) > 0.$$

8.3.11. After the transformation $z = \frac{1}{2}c(w + w^{-1})$ the problem is reduced to an analogous problem for the outside of $K(0; R)$ $(c = \sqrt{a^2-b^2}, a = \frac{1}{2}c(R+R^{-1})$, $b = \frac{1}{2}c(R-R^{-1})$, $R = (a+b)/c > 1)$. The solution of the latter problem is: $\log(|w|/R)+A$. Hence

$$u(z) = \log|z + \sqrt{z^2-c^2}| - \log(a+b) + A;$$

if $A = \log[(a+b)/2]$, then $u(z) - \log|z| = o(1)$ as $z \to +\infty$. The uniqueness follows from the maximum principle.

8.3.12. The mapping $w = 2(z^2 + \frac{1}{2})$ carries the given domain into that considered in Exercise 8.3.10. Hence

$$u(z) = \pi^{-1}\operatorname{im}\operatorname{Log}(2z^2+1+2z\sqrt{z^2+1}) = \pi^{-1}\operatorname{im}\operatorname{Log}F(z).$$

We have $u_x = 0$ on the real axis, hence

$$u_y = \partial u/\partial n = \pi^{-1}|F'(z)| = 2\pi^{-1}(x^2+1)^{-1/2}.$$

8.4.1. $\operatorname{re}[(1+z)/(1-z)] = (1-|z|^2)|1-z|^{-2}$.

8.4.2. (i) Circular arcs with end points a, b situated in the upper half-plane (cf. Ex. 8.3.5);

(ii) circular arcs with end points $e^{i\theta_1}$, $e^{i\theta_2}$, situated inside $K(0; 1)$.

8.4.3. $\omega(z; l_2, G) = \alpha^{-1}\operatorname{Arg} z$, $\omega(z; l_2, G) = 1 - \omega(z; l_1, G)$.

8.4.4. If $\theta(z) = \operatorname{im}\operatorname{Log}[(z-1)/(z+1)]$, then

$$\omega(z; l, G) = 2\pi^{-1}\theta(z) - 1; \quad \omega(z; \gamma, G) = 2[1 - \pi^{-1}\theta(z)];$$

circular arcs with end points $1, -1$ situated inside G.

8.4.5. If $z = x + iy$, then

$$\omega(z; \gamma, G) = 2\pi^{-1}[\pi - \theta(z)] = 2\pi^{-1}[\arctan(y/(r-x)) + \arctan(y/(r+x))]$$
$$= 4y(\pi r)^{-1}(1 + o(1)).$$

8.4.6. The mapping $z = w^2$ carries G_0, γ_0 into G, γ and due to the conformal invariance of ω we have:

$$\omega(z; \gamma, G) = 2\pi^{-1}\operatorname{Arg}[(z-1)/(z+1)] = 2\pi^{-1}\operatorname{Arg}[(w^2-1)/(w^2+1)]$$

8.4.7. We have

$$\omega(r, \theta) = \lambda(1 - \log r/\log R) + \sum_{n=1}^{\infty}[(a_n r^n + b_n r^{-n})\cos n\theta + (c_n r^n + d_n r^{-n})\sin n\theta],$$

where

$$a_n = [\pi n(1-R^{2n})]^{-1}\sin 2\pi n\lambda, \quad b_n = [\pi n(1-R^{-2n})]^{-1}\sin 2\pi n\lambda,$$
$$c_n = [\pi n(1-R^{2n})]^{-1}(1-\cos 2\pi n\lambda), \quad d_n = [\pi n(1-R^{-2n})]^{-1}(1-\cos 2\pi n\lambda).$$

8.4.8. If $\theta(z) = \mathrm{Arg}[(z-b)/(z-a)]$ and $\varphi_k = \theta(\zeta)$, $\zeta \in \gamma_k$, then
$$\omega(z; \gamma_k, G) = [\theta(z)-\varphi_l]/(\varphi_k-\varphi_l) \quad (k, l = 1, 2, \; k \neq l).$$

8.4.9. (i) $\omega(z; \gamma_2, G) - \omega(z; \gamma_1, G)$ is a bounded harmonic functions with nonnegative boundary values, hence it is nonnegative in G;

(ii) consider $\omega(z; \gamma, G_2) - \omega(z; \gamma, G_1)$, $z \in G_1$; again the difference has nonnegative boundary values.

8.4.10. The function
$$h(z) = \log|f(z)| - \omega(z; \alpha, G)\log m - \omega(z; \beta, G)\log M$$
is harmonic except possibly at zeros z_k of f and bounded from above. If $z \to z_k$, then $h \to -\infty$; if $z \to \zeta \in \mathrm{fr}\, G$, $\overline{\lim}\, h(z) \leq 0$ except possibly for end points of α. Hence $h(z) \leq 0$ in G by the maximum principle.

8.4.11. If $z = x+iy$ is fixed, then by Exercise 8.4.10 (with $m = 1$, $\alpha = (-r, r)$, $\beta = C(0; r) \cap H_+$ and $G = K(0; r) \cap H_+$) and by Exercise 8.4.5 we have:
$$\log|f(z)| \leq \omega(z; \beta, G)\log M(r) = 4y(\pi r)^{-1}\log M(r)(1+o(1)).$$
Suppose that $\lim_{r \to +\infty} r^{-1}\log M(r) = \sigma \leq 0$; then there exists a sequence $\{r_n\}$, $r_n \to +\infty$, such that $r_n^{-1}\log M(r_n) \to \sigma \leq 0$, and consequently
$$\log|f(z)| \leq 4y\pi^{-1}\sigma \leq 0, \quad \text{i.e.} \quad |f(z)| \leq 1.$$

8.5.1. If $h(z, z_0)$ is the solution of the Dirichlet problem with boundary values $\log|\zeta-z_0|$, then
$$g(z, z_0; G) = h(z, z_0) - \log|z-z_0|.$$

8.5.2. Suppose $t = f(z)$ maps 1:1 conformally G onto $K(0; 1)$. Then
$$g(z, z_0; G) = \log[|1-\overline{f(z_0)}f(z)|/|f(z)-f(z_0)|].$$

8.5.3. If $\mathrm{dist}(w, \mathrm{fr}\, G_w) \to 0$, then $\mathrm{dist}(z(w), G_z) \to 0$, hence $g(w, w_0; G) \to 0$; if $w \to w_0$, then
$$g(z(w), z(w_0); G_z) = -\log|z(w)-z(w_0)|+O(1)$$
$$= -\log|w-w_0| - \log[|z(w)-z(w_0)|/|w-w_0|] + O(1)$$
$$= -\log|w-w_0| + O(1).$$

8.5.4. $u+iv = \log \Phi(z)$ is analytic in any sufficiently small neighborhood of a boundary point, moreover $u = g(z; z_0; G)$. By Cauchy–Riemann equations: $\partial g/\partial n = \partial u/\partial n = -\partial v/\partial s = |\partial \theta/\partial s|$ because $v = \theta = \arg \Phi$ for $z \in \Gamma$. The image of Γ under Φ is $C(0; 1)$, where θ and the arc length σ coincide; hence

$\partial g/\partial n = d\sigma/ds = |\Phi'(z)|$, because the ratio of arc length elements $d\sigma$ on $C(0; 1)$ and ds on Γ is equal to $|\Phi'|$.

8.5.5. $g(z, z_0; K(0; R)) = \log[|R^2 - z\bar{z}_0|/|R(z-z_0)|]$, hence by Exercise 8.5.4:
$$\partial g/\partial n_\zeta = R^{-1}(R^2 - |z_0|^2)|\zeta - z_0|^{-2}$$
$$= R^{-1}(R^2 - r^2)(R^2 + r^2 - 2Rr\cos(\theta - \varphi))^{-1},$$
$\zeta = Re^{i\theta}$, $z_0 = re^{i\varphi}$; note that $ds = Rd\theta$.

8.5.6. $g(z, z_0; G) = \log[|z - \bar{z}_0|/|z - z_0|]$;
$\partial g/\partial n_\zeta = 2y_0|\zeta - z_0|^{-2} = 2y_0[(x - x_0)^2 + y^2]^{-1}$; $ds = dx$.

8.5.7. (i) $\log|z|$;
(ii) $\log|z| + \log|1 + \sqrt{1 - z^{-2}}|$;
(iii) $-\log(a+b) + \log|z| + \log|1 + \sqrt{1 - (a^2 - b^2)z^{-2}}|$; $(g(z; G) = \log|\Phi(z)|$, where $\Phi\colon G \to K(\infty; 1)$ is such that $\Phi(\infty) = \infty$).

8.5.8. (i) If $\arg w$ increases by $2k\pi$, the increment of $(2\pi i)^{-1}(\log w \mp \log c)$ is an integer, hence $F(w)$ remains unchanged by Exercise 7.6.13 (ii) and consequently F is single-valued;

(ii) F has zeros at ch^{2n} and poles at $c^{-1}h^{2n}$ ($n = 0, \mp 1, \mp 2, \ldots$), hence there is one zero $w = c$ and no poles in A;

(iii) we have $\overline{\vartheta(\bar{z})} \equiv \vartheta(z)$, hence $|\vartheta(\bar{z})| = |\vartheta(z)|$ and this implies
$$|F(e^{i\theta})| = |\vartheta[(2\pi)^{-1}(\theta + i\log c)]| : |\vartheta[(2\pi)^{-1}(\theta - i\log c)]| = 1;$$
moreover,
$$|F(he^{i\theta})| = |\vartheta(z)| : |\vartheta(z + \tau)|,$$
where $z = (2\pi)^{-1}(i\log h - i\log c + \theta)$ and consequently
$$|F(he^{i\theta})| = |\exp[\pi i(2z + \tau)]| = |\exp(\log c + \theta i)| = c.$$

8.5.9. In order to obtain Green's function $g(w, c; A)$ we add to $-\log|F(w)|$ a term $\lambda \log|w|$ such that the sum is equal to 0 on $\operatorname{fr} A$. This gives
$$g(w, c; A) = \log c \log|w|/\log h - \log|F(w)|,$$
where F is the function considered in Exercise 8.5.8.

8.5.10. We have: $\log \Phi(w) = g(w) + ih(w)$, where $|\Phi(w)| = c$ on $C(0; h)$. Thus
$$\partial g/\partial n = \partial h/\partial s = c^{-1}\partial(c \arg \Phi)/\partial s = ie^{i\theta} c^{-1} d\Phi/dw = |\Phi'(w)/\Phi(w)|;$$
Now, $\Phi = w^{\log c/\log h}/F(w)$, hence
$$\partial g/\partial n = |w^{-1}\log c/\log h - F'(w)/F(w)|$$
where $F'/F = w^{-1}\{-\eta_1\pi^{-2}\log c + (2\pi i)^{-1}[\zeta((2\pi i)^{-1}\log(w/c)) - \zeta((2\pi i)^{-1}\log cw)]\}$.

8.5.11. We may assume that $G = K(0; 1)$ since both ρ and g are invariant under conformal mapping. Moreover, we can take $z_0 = 0$. Then $\rho(z, 0) = \frac{1}{2}\log[(1+|z|)/(1-|z|)]$, $g(z, 0) = -\log|z|$, hence $g = -\log\tanh\rho$.

8.5.12. The function $\log|\zeta-z_1|$ is harmonic w.r.t. z_1, hence $g(z, z_1)$ is harmonic in z_1 by (8.5A); therefore $g(z_1, z_2)-g(z_2, z_1) = h(z_1)$ is also harmonic in z_1 and for $z_1 \to \text{fr}\,G$ we obtain $\overline{\lim}\,h(z_1) \leq 0$ which means that $h(z_1) \leq 0$ for a fixed $z_2 \in G$ and all $z_1 \in G$. Similarly $H(z_2) = g(z_1, z_2)-g(z_2, z_1) \geq 0$ for a fixed z_1 and all $z_2 \in G$. Hence $g(z_1, z_2)-g(z_2, z_1) \equiv 0$. We used the fact that a bounded harmonic function necessarily has a harmonic extension to isolated singularities.

8.5.13. A consequence of the extremum principle for $g(z, z_1; G)-g(z, z_1; G_0)$, $z, z_1 \in G_0$.

8.5.14. We have

$$\omega_k(z_0) = (2\pi)^{-1} \int_{\Gamma_k} \partial g(\zeta, z_0; G)/\partial n_\zeta\, ds$$

$$= -(2\pi)^{-1} \int_{\Gamma_k} \partial h(\zeta, z_0; G)/\partial s_\zeta\, ds = -(2\pi)^{-1} \Delta_{\Gamma_k} h(\zeta, z_0),$$

$\zeta \in \Gamma_k$; $\partial g/\partial n = -\partial h/\partial s$ by Cauchy–Riemann equations.

8.5.15. $\log \Phi = g+ih$, hence for w describing $C(0; 1)$:

$$\Delta \log \Phi = i\Delta h = -2\pi i \omega(c; C(0; 1), A) = -2\pi i\left(1 - \frac{\log c}{\log h}\right)$$

$$= \Delta \log w \log c/\log h - \Delta \log F(w),$$

i.e. $\Delta \log F(w) = 2\pi i$, similarly for w describing $C(0; h)$ we have $\Delta \log F(w) = 0$.

8.5.16. D is a convex domain, hence there exists a half plane H containig D and such that $c \in \mathbf{C}\setminus H$. In view of Exercise 8.5.13 we may assume that $D = H$. Therefore we may take $c = 0$, $D = \{z: \text{re}\, z > 0\}$. By Exercise 8.5.6 $g(a, b; H) = \log\frac{|a+\bar{b}|}{|a-b|}$. If the points a, b are rotated round the origin, $|a-b|$ remains unchanged and $|a+\bar{b}| \leq |a|+|b|$ with the sign of equality in case $a, 0, -\bar{b}$ are collinear.

Hence $\sup_D g(a, b; D) = \log[(|a|+|b|)/|a-b|]$ in case $c = 0$; in the general case:

$$\sup_D g(a, b; D) = \log \frac{|a-c|+|b-c|}{|a-b|}.$$

8.5.17. If $g(z) = g(z, \infty; G)$, then
$$\varphi(z) = \begin{cases} Ag(z)+B & \text{for} \quad z \in G, \\ B & \text{for} \quad z \in H. \end{cases}$$
The uniqueness follows from the maximum principle in the usual manner.

8.6.1. If $f, g \in L_2(G)$, so does $af+bg$ for any complex a, b because
$$|af+bg|^2 \leqslant |a|^2|f|^2+|b|^2|g|^2+|ab|(|f|^2+|g|^2).$$

8.6.2. $|f\bar{g}| \leqslant \frac{1}{2}(|f|^2+|g|^2)$, hence $\iint f\bar{g}$ is finite.

8.6.3. Suppose $\varphi: \mathbf{C} \setminus \Gamma \to K(0; R)$ is such that $\varphi'(\zeta) = 1$; then
$$R = r(\zeta; \mathbf{C} \setminus \Gamma) \quad \text{and} \quad \iint_{\mathbf{C}\setminus\Gamma} |\varphi'|^2 dx dy = \pi R^2$$
(cf. Ex. 8.1.12) is finite. Now, $G \subset \mathbf{C} \setminus \Gamma$ and hence $\varphi' \in L_2(G)$.

8.6.4. If $\text{dist}(z, \mathbf{C} \setminus G) = \rho$, then $|f(z)| \leqslant (M/\pi\rho^2)^{1/2}$.

8.6.5. $|f(a)| \leqslant (M/\pi)^{1/2}(1-|a|)^{-1}$.

8.6.6. The family of functions $f \in L_2(G)$ such that $f(\zeta) = 1$ and $\|f\| \leqslant \sqrt{\pi} R$, where R is defined as in the solution of Exercise 6.8.3, is compact and nonempty.

8.6.7. If $\varepsilon > 0$, $0 \leqslant \theta \leqslant 2\pi$, then $f^* = f_0 + \varepsilon e^{i\theta} g \in L_2(G)$ for any $g \in L_2(G)$. We have $\|f_0\| \leqslant \|f^*\|$, hence
$$(f_0, f_0) \leqslant (f_0, f_0) + 2\varepsilon \,\text{re}[e^{i\theta}(g, f_0)] + \varepsilon^2(g, g),$$
i.e. $0 \leqslant 2\text{re}[e^{i\theta}(g, f_0)] + \varepsilon(g, g)$ for any $\varepsilon > 0$ and any $\theta \in [0, 2\pi]$ which implies $(g, f_0) = 0$.

8.6.8. Any function analytic in G has a primitive, hence we need to find φ analytic in G and such that $\varphi(\zeta) = 0$, $\varphi'(\zeta) = 1$ and $\iint_G |\varphi'|^2$ has a minimum. In view of Exercise 8.1.12 the extremal function $f_0 = \varphi'$ is equal to $w'(z)/w'(\zeta)$; if $R = r(\zeta; G)$, then $\|f_0\|^2 = \iint_G |\varphi'|^2 = \pi R^2 = \pi |w'(\zeta)|^{-2}$ by Exercise 8.1.8 and hence
$$k(z, \zeta) = \pi^{-1}|w'(\zeta)|^2 w'(z)/w'(\zeta) = \pi^{-1} w'(z)\overline{w'(\zeta)}.$$

8.6.9. (i) $\pi^{-1}R^2(R^2-z\bar{\zeta})^{-2}$; (ii) $\pi^{-1}(z-\bar{\zeta})^{-2}$.

8.6.10. $w'(z) = \sqrt{\pi/k(\zeta, \zeta)} k(z, \zeta)$.

8.6.11. If $g(\zeta) = 0$, then $(g, f_0) = 0$ (Ex. 8.6.7) and also $(g, k) = 0$, i.e.
$$g(\zeta) = \iint_G f(z)\overline{k(z, \zeta)} dx dy = 0;$$

8. THE DIRICHLET PROBLEM

if $f(\zeta) \neq 0$, then $g(z) = f(z)/f(\zeta) - f_0(z, \zeta)$ vanishes for $z = \zeta$, hence

$$\iint_G [f(z)/f(\zeta) - f_0(z, \zeta)]\overline{f_0(z, \zeta)}\, dx\, dy = 0$$

and consequently

$$\iint_G f(z)\overline{f_0(z, \zeta)}\, dx\, dy = f(\zeta)\|f_0\|^2.$$

8.6.12. Put $f = k(z, \zeta)$ in Exercise 8.6.11 and use Schwarz's inequality.

8.6.13. Suppose that $f(\zeta) = (f, k_1) = (f, k_2)$; then $(f, k_1 - k_2) = 0$ for any $f \in L_2(G)$ and also for $f = k_1 - k_2$. This implies $\|k_1 - k_2\| = 0$, i.e. $k_1 = k_2$.

8.6.14. If $s_n = \sum_{k=0}^{n} a_k \varphi_k$, $\sigma_n = \sum_{k=0}^{n} b_k \varphi_k$, then from the minimal property of Fourier coefficients and from the density property of $\sum_{k=0}^{n} c_k \varphi_k$ it follows that $\|f - s_n\| \to 0$, $\|g - \sigma_n\| \to 0$. Hence

$$(f - s_n, g - \sigma_n) = \left|(f, g) - \sum_{k=0}^{n} a_k \bar{b}_k\right| \leq \|f - s_n\| \cdot \|g - \sigma_n\| \to 0.$$

8.6.15. From Exercise 8.6.11 it follows that $k(\zeta, \eta) = (k_\eta, k_\zeta)$, where $k_\zeta = k(z, \zeta)$, $k_\eta = k(z, \eta)$; moreover, $a_n = (k_\zeta, \varphi_n)$ and hence by Exercise 8.6.11: $\bar{a}_n = (\varphi_n, k_\zeta) = \varphi_n(\zeta)$; similarly $b_n = (k_\eta, \varphi_n) = \overline{\varphi_n(\eta)}$. Using Exercise 8.6.14 we obtain:

$$(k_\eta, k_\zeta) = \sum_{n=0}^{\infty} \overline{\varphi_n(\eta)} \varphi_n(\zeta) = k(\zeta, \eta).$$

8.6.16. Using polar coordinates we easily verify that $\{\varphi_n\}$ form an orthonormal system; the completeness can be proved by verifying Parseval's identity, the integral $\iint |f|^2$ being expressed by Laurent's coefficients.

8.6.17. $k(z, \zeta) = (2\pi \log(1/h))^{-1}(z\bar{\zeta})^{-1} + \pi^{-1} \sum_{n=-\infty}^{+\infty} (1 - h^{2n})^{-1} n(z\bar{\zeta})^{n-1}$.

8.6.18. Both series $\sum_{n=0}^{\infty} (n+1)^{-1}|b_n|^2$, $\sum_{n=2}^{\infty} (n-1)^{-1}|b_{-n}|^2 h^{-2n}$ should be convergent, cf. Exercise 4.2.5.

8.6.19. If $z = z(t)$, $\zeta = z(\tau)$, then after the change of variables in the formula of Exercise 8.6.11 we obtain:

$$f(z(\tau)) = \iint_{G_1} f(z(t))\overline{k(z(t), z(\tau))}\,|z'(t)|^2\, du\, dv$$

and due to uniqueness of k we obtain: $k_1(t, \tau) = k(z, \zeta)|dz/dt|^2$. Hence $\sqrt{k_1(t, t)}\,|dt| = \sqrt{k(z, z)}\,|dz|$.

CHAPTER 9
Two-Dimensional Vector Fields

9.1.1. $w = \overline{f'(z)}$; $|w| = |f'(z)|$.

9.1.2. $g = u_x - iu_y$, $G = U_x - iU_y$ are analytic in D and we may assume that g and G are not identically 0; then g, G do not vanish on $D\setminus H$, $H \in D$ being an isolated set. If $z \in H \cap \gamma$, then the normal vector of γ has components u_x, u_y, or U_x, U_y, and consequently $\arg((U_x - iU_y)/(u_x - iu_y)) = 0$, hence the quotient G/g must reduce to a constant (which is real).

9.1.3. The lines of flow are circles $\mathrm{im}(i/z) = \mathrm{const}$, hence $f(z) = ki/z$; $w = \overline{f'(z)}$, hence $|w_1|/|w_2| = 1/2$.

9.1.4. The lines of flow can be represented in the form $\mathrm{im}(iz^{-2}) = \mathrm{const}$, hence $f(z) = kiz^{-2}$ (k is a real constant); $|w_1|/|w_2| = 2^{-3/2}$.

9.1.5. Circles $C(0; r)$.

9.1.6. (i) Equipotential lines: straight lines $x = \mathrm{const}$; lines of flow: $y = \mathrm{const}$; velocity: $w = a$;
  (ii) $y = C_1$, $x = C_2$, $w = -ai$;
  (iii) $x^2 + y^2 = C_1 x$, $x^2 + y^2 = C_2 y$, $w = \bar{z}^{-2}$;
  (iv) $|z-b|/|z-c| = C_1$, circles through b, c,
$$\bar{w} = (b-c)[(z-b)(z-c)]^{-1};$$
  (v) $x^2 - y^2 = C_1$, $xy = C_2$, $w = 2\bar{z}$;
  (vi) $x^2 + y^2 = C_1$, $y = C_2 x$, $w = a/\bar{z}$;
  (vii) $y = C_1 x$, $x^2 + y^2 = C_2$, $w = -ai/\bar{z}$.

9.1.7. If w is the velocity and $f = u + iv$ is the complex potential of flow, then $\bar{w} = f' = u_x - iu_y$ and this means that u_x, u_y are components of w. Hence
$$\int_\gamma f'(z)\,dz = \int_\gamma (u_x - iu_y)(dx + i\,dy) = \int_\gamma u_x\,dx + u_y\,dy + i\int_\gamma u_x\,dy - u_y\,dy$$
$$= \int_\gamma w_s\,ds + i\int_\gamma w_n\,ds = \Gamma + iQ.$$

(w_n denotes the normal component of w).

9.1.8. The points $\mp(1\mp i)/\sqrt{2}$ are sources of intensity 2π, the points 0, ∞ are sinks of intensity 4π.

9.1.9. Equipotential lines: equilateral hyperbolas through $\mp a$ with center 0; lines of flow: lemniscates with foci $\mp a$, $\Gamma = -4\pi$.

9.1.10. $Q = 6\pi$, $\Gamma = 0$.

9.1.11. $f(z) = Qi\pi^{-1}\text{Arc}\sin z$; according to Bernoulli law the pressure
$$p = C - \tfrac{1}{2}|w'|^2 = C - Q^2/[2\pi^2(1-x)^2],$$
$C = \text{const} = p_\infty$.

9.1.12. $f(z) = w_\infty(z+z^{-1})$, $w(i) = 2w_\infty$, $w(2) = \tfrac{3}{4}w_\infty$, $p(2)-p(i) = \tfrac{55}{32}w_\infty^2$.

9.1.13. The complex potential of flow f_1 outside the circular cylinder $|\zeta| > R$ has the form $f_1(\zeta) = w(e^{-i\alpha}\zeta + R^2 e^{i\alpha}\zeta^{-1})$, the velocity at ∞ being $w_\infty e^{i\alpha}$. Under the mapping $z = \tfrac{1}{2}c(k\zeta + 1/k\zeta)$ where $kc = 2$, the domain $\{\zeta: |\zeta| > R\}$, $R = \tfrac{1}{2}(a+b)$, is mapped onto the outside of the given ellipse so that $(dz/d\zeta)_\infty = 1$. Hence $f(z) = f_1(\zeta(z))$, where $\zeta(z) = \tfrac{1}{2}(z + \sqrt{z^2-c^2})$.

9.1.14. (i) Map the outside of the wedge, or segment onto the outside of $K(0; R)$ so that $(d\zeta/dz)_\infty = 1$; then $w(z) = f_1(\zeta(z))$ similarly as in Exercise 9.1.13.

9.1.15. $f(z) = (2\pi)^{-1}Q\log(z-a)$.

9.1.16. $f(z) = (2\pi)^{-1}Q\log((z+1)/z)$.

9.1.17. $f(z) = (2\pi)^{-1}Q\log(1+4z^{-4})$.

9.1.18. $f(z) = -p/\pi z$; circles $C(0; r)$.

9.1.19. $f(z) = (2\pi i)^{-1}\Gamma\log(z-a)$.

9.1.20. $f(z) = w_\infty z + (2\pi)^{-1}Q\log(z^2+a^2)$; $w(0) = w_\infty$.

9.1.21. (i) $f(z) = -\pi^{-1}[p(z-ai)^{-1} + p(z+ai)^{-1}] + w_\infty z$;
(ii) $f(z) = (2\pi i)^{-1}\Gamma\log((z-ai)/(z+ai)) + (2\pi)^{-1}Q\log(z^2+a^2) + w_\infty z$.

9.1.22. $f(z) = (2\pi i)^{-1}\Gamma\log\{(z-ai)(z+i/a)[(z+ai)(z-i/a)]^{-1}\}$.

9.1.23. $f(z) = (2\pi)^{-1}Q\log(z+z^{-1}-a-a^{-1})$.

9.1.24. $f(z) = (2\pi i)^{-1}\Gamma\log((z-a)/(z-a^{-1}))$.

9.1.25. We have $w = ie^{i\theta}[(2\pi R)^{-1}\Gamma - 2a\sin\theta]$ at $z = Re^{i\theta}$, hence $C(0; R)$ is a line of flow; moreover, $w(iR) = 0$, if $a = \Gamma/4\pi R = w$; the force $F = -iw_\infty\Gamma$.

9.2.1. Suppose that the charged wire coincides with the OZ axis of the rectangular system $OXYZ$ of coordinates and the unit charge is placed at the point a in the XY-plane. The force acting at a due to two linear charges $q\,dZ$ situated at

$Z, -Z$ has the direction Oa and its magnitude is $2q(Z^2+r^2)^{-1}\cos\alpha\, dZ = 2qr(Z^2+r^2)^{-3/2}dZ$, where $r = |Oa|$ and $\alpha = \measuredangle\,(aOZ)$. Hence the magnitude of the resulting force is equal to $2rq\int_0^{+\infty}(Z^2+r^2)^{-3/2}dZ = 2q/r$.

9.2.2. The components of the force w are equal to

$$2qx(x^2+y^2)^{-1} = \frac{\partial}{\partial x}q\log(x^2+y^2), \quad 2qy(x^2+y^2)^{-1} = \frac{\partial}{\partial y}q\log(x^2+y^2),$$

hence the (real) electrostatic potential $\psi(z) = -2q\log|z|$ and the complex potential $f(z) = -2qi\log z$.

9.2.3. $f(z) = -2iq\log(z+h)(z-h)$; lemniscates with foci $\mp h$.

9.2.4. $f(z) = 2Miz^{-1}$ if the positive charge is placed at $h > 0$; equipotential lines: $x^2+y^2-C_1 x = 0$, lines of force; $x^2+y^2-C_2 y = 0$; $w = -i\overline{f'(z)} = 2M\bar{z}^{-2}$.

9.2.5. Consider a portion Δ of the surface of the conductor corresponding to the arc ds on its intersection with the plane of reference and let S be the closed surface being the boundary of a solid formed by shifting Δ both sides along the normal. According to Gauss theorem:

$$\iint_S w_n\, dS = |\Delta|w_n = 4\pi|\Delta|\sigma,$$

hence

$$\sigma = (4\pi)^{-1}w_n = -(4\pi)^{-1}\partial\psi/\partial n.$$

9.2.6. $q = \int_\Gamma \sigma\, ds = -(4\pi)^{-1}\int(\partial\psi/\partial n)\, ds$.

9.2.7. The total charge on the outer coating

$$q_0 = \int_{\Gamma_0}\sigma_0\, ds = -(4\pi)^{-1}\int_{\Gamma_0}(\partial\psi/\partial n)\, ds,$$

similarly

$$q_1 = -(4\pi)^{-1}\int_{\Gamma_1}(\partial\psi/\partial n)\, ds.$$

The Green's formula with $h = 1$ gives the result.

9.2.8. The electrostatic potential of a condenser has the form

$$\psi(x,y) = \psi(z) = \psi_0 + (\psi_1-\psi_0)\omega(z;\Gamma_1,G),$$

hence

$$c = [4\pi(\psi_1-\psi_0)]^{-1}\int_{\Gamma_1}(\psi_1-\psi_0)(-\partial\omega/\partial n)\, ds = -(4\pi)^{-1}\int_{\Gamma_0+\Gamma_1}\omega(\partial\omega/\partial n)\, ds$$

9. TWO-DIMENSIONAL VECTOR FIELDS

since $\omega = j$ on $\Gamma_j, j = 0, 1$. The application of Green's formula gives the result ($\omega_{xx}+\omega_{yy} = 0$).

9.2.9. After a change of variables:
$$\omega_u^2+\omega_v^2 = (\omega_x^2+\omega_y^2)\,\partial(x,y)/\partial(u,v).$$

9.2.10. If $\psi_0 = 0$, $\psi_1 = 1$, then $f(z) = (i\log(z/R))\colon \log(r/R)$; introduce polar coordinates and apply Exercise 9.2.8.

9.2.11. Suppose that the cross-sections of wires are disks $K(\mp\tfrac{1}{2}b;a)$; after the linear transformation $w = (z-d)(z+d)^{-1}$, where $\mp d = \mp\sqrt{b^2-4a^2}/2$ ($\mp d$ are symmetric w.r.t. either circle) the problem is reduced (cf. Ex. 9.2.9) to Exercise 9.2.10.

9.2.12. If $d = 7-4\sqrt{3}$ and $w = (z+d)(1+zd)^{-1}$, we obtain the concentric cylinder $1 \leqslant |w| \leqslant 2+\sqrt{3}$; hence
$$f(z) = -(i\log(w/(2+\sqrt{3})))\colon \log((2+\sqrt{3})^{-1}), \quad c = \tfrac{1}{2}[\log(2+\sqrt{3})]^{-1}.$$

9.2.13. The complex potential can be evaluated similarly as in Exercise 9.2.12.; $|\sigma| = (4\pi)^{-1}|f'(z)|$, hence
$$(96\pi\log 2)^{-1} \leqslant |\sigma| \leqslant 3(32\pi\log 2)^{-1}.$$

9.2.14. $f(z) = 2i\lambda\log[(z+ih)/(z-ih)]$, $\sigma = \pi^{-1}\lambda h(x^2+h^2)^{-1}$ (z-plane is intersected by both wires at a right angle, the intersection points being $\mp ih$).

9.2.15. If the density of charge per unit length of $\Gamma = \bigcup_{k=1}^{n}\Gamma_k$ is equal to $\sigma(\tau)$, then the real potential
$$\psi(z) = 2\int_\Gamma \sigma(\tau)\log|z-\tau|^{-1}ds_\tau$$
$$= 2\log|z|^{-1}\int_\Gamma \sigma(\tau)\,ds_\tau + 2\int_\Gamma \sigma(\tau)\log|1-\tau/z|^{-1}ds_\tau = 2\lambda\log|z|^{-1}+o(1)$$
as $z \to \infty$. We have, moreover, $g(z)-\log|z| = \gamma+o(1)$ since a bounded harmonic function with an isolated singularity has a harmonic extension to this singularity. We have also
$$2\lambda\log|z|^{-1}+2\lambda g(z)-2\lambda\gamma = o(1), \quad \text{or} \quad \psi(z)+2\lambda g(z)-2\lambda\gamma = o(1).$$
The left-hand side has a constant value on Γ and vanishes at ∞, hence it is identically 0. Therefore $\psi(z) = -2\lambda g(z)+2\lambda\gamma$, which gives $\psi = 2\lambda\gamma$ on Γ.

9.2.16. By Exercise 8.5.7 (iii),
$$g(z) = \log|z|-\log(a+b)+\log|1+\sqrt{1-(a^2-b^2)/z^2}|,$$
hence $g(z)-\log|z|+o(1) = -\log[(a+b)/2] = \gamma$.

CHAPTER 10

Univalent Functions

10.1.1. If $\int_a^b f\,dg = 0$, the inequality is obvious. If $\int_a^b f\,dg \neq 0$, take real α such that $e^{i\alpha}\int_a^b f\,dg = |\int_a^b f\,dg|$. It is easy to verify that $c\int_a^b f\,dg = \int_a^b cf\,dg$ for any complex c, hence

$$\left|\int_a^b f\,dg\right| = e^{i\alpha}\int_a^b f\,dg = \int_a^b e^{i\alpha}f\,dg$$

$$= \mathrm{re}\left\{\int_a^b e^{i\alpha}f\,dg\right\} = \int_a^b \mathrm{re}[e^{i\alpha}f]\,dg \leqslant V_a^b(g)\max_t |f|.$$

10.1.2. Obviously f is continuous, hence by a theorem of Helly (cf. [23], p. 252): $\int_a^b f\,dg_n \to \int_a^b f\,dg$. Moreover,

$$\left|\int_a^b (f_n-f)\,dg_n\right| \leqslant \max|f_n-f| \cdot K \to 0$$

as $n \to +\infty$, if $V_a^b(g_n) \leqslant K$ for all n. Therefore

$$\left|\int_a^b f_n\,dg_n - \int_a^b f\,dg\right| \leqslant \left|\int_a^b (f_n-f)\,dg_n\right| + \left|\int_a^b f\,dg_n - \int_a^b f\,dg\right| \to 0.$$

10.1.3. $\mu_n(t) = (2\pi)^{-1}\int_0^t u(R_n e^{i\theta})\,d\theta$ is a continuously differentiable function of $t \in [0, 2\pi]$, hence for any $z \in K(0; R_n)$

$$f(z) = \frac{1}{2\pi}\int_0^{2\pi} \frac{R_n e^{i\theta}+z}{R_n e^{i\theta}-z} u(R_n e^{i\theta})\,d\theta = \int_0^{2\pi} \frac{R_n e^{i\theta}+z}{R_n e^{i\theta}-z}\,d\mu_n(t).$$

Since $\mu_n' > 0$, μ_n is strictly increasing for any fixed n, moreover $\mu_n(0) = 0$, $\mu_n(2\pi) = u(0) = 1$. By Helly's selection principle (cf. [23], p. 241), we may choose a subsequence $\{\mu_{n_k}\}$ convergent in $[0, 2\pi]$ to μ which satisfies $\mu(0) = 0$, $\mu(2\pi) = 1$.

10. UNIVALENT FUNCTIONS

Obviously $(R_n e^{it}+z)/(R_n e^{it}-z)$ tends uniformly in $[0, t]$ to $(Re^{it}+z)/(Re^{it}-z)$ for any fixed $z \in K(0; R)$. Apply now Exercise 10.1.2.

10.1.4. The linear transformation $w = (1+z)/(1-z)$ maps in a univalent manner $K(0; 1)$ onto $\{w: \operatorname{re} w > 0\}$, and the image of $C(0; 1)$ is the imaginary axis. Hence $u(e^{i\theta}) \equiv 0$ and the Schwarz formula gives 0. We can take $\mu(0) = 0$, $\mu(t) = 1$ for $t \in (0, 2\pi]$, $R = 1$, in Herglotz formula.

10.1.5. If $w \in \operatorname{conv} \Gamma$, H being continuous we can find $t_k \in [a, b]$, $k = 1, 2$, such that $w = \lambda H(t_1)+(1-\lambda)H(t_2)$, $0 \leq \lambda \leq 1$, $t_1 < t_2$. Take now $\mu(t)$ as a step function with jumps λ, $1-\lambda$ at t_1, t_2 resp., which gives $w = \int_a^b H(t)\,d\mu(t)$. Conversely, each integral sum of $\int_a^b H(t)\,dt$ has the form: $w_n = \sum_{k=1}^n H(t_k) \Delta\mu_k$ with $\Delta\mu_k \geq 0$, $\sum_{k=1}^n \Delta\mu_k = 1$, hence $w_n \in \operatorname{conv} H$ and $\operatorname{conv} H$ being closed the result follows.

10.1.6. $c_n = \int_0^{2\pi} 2e^{-int}\,d\mu(t)$; the circle $C(0; 2)$ is the curve Γ of Exercise 10.1.5 for any positive integer n.

10.1.7. The curve Γ of Exercise 10.1.5 is the circle $(1+ze^{-it})/(1-ze^{-it})$, $0 \leq t \leq 2\pi$, whose diameter is the segment $[(1-r)(1+r)^{-1}, (1+r)(1-r)^{-1}]$. Therefore $\operatorname{conv} \Gamma = \bar{K}(z_0; \rho)$, where $z_0 = (1+r^2)(1-r^2)^{-1}$, $\rho = 2r(1-r^2)^{-1}$, $r = |z|$.

10.1.8. $\varphi(z)$ has the form: $(w-w_0)/(w+\bar{w}_0)$, where $\operatorname{re} w > 0$, $\operatorname{re} w_0 > 0$, hence $|\varphi(z)| < 1$. Moreover, $\varphi(0) = 0$, hence by Schwarz lemma $|\varphi'(0)| \leq 1$ which gives

$$(1-|z_0|^2)|f'(z_0)| \leq f(z_0)+\overline{f(z_0)} = 2\operatorname{re} f(z_0) = 2|f(z_0)|\cos\alpha.$$

10.1.9. We have:

$$\frac{d}{dr}\log|f(re^{i\theta})| = |f(re)^{i\theta}|^{-1} \cdot \frac{d}{dr}|f(re^{i\theta})|$$

$$\leq |f(re^{i\theta})|^{-1}|f'(re^{i\theta})| \leq \frac{d}{dr}\log\psi(r),$$

hence $\dfrac{d}{dr}\log\{|f(re^{i\theta})|/\psi(r)\} \leq 0$. A nonnegative, decreasing function has always a finite limit as $r \to R-$.

10.1.10. By the inequality of Exercise 10.1.8:

$$|f'(re^{i\theta})|/|f(re^{i\theta})| \leqslant \frac{d}{dr} \log[(1+r)(1-r)^{-1}],$$

hence

$$\lim_{r \to 1-} |f(re^{i\theta})|(1-r)(1+r)^{-1} = \tfrac{1}{2} \lim_{r \to 1-} (1-r)|f(re^{i\theta})|$$

exists for any real θ.

10.1.11. Suppose that $\varepsilon > 0$ is arbitrary. Choose $\delta > 0$ such that $\mu(\theta_0+\delta) - \mu(\theta_0-\delta) < \tfrac{1}{4}\varepsilon$. Split now the integral I of the Herglotz formula into 3 integrals: $\int_0^{\theta_0-\delta}, \int_{\theta_0+\delta}^{2\pi}, \int_{\theta_0-\delta}^{\theta_0+\delta}$; the integrand in two initial terms is $O(1)$ as $z \to e^{i\theta_0}$, hence both integrals multiplied by $1-r$ tend to 0 as $r \to 1-$. The third integral I_3 has for $z = re^{i\theta_0}$ an upper estimate $2(1-r)^{-1} V_{\theta_0-\delta}^{\theta_0+\delta}(\mu) = \tfrac{1}{2}\varepsilon(1-r)^{-1}$, i.e. $(1-r)I_3 < \tfrac{1}{2}\varepsilon$. Finally $(1-r)|I| < \varepsilon$ for all r sufficiently close to 1.

10.1.12. Suppose that $h(t) = \mu(\theta_0-)$ for $0 \leqslant t < \theta_0$, and $h(t) = \mu(\theta_0+)$ for $\theta_0 < t \leqslant 2\pi$ and $h(\theta_0) = \mu(\theta_0)$. Obviously $\gamma(t) = \mu(t)-h(t)$ is continuous at $t = \theta_0$ and increasing in $[0, 2\pi]$. Hence

$$f(z) = \int_0^{2\pi} (e^{it}+z)(e^{it}-z)^{-1} d\gamma(t) + h(e^{i\theta_0}+z)(e^{i\theta_0}-z)^{-1}.$$

Similarly as in Exercise 10.1.11 we show that

$$(1-r) \int_0^{2\pi} (e^{it}+re^{i\theta_0})(e^{it}-re^{i\theta_0})^{-1} d\gamma(t) \to 0 \quad \text{as} \quad r \to 1-$$

and clearly

$$\lim_{r \to 1-} (1-r)h(e^{i\theta_0}+re^{i\theta_0})(e^{i\theta_0}-re^{i\theta_0})^{-1} = 2h.$$

10.1.13. If μ is continuous at $2\pi-\alpha$ and $\nu(\varphi) = \mu(\varphi+2\pi)-\mu(2\pi-\alpha)$ for $-\alpha \leqslant \varphi \leqslant 0$ and $\nu(\varphi) = \mu(\varphi)+1-\mu(2\pi-\alpha)$ for $0 < \varphi \leqslant 2\pi-\alpha$, then $\nu(-\alpha) = 0$, $\nu(2\pi-\alpha) = 1$, moreover

$$f(z) = \int_{-\alpha}^{-\alpha+2\pi} (e^{i\varphi}+z)(e^{i\varphi}-z)^{-1} d\nu(\varphi),$$

hence

$$\lim_{r \to 1-} (1-r)f(r) = 2[\nu(0+)-\nu(0-)] = 2[\mu(0+)+1-\mu(2\pi-)].$$

10.1.14. The function $\mu(t)$ is increasing and hence it has at most a countable number of discontinuities θ_k ($\neq 0, 2\pi$) with jumps h_k; we have $\alpha(\theta_k) = 2h_k$, whereas for $\theta_0 = 0, 2\pi$ we have

$$\alpha(\theta_0) = 2[\mu(0+)+1-\mu(2\pi-)].$$

Hence

$$\sum_{k=0}^{\infty} \alpha(\theta_k) = 2\left[\sum_{k=1}^{\infty} h_k + \mu(2\pi) - \mu(2\pi-) + \mu(0+) - \mu(0)\right] \leqslant 2.$$

For $\theta \neq \theta_k$ we have $\alpha(\theta) = 0$ by Exercise 10.1.11.

10.2.1. $\log f(z) = \log|f(z)| + i \arg f(z)$ is analytic in the annulus slit along a radial segment. If $z = re^{i\theta}$, then after differentiating both sides w.r.t. θ we obtain:

$$ire^{i\theta}f'/f = \frac{\partial}{\partial \theta}\log|f| + i\frac{\partial}{\partial \theta}\arg f;$$

now compare the imaginary parts of both sides.

10.2.2. If $\gamma(r)$ is the image curve of $C(0; r)$, $0 < r < R$, and $p(z) = zf'(z)/f(z)$, then the index

$$n(\gamma(r), 0) = (2\pi i)^{-1} \int_{\gamma(r)} w^{-1}dw = (2\pi i)^{-1} \int_{C(0;r)} z^{-1}p(z)dz = 1.$$

Hence $\Phi = \arg f(re^{i\theta})$ is a strictly increasing function of θ which increases by 2π as θ ranges over $[0, 2]$ and therefore $\gamma(r)$ has a representation in polar coordinates: $R = R(\Phi)$, $\alpha \leqslant \Phi \leqslant \alpha + 2\pi$. If $w_0 = R_0 e^{i\Phi_0}$ is not on $\gamma(r)$ and $R_0 < R(\Phi_0)$, then the segment $[0, w_0]$ does not meet $\gamma(r)$ and consequently $n(\gamma(r), w_0) = n(\gamma(r), 0) = 1$ which means (argument principle) that the equation $f(z) = w_0$ has 1 root in $K(0; r)$. Similarly, if $R_0 > R(\Phi_0)$ then the ray with origin at w_0 being the prolongation of $[0, w_0]$ does not meet $\gamma(r)$, hence $n(\gamma(r), w_0) = n(\gamma(r), \infty) = 0$ which means that the equation $f(z) = w_0$ has no roots in $K(0; r)$; f being univalent in $K(0; r)$ for any $r < R$ is univalent in $K(0; R)$.

10.2.3. (i) We have $\mathrm{re}[zf_1'/f_1] = \mathrm{re}[(1+2z)/(1-z)] > 0$ in $K(0; \frac{1}{2})$. In fact, the image curves of $C(0; r)$ under $(1+2z)/(1-z)$ are Apollonius circles with limit points $-2, 1$ situated in the right half-plane for $r < \frac{1}{2}$;

(ii) $\mathrm{re}(zf'/f) > \frac{1}{2}(2-\alpha)$ in $K(0; 1)$.

10.2.4. From the solution of Exercise 10.2.2 it follows that the conditions are sufficient for D_f to be starlike. We now prove the converse. Suppose D_f is starlike, f being univalent with $f(0) = 0$, $f'(0) = 1$. We first prove that also $f[K(0; r)]$ is starshaped for any $r \in (0, 1)$. Obviously $\omega(z) = f^{-1}(tf(z))$ satisfies the conditions of Schwarz's lemma for any $0 < t < 1$. Hence $|\omega(z)| \leqslant |z|$. If $w_1 \in f[K(0;r)]$ and $w_1 = f(z_1)$, $|z_1| < r$, then $|\omega(z_1)| = |f^{-1}(tw_1)| \leqslant |z_1| < r$. Therefore $tw_1 = f(z_2)$ with $z_2 \in K(0;r)$ which means that $f[K(0;r)]$ is starshaped. Hence $f[C(0;r)]$ is a curve starlike w.r.t. the origin and so $\arg f(re^{i\theta})$ increases with θ which yields $\operatorname{re}[zf'(z)/f(z)] \geqslant 0$ as in Exercise 10.2.1. The case $\operatorname{re}[zf'(z)/f(z)] = 0$ in $K(0;1)$ is excluded by the maximum principle.

10.2.5. $f'(z)/f(z) - z^{-1} = z^{-1}[p(z)-1] = 2\int_0^{2\pi} (e^{i\theta}-z)^{-1}d\mu(\theta)$ and the integration gives

$$f(z) = z\exp\left\{-2\int_0^{2\pi} \operatorname{Log}(1-ze^{-i\theta})d\mu(\theta)\right\}.$$

10.2.6. $|f(z)/z| \leqslant \exp\{2\log(1-|z|)^{-1}\} = (1-|z|)^{-2}$, hence $|f(z)| \leqslant |z|(1-|z|)^{-2}$ and similarly $(1+|z|)^{-2}|z| \leqslant |f(z)|$. Hence using the estimates

$$(1-|z|)(1+|z|)^{-1} \leqslant |zf'(z)/f(z)| \leqslant (1+|z|)(1-|z|)^{-1},$$

cf. Exercise 10.1.7, the estimates of f' easily follow:

$$(1-|z|)(1+|z|)^{-3} \leqslant |f'(z)| \leqslant (1+|z|)(1-|z|)^{-3}.$$

10.2.7. $\operatorname{Log}[z/f(z)]^{1/2} = \int_0^{2\pi} \operatorname{Log}(1-ze^{-i\theta})d\mu(\theta)$ by Exercise 10.2.5. Using Exercise 10.1.5 and Exercise 2.6.8 we see that $\operatorname{Log}[z/f(z)]^{1/2}$ ranges over a convex domain being the image of $\overline{K}(1;r)$ under Log. Hence $[z/f(z)]^{1/2}$ ranges over $\overline{K}(1;r)$ for varying $f \in S^*$ and fixed z, $|z| = r$.

10.2.8. If f, z range over S^* and $K(0;1)$ resp., then $[z/f(z)]^{1/2}$ ranges over $\bigcup_{r\in(0,1)} \overline{K}(1;r) = K(1;1)$, i.e. $[f(z)/z]^{1/2}$ ranges over $\{w: \operatorname{re} w > \tfrac{1}{2}\}$.

10.2.9. Suppose that $|z_1| \leqslant |z_2| < r$ and $w_k = f(z_k)$, $k = 1, 2$; by Schwarz's lemma $|\psi(z)| \leqslant |z|$ and for $z = z_2$ we have

$$|f^{-1}[tw_1+(1-t)w_2]| \leqslant |z_2| < r;$$

if $z_0 = f^{-1}[tw_1+(1-t)w_2]$, then $|z_0| < r$ and $f(z_0) = tw_1+(1-t)w_2$ which means that B_r is convex since $t \in (0,1)$ can be arbitrary.

10.2.10. If $zf' \in S^*$, then the tangent vector of $f[C(0;r)]$ turns monotonically and its argument increases by 2π as z describes $C(0;r)$. Hence $f[C(0;r)]$ is a convex curve whose interior domain is convex. A converse can be proved in an analogous manner.

10.2.11. $f' = F/z$ with $F \in S^*$ (cf. Ex. 10.2.8).

10.2.12. We have $\operatorname{re} F(z) > 0$ for $z \in K(0; 1)$ and also for $z \in [z_1, z_2]$, hence

$$0 < \int_0^1 \operatorname{re} F[z_1 + t(z_2 - z_1)] dt = \operatorname{re}\left\{(z_2 - z_1)^{-1} \int_{z_1}^{z_2} (1+z)(1-z)^{-1} dz\right\}$$
$$= \operatorname{re}\{2(z_2 - z_1)^{-1} \operatorname{Log}[(1-z_1)(1-z_2)^{-1}] - 1\}.$$

10.2.13. $p(z) = 2\sqrt{f'(z)} - 1 \in \mathscr{P}$ by Exercise 10.2.11. Hence by Exercise 10.1.3:

$$p(z) = \int_0^{2\pi} (e^{i\theta} + z)/(e^{i\theta} - z) d\mu(\theta)$$

and this implies

$$f'(z) = \int_0^{2\pi}\int_0^{2\pi} e^{i\varphi} e^{i\psi} (e^{i\varphi} - z)^{-1}(e^{i\psi} - z)^{-1} d\mu(\varphi) d\mu(\psi)$$

(where μ is increasing, $\mu(0) = 0$, $\mu(2\pi) = 1$). Integrating w.r.t. z we obtain:

$$z^{-1} f(z) = \int_0^{2\pi}\int_0^{2\pi} (ze^{-i\psi} - ze^{-i\varphi})^{-1} \operatorname{Log} \frac{1 - e^{-i\varphi} z}{1 - e^{-i\psi} z} d\mu(\varphi) d\mu(\psi)$$

and hence by Exercise 10.2.12: $\operatorname{re}[z^{-1} f(z)] > \tfrac{1}{2}$.

10.2.14. From Exercise 10.2.13 it follows that $|z/f(z) - 1| < 1$. Now apply Schwarz's lemma.

10.2.15. $g \in S^c$, hence

$$\operatorname{re}\{\zeta f'(\zeta)/f(\zeta)\} = \operatorname{re}\{(1-|\zeta|^2)^{-1} \zeta/g(\zeta)\} \geq (1-|\zeta|^2)^{-1}(1-|\zeta|) = (1+|\zeta|)^{-1} > \tfrac{1}{2}$$

since $\operatorname{re}\{\zeta/g(\zeta)\} \geq 1 - |\zeta|$ (cf. Ex. 10.2.14).

10.2.16. By Exercise 10.2.15 $zf'/f \prec (1-z)^{-1}$, hence zf'/f takes inside $\overline{K}(0; r)$ the values contained in the disk with diameter $[1-r, 1+r]$.

10.2.17. By Exercise 10.2.13 $z^{-1} f \prec (1-z)^{-1}$, hence the estimates are the same as in Exercise 10.2.16.

10.3.1. Consider any partition of $[0, 2\pi]$ containing all the end points of monotoneity intervals of $\Phi(\theta)$ and consider a corresponding integral sum for the Stieltjes integral $\int_0^{2\pi} h(R(\theta)) d\Phi(\theta)$. Any ray $\Phi = \varphi_k$ situated inside an angle obtained by the partition intersects Γ at an odd number of points: $R_1^{(k)} > R_2^{(k)} > \ldots > R_{2n_k+1}^{(k)}$ and if $\Delta\Phi$ is the measure of the corresponding angle, then the integral sum is equal to $\sum_k \Delta\Phi_k[h(R_1^{(k)}) - h(R_2^{(k)}) + \ldots + h(R_{2n_k+1}^{(k)})] > 0$ because the orientation is positive. Take now the limit for a normal sequence of partitions.

10.3.2. We have:

$$r \frac{d}{dr} \int_0^{2\pi} g(R) d\theta = r \int_0^{2\pi} \frac{\partial}{\partial r} g(R) d\theta = r \int_0^{2\pi} g'(R) R \frac{\partial}{\partial r} \log R \, d\theta$$

$$= \int_0^{2\pi} g'(R) R \frac{\partial \Phi}{\partial \theta} d\theta = \int_0^{2\pi} R g'(R) d_\theta \Phi(r, \theta).$$

10.3.3. If $g(R) = R^2$, then by Exercise 10.3.2 we obtain

$$r \frac{d}{dr} \int_0^{2\pi} |f(re^{i\theta})|^2 d\theta = r \frac{d}{dr} \int_0^{2\pi} R^2 d\theta = \int_0^{2\pi} 2R^2(r, \theta) d_\theta \Phi(r, \theta)$$

which is $\geqslant 0$ in view of Exercise 10.3.1 ($h(R) = 2R^2$).

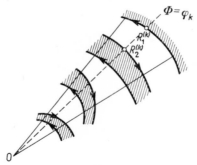

FIG. 7

10.3.4. The values of F do not cover $\mathbf{C}$ completely (Liouville's theorem); if $F(z) \neq w_0$ for any $z \in K(\infty; 1)$, then by the argument principle the images of $C(0; r)$ by $F(z) - w_0$ contain $w = 0$ inside, hence

$$I(r) = \int_0^{2\pi} |F(re^{i\theta}) - w_0|^2 d\theta = 2\pi[r^2 + |b_0 - w_0|^2 + |b_1|^2 r^{-2} + |b_2|^2 r^{-4} + \ldots]$$

is an increasing function of r and consequently $I'(r) \geqslant 0$.

10.3.5. $\sum_{n=1}^{N} n|b_n|^2 r^{-2n-1} \leqslant r$; make first $r \to 1$ which gives $\sum_{n=1}^{N} n|b_n|^2 \leqslant 1$ and then make $N \to +\infty$.

10.3.6. If $f \in S$ and $|z| > 1$, then $F(z) = 1/f(z^{-1})$ belongs to Σ_0. If $F \in \Sigma_0$ and $|\zeta| < 1$, then $f(\zeta) = 1/F(\zeta^{-1})$ belongs to S. The function $F(z) = z(1+z^{-3})^{2/3}$ corresponds in this way to a function $f \in S^*$.

10.3.7. $G(z) = [F(z^2)]^{1/2} = z + \frac{1}{2}b_0 z^{-1} + \ldots \in \Sigma_0$, hence by Exercise 10.3.5 we have: $\frac{1}{2}|b_0| \leqslant 1$. The case of equality occurs only for $G(z) = z + e^{i\alpha}z^{-1}$, or $F(z) = z + 2e^{i\alpha} + e^{2i\alpha}z^{-1}$, with α real.

10.3.8. $F - w_k \in \Sigma_0$, hence $|w_k - b_0| \leqslant 2$ $(k = 1, 2)$ and
$$|w_1 - w_2| \leqslant |w_1 - b_0| + |b_0 - w_2| \leqslant 4.$$

10.3.9. If $f \in S$, then $F(\zeta) = 1/f(\zeta^{-1}) = \zeta - a_2 + \ldots \in \Sigma_0$ so that $|a_2| \leqslant 2$ (cf. Ex. 10.3.7). Equality holds only for $F(\zeta) = \zeta + 2e^{i\alpha} + e^{2i\alpha}\zeta^{-1}$, i.e. for $f(z) = f_\alpha(z) = z(1 + e^{i\alpha}z)^{-2}$.

10.3.10. φ is univalent and analytic in $K(0; 1)$ being a superposition of a linear transformation and $f \in S$; moreover, $\varphi(z) = z + (a_2 + h^{-1})z^2 + \ldots$, hence $|a_2 + h^{-1}| \leqslant 2$, i.e. $|h^{-1}| \leqslant 2 + |a_2| \leqslant 4$. Equality holds only, if $|a_2| = 2$ which means that $f = f_\alpha$.

10.3.11. φ is univalent in $K(0; 1)$, moreover, $\varphi(0) = 0$, $\varphi'(0) = 1$. Hence $\varphi \in S$. Now, $z(t) \neq b$ and this implies that $\varphi(t) \neq [f(b) - f(a)]t'(a)/f'(a) = h$ for any $t \in K(0; 1)$; $t(z)$ denotes here the inverse of $z(t)$. The result follows immediately from Exercise 2.9.21 and Exercise 10.3.10.

10.3.12. (i) Take $b = 0$ which gives
$$|af'(a)/f(a)| \leqslant (1+|a|)(1-|a|)^{-1};$$
(ii) on integrating both sides in (i) we obtain: $|f(a)| \leqslant |a|(1-|a|)^{-2}$; taking $a = 0$ in Exercise 10.3.11 we obtain $|b|(1+|b|)^{-2} \leqslant |f(b)|$;
(iii) $\psi \in S$, hence $|\psi(t)| \leqslant |t|(1-|t|)^{-2}$ by (ii) and putting $t = -z$ we obtain
$$|zf'(z)/f(z)| \geqslant (1-|z|)(1+|z|)^{-1}.$$
Equality can be attained in all cases for $f = f_\alpha$.

10.3.13. Multiply both estimates for $|f(z)|$ and $|zf'(z)/f(z)|$ side by side.

10.3.14. We have
$$f'(b) = \varphi'(0)(1-|b|^2)^{-1}, \quad f'(a) = \varphi'(\zeta_1)(1-|b|^2)(1-a\bar{b})^{-2}$$
with $\zeta_1 = (a-b)(1-a\bar{b})^{-1}$. Hence
$$f'(a)/f'(b) = (1-|b|^2)^2(1-a\bar{b})^{-2}\varphi'(\zeta_1)/\varphi'(0).$$
Now, $[\varphi(\zeta) - \varphi(0)]/\varphi'(0) = \psi(\zeta) \in S$, thus
$$|\psi'(\zeta_1)| \leqslant (1+|\zeta_1|)(1-|\zeta_1|)^{-3}.$$

and finally
$$\left|\frac{f'(a)}{f'(b)}\right| \leqslant \frac{(1-|b|^2)}{(1-|a|^2)}\left(\frac{|1-a\bar{b}|+|a-b|}{|1-a\bar{b}|-|a-b|}\right)^2$$
(cf. Ex. 1.1.8 (iii)). The lower estimate is obtained by interchanging a, b.

10.3.16. (i) $(|z|-1)^2|z|^{-1} \leqslant |F(z)| \leqslant (|z|+1)^2|z|^{-1}$;

(ii) $(|z|-1)(|z|+1)^{-1} \leqslant |zF'(z)/F(z)| \leqslant (|z|+1)(|z|-1)^{-1}$, which is easily obtained by using Exercise 10.3.12 and Exercise 10.3.6.

10.3.17. (i) Take a sequence $\{z_n\}$ such that $|z_n| \to 1$, $F(z_n)$ is convergent and
$$\lim|F(z_n)-b_0| = \varlimsup_{|z|\to 1+}|F(z)-b_0|.$$
Then $w_0 = \lim F(z_n)$ is not a value of F, hence $F-w_0 \in \Sigma_0$. Now, by Exercise 10.3.7, $|b_0-w_0| \leqslant 2$.

(ii) $\varlimsup_{|z|\to 1+}|z(F(z)-z-b_0)|^2 \leqslant \varlimsup_{|z|\to 1+}|F(z)-b_0|+1 \leqslant 3$, moreover, $|z(F(z)-z-b_0)|$ $\to |b_1| \leqslant 1$ as $z \to +\infty$, hence $|z(F(z)-z-b_0)| \leqslant 3$ by the maximum principle.

10.3.18. (i) $|F'(z)-1| \leqslant |z|^-(1\cdot|b_1|+\sqrt{2}|z|^{-1}\sqrt{2}|b_2|+ ...)$
$$\leqslant |z|^{-2}\sqrt{(1-|z|^{-2})^{-2}}\sqrt{|b_1|^2+2|b_2|^2+...} \leqslant (|z|^2-1)^{-1}$$
by Schwarz's inequality and the area theorem.

(ii) $|F'(z)| \leqslant |F'(z)-1|+1 \leqslant |z|^2(|z|^2-1)^{-1}$.

10.3.19. $F(z)$ is a superposition of $a(\zeta+\zeta^{-1})+b$ and $\zeta = i(h^2-1)^{-1/2}(hz-1)$ hence it must be univalent. The latter transformation carries $C(0;1)$ into a circle through ∓ 1 and by Exercise 2.5.2: $\mathbf{C}\setminus F[K(\infty;1)]$ is a circular arc.

10.3.20. (ii) $g \in S$ hence in view of Exercise 10.3.12 (ii) we have
$$|f(z)||1+e^{i\alpha}f(z)/M| \leqslant |z|(1-|z|)^{-2}$$
for an arbitrary real α. Thus
$$|f(z)|(1-|f(z)|/M)^{-2} \leqslant |z|(1-|z|)^{-2} = |w|(1-|w|/M)^{-2},$$
where $w = f_M(z)$. Now, $u(1-u/M)^{-2}$ is an increasing function of $u \in (0, M)$, hence $|f(z)| \leqslant |w| = |f_M(z)|$. A similar proof for the lower estimate.

(i) $M(|z|,f) = |z|+|a_2||z|^2+O(|z|)^3 \leqslant M(|z|,f_M)$
$$= |z|+2(1-M^{-1})|z|^2+O(|z|^3), \text{ hence } |a_2| \leqslant 2(1-M^{-1}).$$

10.3.21. $g(z) = z+\bar{\zeta}-\tfrac{1}{2}(1-|\zeta|^2)f''(\zeta)/f'(\zeta)-\tfrac{1}{6}(1-|\zeta|^2)^2\{f,\zeta\}z^{-1}+ ...$ Now, by the area theorem (Ex. 10.3.5) the coefficient of z^{-1} is at most 1 in absolute value, hence $|\{f, \zeta\}| \leq 6(1-|\zeta|^2)^{-2}$. If h is analytic and univalent in $K(0; 1)$, then $f = ah+b \in S$ for suitably chosen constants a, b and $\{h, z\} = \{f, z\}$ by Exercise 7.2.14. If h is meromorphic and univalent in $K(0; 1)$ and $h(z) \neq w_0$ for any $z \in K(0; 1)$, then $H = (h-w_0)^{-1}$ is analytic and univalent and again $\{H, z\} = \{h, z\}$.

10.4.1. The proofs for Steiner and Pólya symmetrization are similar. For sake of simplicity we give here the proof for Steiner symmetrization Suppose that $\sigma(x_1)$, $\sigma(x_2)$ meet G^* and also G. Then $\sigma(x)$ meets G for any $x \in [x_1, x_2]$ since in the opposite case G would be disconnected. This implies that $[x_1, x_2] \subset G^*$ and obviously two arbitrary points $z_1, z_2 \in G^*$ with $\operatorname{re} z_1 = x_1$, $\operatorname{re} z_2 = x_2$ can be joined by the polygonal line $[z_1, x_1, x_2, z_2]$ in G^*. Hence G^* is arc-wise connected, if G is connected Evidently $\mathbf{C} \setminus G^*$ is also simply connected in this case. Suppose now G is an open set and $z_0 = x_0 + iy_0 \in G^*$. It is easily verified that the linear measure $l(x)$ of $\sigma(x) \cap G$ is a lower semicontinuous function of x for open G (note that $\sigma(x) \cap G$ is an at most countable system of open segments: replace the countable system by a finite system of closed sub-segments with a slightly less total measure and use Heine–Borel theorem for the latter system). Now, in our case $l(x_0) > 2|y_0|$. From lower semicontinuity it follows that for any $y_1 > 0$ such that $l(x_0) > 2y_1 > 2|y_0|$ we can choose δ such that $\frac{1}{2}l(x) > y_1$ for any $x \in (x_0-\delta, x_0+\delta)$. Then the open rectangle $\{x+iy: |x-x_0| < \delta, |y| < y_1\} \subset G^*$ which proves that G^* is open. In case of Steiner symmetrization G^* is always simply connected if G is a domain. In case of Pólya symmetrization we should make an additional hypothesis that G is a simply connected domain.

10.4.2. If G is a domain starshaped w.r.t. the origin 0 and $w_0 \in \mathbf{C} \setminus G$, then the whole ray emanating from w_0 whose prolongation contains 0 belongs to $\mathbf{C} \setminus G$. This implies that the function $l(r)$ introduced in the definition is decreasing which again implies that G^* is starshaped w.r.t. the origin.

10.4.3. Suppose G is a convex domain. It is sufficient to consider bounded, convex domains since each convex domain is a sum of an increasing sequence of bounded convex domains e.g. $G \cap K(0; n)$. Now the boundary of a bounded convex domain consists of the graphs of 2 functions $\varphi_1, \varphi_2, a \leq x \leq b$, and possibly two segments $\operatorname{re} z = a$, $\operatorname{re} z = b$, the functions $\varphi_1, -\varphi_2$ being convex. This implies that $\frac{1}{2}(\varphi_1 - \varphi_2) = l(x)$ is convex and non-positive, hence G^* being bounded by two convex arcs and possibly two segments must necessarily be convex.

$$D = [S \cap K(0; 2)] \cup [S \cap K(0; 4) \cap \{z: \operatorname{re} z > 0\}],$$

where $S = \{z: |\operatorname{lim} z| < 1\}$, is convex, however D^* is not convex.

10.4.4. If $f(z) = a_0 + a_1 z + a_2 z^2 + \ldots$, then

$$\iint_{K(0;1)} |f'(re^{i\theta})|^2 r\, dr\, d\theta = \pi \sum_{n=1}^{\infty} n|a_n|^2 < +\infty.$$

Suppose that $\varepsilon > 0$ is arbitrary and choose $m \geq 2$ such that $\sum_{n=m}^{\infty} n|a_n|^2 < \tfrac{1}{4}\varepsilon^2$. Thus

$$\sum_{n=m}^{\infty} n|a_n| r^{n-1} = \sum_{n=m}^{\infty} \sqrt{n}|a_n| \cdot \sqrt{n} r^{n-1}$$

$$\leq \left(\sum_{n=m}^{\infty} n|a_n|^2\right)^{1/2} \left(\sum_{n=m}^{\infty} n r^{2n-2}\right)^{1/2} < \tfrac{1}{2}\varepsilon(1-r^2)^{-1}$$

and therefore

$$(1-r)\sum_{n=m}^{\infty} n|a_n| r^{n-1} \leq \tfrac{1}{2}\varepsilon(1+r)^{-1} < \tfrac{1}{2}\varepsilon$$

for any $r \in (0, 1)$, if $m = m(\varepsilon)$ is large enough. Choose now r_0, m being fixed as before, such that $(1-r)\sum_{n=1}^{m} n|a_n| < \tfrac{1}{2}\varepsilon$ for all $r \in (r_0, 1)$. Then $(1-r)\sum_{n=1}^{m} n|a_n| r^{n-1} < \tfrac{1}{2}\varepsilon$ and also

$$(1-r)M(r, f') \leq (1-r)\sum_{n=1}^{\infty} n|a_n| r^{n-1} < 2\varepsilon/2 = \varepsilon$$

for all $r \in (r_0, 1)$ which proves our result.

10.4.5. By Exercise 8.1.9: $r(w, G) = (1-|z|^2)|f'(z)|$ where f is univalent in $K(0; 1)$, $w = f(z)$, and $f[K(0;1)] = G$. Hence $r(w; G)$ is continuous. Note that the area $|G|$ is finite and use Exercise 10.4.4.

10.4.6. (i) In view of the maximum principle we need only to estimate the difference $h(z, z_0) - h(z, z_1)$ on the boundary of G. We have for $\zeta \in \operatorname{fr} G$:

$$h(z, z_0) - h(z, z_1) < \log[|\zeta - z_0|/|\zeta - z_1|] = \log|1 + (z_1 - z_0)(\zeta - z_1)^{-1}|$$

$$\leq |(z_1 - z_0)(\zeta - z_1)^{-1}| \leq |z_1 - z_0|/\delta;$$

after interchanging z_0, z_1 we obtain $|h(z, z_0) - h(z, z_1)| < |z_1 - z_0|/\delta$ which gives (i).

The inequality $\log|1+z| \leq |z| = \log e^{|z|}$ is equivalent to another, quite obvious inequality:

$$|1+z| \leq 1 + |z| \leq 1 + |z| + |z|^2/2! + \ldots = e^{|z|}.$$

(ii) the inner radius $r(z_0; G)$ satisfies the equality: $\log r(z_0; G) = h(z_0, z_0)$, hence

$$|\log r(z_0; G) - \log r(z_1; G)| = |h(z_0, z_0) - h(z_0, z_1) + h(z_1, z_0) - h(z_1, z_1)|$$
$$< 2|z_1 - z_0|/\delta$$

in view of (i). We used the symmetry of h which is equivalent to the symmetry of the Green function.

10.4.7. Suppose that G is the union $\bigcup_{n=1}^{\infty} G_n$ of an increasing sequence of bounded domains each being regular with respect to the Dirichlet problem. If for some m:

$$\text{dist}(z'; \text{fr}\, G_m) \geqslant \delta, \quad \text{dist}(z''; \text{fr}\, G_m) \geqslant \delta,$$

then the same is true for all $n > m$ and also

$$|\log r(z'; G_n) - \log r(z''; G_n)| \leqslant 2|z' - z''|/\delta$$

for all $n \geqslant m$. Hence either both sequences $r(z'; G_n)$ tend to $+\infty$, or both tend to a finite limit. In the limiting case we have:

$$|\log r(z'; G) - \log r(z''; G)| \leqslant 2|z' - z''|/\delta$$

which proves continuity of $r(z; G)$. We have proved even more: $r(z; G)$ is Lipschitzian on compact subsets of G.

10.4.8. $r(z; H)$ attains a finite maximum at some point $\zeta \in H$ in view of Exercise 10.4.5. Consider now Steiner symmetrization w.r.t. a line through ζ parallel to the major axis. This gives $r(\zeta_1; H) \geqslant r(\zeta; H)$, ζ_1 being the projection of ζ on the major axis. Another Steiner symmetrization w.r.t. a line through ζ_1 parallel to the minor axis shows that $r(0; H) \geqslant r(\zeta; H)$, 0 being the center of H.

10.4.9. If $G \in \mathscr{G}$ and the positive, real axis is the half-line of Pólya symmetrization, then the symmetrized domain $G^* \in \mathscr{G}$. Moreover, if $G_0 = K(0; R) \setminus (-R, -r]$, then $G^* \subset G_0$. Hence

$$r(0; G) \leqslant r(0; G^*) \leqslant r(0; G_0) = 4rR^2(R+r)^{-2}$$

(cf. Ex. 2.9.19, 8.1.9). The l.u.b. is actually attained because $G_0 \in \mathscr{G}$.

10.4.10. δ is the root of the equation $4R^2\delta(R+\delta)^{-2} = 1$, i.e. $\delta = R[2R-1 - 2\sqrt{R(R-1)}]$. For suppose, contrary to this, that there exists $G_0 \in \mathscr{G}(R)$ such that $\text{dist}(0; \Gamma) = \delta_0 < \delta$, Γ being the component of $\mathbf{C} \setminus G$ containing ∞. If we symmetrize G_0 w.r.t. the positive real axis and replace G_0^* by $G_1 = K(R) \setminus (-R, -\delta_0]$ which contains G_0^*, then

$$r(0; G_0) = 1 \leqslant r(0; G_0^*) \leqslant r(0; G_1)$$

which is a contradiction because $\delta_0 < \delta$ implies $r(0; G_1) < 1$.

10.4.11. We may suppose that $G = f[K(0; 1)]$ and $f \in S$. We have:
$$r(w_0; G) = (1-|z_0|^2)|f'(z_0)|,$$
where $w_0 = f(z_0)$; hence
$$|f'(z_0)/f(z_0)| = r(w_0; G)\,[|w_0|(1-|z_0|^2)]^{-1} \leqslant (1+|z_0|)\,[|z_0|(1-|z_0|)]^{-1}$$
by Exercise 10.3.12 (i). Hence
$$r(w_0; G) \leqslant |w_0|(1+|z_0|)^2|z_0|^{-1} = 4|w_0|+|w_0|(1-|z_0|)^2|z_0|^{-1} \leqslant 4|w_0|+1,$$
(cf. Ex. 10.3.2 (ii)).

10.4.12. If $0 \in G$ and $w_0 \neq \infty$ is such that $|w_0| = \operatorname{dist}(0; \mathbf{C}\setminus G)$, then $(-\infty, -|w_0|] \subset \mathbf{C}\setminus G^*$ and from the definition of circular symmetrization it follows that $\mathbf{C}\setminus G^*$ is arcwise connected. Hence G^* is of hyperbolic type.

If $G = \{w: \operatorname{im} w > -\tfrac{1}{2}\}$ and the real axis is the line of Steiner symmetrization then $G^* = \mathbf{C}$ which is not of hyperbolic type.

10.4.13. From the conformal invariance of $|dw|/r(w; G)$ (cf. Ex. 8.1.11) it follows that
$$|dw|/r(w; G) = |dz|/r(z; K(0; 1)) = (1-|z|^2)^{-1}|dz|,$$
hence
$$\int_\Gamma [r(w; G)]^{-1}|dw| = \rho(w_1, w_2; G),$$
Γ being the h-segment joining w_1 to w_2. We now associate with Γ a curve $\gamma \subset G^*$ joining w_1 to w_2 in the following manner. Suppose that $w = Re^{i\phi} \in \Gamma$. Then the point $\zeta \in \gamma$ associated with w is the point $\zeta = R$. From $|dw|^2 = dR^2 + R^2 d\phi^2$ it follows that $|d\zeta| \leqslant |dw|$. We consider a ray $0w$, $w \in \Gamma$ and the symmetrized domain G_w^* of G w.r.t. ray $0w$. We have: $r(w; G_w^*) \geqslant r(w; G)$ due to Pólya–Szegö theorem. However, G_w^*, G^* are congruent and a rotation round the origin by the angle Φ carries G^* into G_w^*. Since the inner radius is invariant under motion, $r(w; G_w^*) = r(\zeta; G^*)$. Thus $|d\zeta|/r(\zeta; G^*) \leqslant |dw|/r(w; G)$ and after integration we obtain
$$\int_\gamma [r(\zeta; G^*)]^{-1}|d\zeta| \leqslant \int_\Gamma [r(w; G)]^{-1}|dw| = \rho(w_1, w_2; G).$$
Observe that γ joins w_1 to w_2 in G^* and
$$\rho(w_1, w_2; G^*) = \inf_C \int_C [r(\zeta; G^*)]^{-1}|d\zeta|,$$
the g.l.b. being taken w.r.t. all regular curves joining w_1 to w_2 in G^*.

10.4.14. Observe that $g(w_1, w_2; G) = -\log\tanh\rho(w_1, w_2; G)$, (cf. Ex. 8.5.11) and use the result of Exercise 10.4.13.

10.4.15. f takes real values on the real axis which follows from the uniqueness and symmetry considerations. Suppose that $w = f(z_0)$, $z_0 \neq 0$, and $|w_0| = M(|z_0|, f)$. If we symmetrize G w.r.t. the ray $0w_0$, we obtain the domain G_0 which arises from G after rotation by the angle $\arg w_0$. In view of Exercise 10.4.14:

$$g(0, w_0; G) = -\log|z_0| \leq g(0, w_0; G_0) = g(0, |w_0|; G) = -\log r,$$

where $f(r) = |w_0|$ ($f(x)$ takes on $(0, 1)$ all values between 0 and $\sup_{|z|<1}|f(z)|$, hence it takes the value $|w_0|$). This means that $|f(z_0)| = f(r) = |w_0|$, while $|z_0| \geq r$. Now, f strictly increases on the real axis, hence

$$f(|z_0|) \geq f(r) = |f(z_0)| = M(|z_0|, f).$$

10.4.16. Suppose that $z_0 \neq 0$, $|f(z_0)| = M(|z_0|, f)$ and $w_0 = f(z_0)$. The symmetrized domain G_0^* obtained from G by symmetrizing w.r.t. the ray $0w_0$ becomes G^* after a rotation hence $w_0 = e^{i\alpha}f^*(r^*)$. Now, $g(0, w_0; G_0^*) = -\log r^*$, $g(0, w_0; G) = -\log r$, hence by the result of Exercise 10.4.14: $r^* \leq r$. Thus

$$M(r, f^*) \geq M(r^*, f^*) = f(r) = |w_0| = |f(z_0)| = M(r, f).$$

10.4.17. Suppose that $f \in S$ and $D = f[K(0; 1)]$. If $he^{i\alpha} \in \mathbf{C} \setminus D$ for some $h \in (0, \frac{1}{4})$, then $\mathbf{C} \setminus D$ being connected, any circle $C(0; r)$, $r \geq h$, meets $\mathbf{C} \setminus D$ and this implies that $\mathbf{C} \setminus D^*$ contains the ray $(-\infty, -h]$. For $D_0 = \mathbf{C} \setminus (-\infty, -h]$ we have $r(0; D_0) = 4h < 1$. On the other hand $1 = r(0; D) \leq r(0; D^*) \leq r(0; D_0) = 4h < 1$ which is a contradiction.

10.5.1. h is harmonic in $K(0; 1)$ except for the points z such that $f(z) = a_0$. If $|z| \to 1$, then $\varlimsup h(z) \geq 0$; if $z \to z'$ ($z' \neq 0, f(z') = a_0$), then $h(z) \to +\infty$ and finally if $z \to 0$, then

$$h(z) = -\log|f(z) - a_0| + \log r(a_0; G) + \log|z| + o(1)$$
$$= -\log|a_1| + \log r(a_0; G) + o(1)$$

which implies that $z = 0$ is a removable singularity.

The maximum principle yields now $h(z) \geq 0$ and consequently

$$-\log|a_1| + \log r(a_0; G) \geq 0, \quad \text{i.e.} \quad |a_1| = |f'(0)| \leq r(a_0; G).$$

In case G is a simply connected domain the result follows by subordination.

10.5.2. We may assume that $r(a_0; G)$ is finite. Choose ρ such that $f'(\rho e^{i\theta}) \neq 0$ for all $\theta \in [0, 2\pi]$ and consider the domain $G(\rho) = f[K(0; \rho)]$ whose boundary consists of a finite number of arcs on $\Gamma(\rho) = f[C(0; \rho)]$. We can add at possible singular points of $\partial G(\rho)$ small disks so that the domain $\tilde{G}(\rho)$ so obtained possesses the classical Green function and $G(\rho) \subset \tilde{G}(\rho) \subset G$. The result of Exercise 10.5.1 as applied to $f(\rho z)$ and the domain $\tilde{G}(\rho)$ gives

$$\rho|f'(0)| \leqslant r(a_0; \tilde{G}(\rho)) \leqslant r(a_0; G);$$

make now $\rho \to 1$.

10.5.3. $|f'(0)| \leqslant r(a_0; D_f)$; moreover, Pólya–Szegö theorem gives $r(a_0; D_f) \leqslant r(a_0; D^*)$. On applying the monotoneity property of the inner radius, the result follows.

10.5.4. We may assume that $a_0 > 0$ (otherwise consider $e^{-i\lambda}f(z)$, where $\lambda = \arg a_0$). The domain D^* obtained from D_f by circular symmetrization w.r.t. the positive real axis is contained in $D_0 = \{w: |\arg w| < \tfrac{1}{2}\alpha\pi\}$. The function

$$f(z) = a_0[(1+z)/(1-z)]^\alpha = a_0(1+2\alpha z + \ldots)$$

maps 1:1 conformally $K(0; 1)$ onto D_0 with $f(0) = a_0$, $r(a_0; D_0)$ being equal to $2\alpha a_0$. Now apply Exercise 10.5.3.

10.5.5. We may assume that $a_0 > 0$. We symmetrize D_f w.r.t. the positive real axis and obtain a domain D contained in $D_0 \setminus (-\infty, -R]$. Now, f_0 maps $K(0; 1)$ onto D_0 so that $f_0(0) = a_0$, hence

$$r(a_0; D_0) = f_0'(0) = 4(a_0 + R).$$

Use now Exercise 10.5.3.

10.5.6. By Exercise 10.5.5 we have $1 = |a_1| \leqslant 4R_f$.

10.5.7. If $f \in S$, then $\hat{\mathbf{C}} \setminus D_f$ is connected, hence $C(0; \rho) \subset D_f$ implies $K(0; \rho) \subset D_f$. Use now the result of Exercise 10.5.6.

10.5.8. g omits the values $\exp[2\pi h^{-1}(u+iv)]$, hence on each circle $C(0; \rho)$ there exists a value omitted by g and therefore $R_g = 0$ (Ex. 10.5.5.). Now

$$g(z) = \exp(2\pi a_0/h)(1 + 2\pi a_1 h^{-1}z + \ldots)$$

and the inequality $|2\pi a_1 h^{-1}| \leqslant 4$ follows from Exercise 10.5.5. Equality holds for

$$f(z) = a_0 + h\pi^{-1}\log[(1+z)(1-z)^{-1}].$$

10.5.9. Write the inequality

$$|\varphi'(0)| \leqslant 4(|\varphi(0)| + R)$$

for $\varphi(\zeta) = f[(z_0+\zeta)(1+\bar{z}_0\zeta)^{-1}]$ and $R = 0$.

10.5.10. We have $|f'(z)/f(z)| \leqslant 4(1-|z|^2)^{-1}$ (Ex. 10.5.9) and after integration we obtain:

$$|\log[f(z)/f(0)]| = \left|\int_0^z f'(z)[f(z)]^{-1}dz\right| \leq 4\int_0^{|z|}(1-r^2)^{-1}dr$$
$$= \log[(1+|z|)^2(1-|z|)^{-2}]$$

and this implies the right-hand inequality; the left-hand one is obtained by applying the right inequality to $[f(z)]^{-1}$.

10.5.11. We may assume that $a_0 = f(0) > 0$. On symmetrizing $D_\varphi = \varphi[K(0;1)]$ w.r.t. the positive real axis we obtain a set D^* contained in the right half-plane. Write now the inequality $|f'(0)| \leq r(a_0; D_0)$ of Exercise 10.5.3 for $f(\zeta) = \varphi[(z_0+\zeta)(1+\bar{z}_0\zeta)^{-1}]$ and note that $r(a_0; D_0) = 2|a_0| = 2|f(0)|$.

10.5.12. (i) Suppose that f is not identically 0. Then

$$f(z) = a_n z^n + a_{n+1}z^{n+1} + \ldots \quad \text{with} \quad a_n \neq 0.$$

By the local mapping theorem (cf. [1], p. 131) there exists $\delta > 0$ such that $K(0;\delta) \subset f[K(0;1)]$. Suppose now that there exists $z_1 \in K(0;1)$ such that $f(z_1) = Re^{i\alpha}$ with $R > \delta^{-1}$. Then for some $z_2 \in K(0;1)$ we have $f(z_2) = R^{-1}e^{-i\alpha} \in K(0;\delta)$, i.e. $f(z_1)f(z_2) = 1$ which is a contradiction.

(ii) From $[1+f(z_1)][1-f(z_1)]^{-1} = -[1+f(z_2)][1-f(z_2)]^{-1}$ $(z_1 \neq z_2)$ it follows that $1 = f(z_1)f(z_2)$, which is a contradiction.

10.5.13. (i) Write the inequality obtained in Exercise 10.5.11 for

$$\varphi(z) = [1+f(z)][1-f(z)]^{-1}.$$

(ii) Put $z = 0$ in (i).

10.5.14. The transformation $W = W(Z)$, where $W = \rho z(z+\rho)(1+\rho z)^{-1}$, $t = \frac{1}{2}(z+z^{-1}+\rho+\rho^{-1}) = 2(\rho-1)^2\rho^{-1}Z(1+Z)^{-2}$, maps 1:1 conformally $\{Z: |Z| < 1\}$ onto the W-plane slit along the ray $[\rho, +\infty)$ and the circular arc: $|W| = \rho$, $|\arg W| \leq 2\arcsin\rho^{-1}$, (cf. Ex. 2.9.22). We have: $W = W(Z) = 4\rho^3(1+\rho)^{-2}Z + \ldots$, hence $w = \frac{1}{4}\rho^{-3}(1+\rho)^2 W(Z) = Z + \ldots = f_0(Z)$ is a function which belongs to S and maps $K(0;1)$ onto the domain $G(\rho)$ such that $\mathbf{C}\setminus G(\rho)$ consists of the ray $(r, +\infty)$ and the circular arc $w = r = (1+\rho)^2/4\rho^2$, $|\arg w| \leq 2\arcsin\rho^{-1}$. We have $r(0; G(\rho)) = 1$. Suppose now that $L(r, f) > L(r, f_0)$ for some $f \in S$. We symmetrize $f[K(0;1)]$ w.r.t. the negative real axis and obtain a domain D^* such that $f_0[K(0;1)] \subset D^*$ and $D^* \setminus f_0[K(0;1)] \neq \emptyset$. Thus we have $r(0; D^*) > r(0, f_0[K(0;1)]) = 1$ which contradicts the inequality $r(0; f[K(0;1)]) = 1 \leq r(0; D^*)$.

10.5.15. Obviously $f(z) = F[(t-\bar{t}_0 z)/(1-z)]$, where $\sqrt{t_0} = \frac{1}{2}\theta + i\sqrt{1-\frac{1}{4}\theta^2}$. Similarly as in Exercise 10.5.14 we can show that the extremal domain is the image domain of $K(0;1)$ under the mapping $g(z) = f(z)/f'(0)$ and arises from $H(\theta)$ by a similarity with the ratio $r = |f'(0)|^{-1}$. The set of values not taken by g and situated on $C(0; r)$ is a circular arc subtending the angle $(2-\theta)\pi = \pi\varphi(r)$, where $\pi\varphi(r)$ is the maximal value to be evaluated.

We have: $|f'(0)| = \frac{1}{4}[(2+\theta)^{2+\theta}(2-\theta)^{2-\theta}]^{1/2}$, hence $\varphi(r) = 2-\theta$ satisfies the equation:

$$r = 4[(4-\varphi)^{4-\varphi}\varphi^{\varphi}]^{-1/2}.$$

Bibliography

1. BASIC TEXTBOOKS

1. L. V. Ahlfors, *Complex Analysis*. McGraw-Hill, New York, 1966.
2. H. Behnke, und F. Sommer, *Theorie der analytischen Funktionen einer komplexen Veränderlichen*. Springer, Berlin–Göttingen–Heidelberg, 1955.
3. C. Carathéodory, *Funktionentheorie*, Vols. I and II. Birkhäuser, Basel, 1950.
4. B. A. Fuchs, V. I. Levin, and B. V. Shabat, *Functions of a Complex Variable and Some of Their Applications*, Vols. I and II. Pergamon, Oxford, 1961.
5. M. Heins, *Complex Function Theory*. Academic Press, New York, 1968.
6. E. Hille, *Analytic Function Theory*, Vols. I and II. Ginn, Boston, 1962.
7. A. Hurwitz, und R. Courant, *Vorlesungen über allgemeine Funktionentheorie und elliptische Funktionen*. Springer, Berlin–Göttingen–Heidelberg, 1964.
8. G. W. Mackey, *Lectures on the Theory of Functions of a Complex Variable*. Van Nostrand, Princeton, 1967.
9. R. Nevanlinna, und V. Paatero, *Einführung in die Funktionentheorie*. Birkhäuser, Basel, 1965.
10. S. Saks and A. Zygmund, *Analytic Functions*. PWN, Warszawa, 1965.
11. E. C. Titchmarsh, *The Theory of Functions*. Oxford University Press, 1947.

2. COLLATERAL READING

12. S. Bergman, *The Kernel Function and Conformal Mapping*. Amer. Math. Soc., New York 1950.
13. R. P. Boas, *Entire Functions*. Academic Press, New York, 1954.
14. R. Courant, *Dirichlet's Principle, Conformal Mapping and Minimal Surfaces*. Interscience Publ., New York, 1950.
15. W. H. J. Fuchs, *Topics in the Theory of Functions of One Complex Variable*. Van Nostrand, Princeton, 1967.
16. G. M. Golusin, *Geometrische Funktionentheorie*. VEB Deutscher Verlag der Wissenschaften, Berlin, 1957.
17. W. K. Hayman, *Multivalent Functions*. Cambridge University Press, 1958.
18. W. K. Hayman, *Meromorphic Functions*. Oxford University Press, 1964.
19. M. Heins, *Selected Topics in the Classical Theory of Functions of a Complex Variable*. Holt, Rinehart and Winston, New York, 1962.
20. J. A. Jenkins, *Univalent Functions and Conformal Mapping*. Springer, Berlin–Göttingen–Heidelberg, 1958.
21. G. Julia, *Exercices d'analyse*, Vol. II. Gauthier-Villars, Paris, 1969.
22. J. E. Littlewood, *Lectures on the Theory of Functions*. Oxford University Press, 1944.

23. I. P. Natanson, *Theorie der Funktionen einer reellen Veränderlichen*. VEB Deutscher Verlag der Wissenschaften, Berlin, 1954.
24. Z. Nehari, *Conformal Mapping*. McGraw-Hill, New York, 1952.
25. R. Nevanlinna, *Eindeutige analytische Funktionen*. Springer, Berlin–Göttingen–Heidelberg, 1953.
26. I. I. Priwalow, *Randeigenschaften analytischer Funktionen*. VEB Deutscher Verlag der Wissenschaften, Berlin, 1956.
27. M. Tsuji, *Potential Theory in Modern Function Theory*. Maruzen, Tokyo, 1959.

3. PROBLEMS AND EXERCISES

28. M. A. Evgrafov, Yu. V. Sidorov, M. V. Fedoryuk, M. I. Shabunin and K. A. Bezhanov, *Collection of Problems on the Theory of Analytic Functions* (in Russian). Moscow, 1969.
29. K. Knopp, *Problem Book in the Theory of Functions*. Dover, New York, 1948.
30. G. Pólya und G. Szegö, *Aufgaben und Lehrsätze aus der Analysis,* Vols. I and II. Springer, Berlin, 1925.
31. M, R. Spiegel, *Theory and Problems of Complex Variables*. Schaum, New York, 1964.
32. L. I. Volkovyskii, G. L. Lunts, and I. G. Aramanovich, *A Collection of Problems on Complex Analysis*. Pergamon, Oxford, 1965.

Index

Abel's limit theorem, 65
Almost uniformly convergent sequence of functions, 57
Almost uniformly convergent series of functions, 57
Analytic continuation of a function, 95
Analytic element, 95
Analytic function, 19
    univalent in a domain, 23
Area of a spherical triangle, 10
Area theorem, 131
Argument principle, 54

Bergman kernel function, 119
Bernoulli numbers, 63
Bessel function, 68
Biberbach–Eilenberg function, 137

Cauchy's coefficient formula, 61
Cauchy's integral formula, 38
Cauchy's theorem
    for a rectangle, 38
    for simply connected domains, 38
    homological version of, 38
Cauchy–Hadamard formula, 59
Cauchy–Riemann equations, 19
Center of Taylor's series, 60
Chain, 33
Circular symmetrization, 134
Circulation along a contour, 122
Close-to-convex function, 104
Compact family of analytic functions, 75
Complete analytic function, 95
Complete Legendre elliptic integral, 105
Complex function, 19

Complex potential
    of an electrostatic field, 124
    of a flow, 121
Conformal mapping, 23
Conjugate harmonic functions, 21
Continuation of a function
    analytic, 95
    direct analytic, 95, 97
Contour, 34
Convergence
    of an infinite product, 82
    disk of, 59
Convergent infinite product, 82
Convex hull of a set, 6
conv $\{z_1, z_2, \ldots, z_n\}$, 6
Criterion
    Marty's, for normality, 75
    Montel's, 75
Curve
    Jordan, 34
    regular, 17, 18, 33
    starshaped image, 94
Cycle 33
    homologous to zero, 38

Decomposition of an entire function, 84
Derivative
    of a complex domain, 19
    Schwarzian, 98
Differentiable complex function, 19
Direct analytic continuation of a function, 95, 97
Disk of convergence, 59
Domain
    Jordan, 110
    of hyperbolic type, 135
    regular with respect to the Dirichlet problem, 113

Eilenberg–Biberbach function, 137
Elliptic linear transformation, 15
Elliptic modular function, 99
Electrostatic potential, 124
Entire function, 76
    of finite order, 88
Equipotential lines, 121
Essential isolated singularity, 41

Fibonacci sequence, 62
Field
    of force, 121
    stationary, 121
    two-dimensional electrostatic, 124
Flow
    complex potential of, 121
    function, 121
    stationary two-dimensional, 121
Formula
    Cauchy's coefficients, 61
    Cauchy's integral, 38
    Cauchy–Hadamard, 59
    for nth derivative, 39
    Green's, 125
    Jensen's, 80, 81
    Legendre's, 107
    Poisson's, 111
    Pringsheim's interpolation, 85
    Taylor's 42
Formulas, Schwarz–Christoffel, 100
Fourier series representation, 112
Function
    analytic, 19
    analytic, univalent in a domain, 23
    Bergman kernel, 119
    Bessel, 68
    Biberbach–Eilenberg, 137
    close-to-convex, 104
    cn, 105
    complete analytic, 95
    complex, 19
    differentiable complex, 19
    dn, 105
    elliptic modular, 99

Function
    entire, 76
    entire, of finite order, 88
    flow, 121
    gamma, 73
    Green's, 114
    harmonic, 21
    holomorphic, 19
    integrable complex-valued, of real variable, 33
    Koebe's, 131, 166
    Pick's 133
    regular, 19
    subordinate, 92
    Weierstrass, ζ, 106
    Weierstrass, $\wp$ 106
    Weierstrass, σ 106
Function element, 95
Functions, conjugate harmonic, 21

Gamma function, 73
Gauss and Lucas theorem, 6
Gauss theorem, 125
Genus of a sequence, 84
Goursat lemma, 38
Green's formula, 125
Green's function, 114

Hadamard's three circles theorem, 90
h-area of a regular domain, 18
Harmonic function, 21
Harmonic measure of a system of arcs, 114
h-boundary rotation, 17
h-distance, 18
h-length of a regular curve, 18
h-line at infinity, 17
h-parallels, 17
h-rotations, 17
h-segment, 17
h-translation, 17
h-triangle, 18
Helly's selection principle, 260
Herglotz theorem, 127
Holomorphic function, 19
Homotopic paths, 98

INDEX

de l'Hospital's rule, 43
Hurwitz theorem, 76
Hyperbolic distance, 17, 18
Hyperbolic length of regular curve, 18
Hyperbolic linear transformation, 15
Hyperbolic metric, 111
Hyperbolic motions, 16
Hyperbolic straight lines, 16

Index of a point, 36
Infinite product, 82
Inner radius
    of a domain at a point, 111
    of a simply connected domain, 133
Isolated singularity, 41
Intensity, 122
Invariant point of a linear transformation, 14
Involution, 14
Integrable complex-valued function of a real variable, 33
Integral
    complete Legendre elliptic, 105
    line, of a complex-valued function, 33
    Schwarz–Christoffel, 105
    Stieltjes, 127
    unoriented line, 34

Jensen's formula, 80, 81
Jordan curve, 34
Jordan domain, 110
Jordan's theorem for curves starlike w.r.t. an origin, 36

Koebe function, 131, 166
Koebe one quarter theorem, 136

Luarent coefficients, 67
Laurent expansion, 45
Laurent series, 66
    principal part of, 66, 67
Legendre formula, 107
Line integral of a complex valued function, 33

Lines
    equipotential, 121
    of flow, 121
Liouville's theorem, 40
Lucas and Gauss theorem, 6

Marty's criterion, 75
Mercator projection, 163
Mittag–Leffler representation 77
Montel's compactness condition, 75
Montel's criterion, 75
M-test of Weierstrass, 57

Nevanlinna's characteristic, 81
Normal family of analytic functions, 75
n-sheeted unit disk, 98

Order
    of an entire function, 88
    of a pole, 41
Osgood–Taylor–Carathéodory theorem, 110

Parabolic linear transformation, 15
Parallel axiom, 16
Parseval's identity, 59, 119
Pentagram, 103
Phragmén–Lindelöf theorem, 115
Pick function, 133
Point
    regular, 61
    singular, 61
Points of a hyperbolic plane, 16
Poisson's formula, 111
Poisson's kernel, 112
Pole, 41
Polynomial represented as a product, 84
Potential, 124
    electrostatic, 124
    complex, of an electrostatic field, 124
    velocity, 121
Power series, 58

Pólya symmetrization, 134
Principal part of the Laurent series, 66, 67
Pringsheim's interpolation formula, 85
Product
    convergent infinite, 82
    infinite, 82
    Weierstrass, 84

Radius
    inner, 133
    of a domain, 111
    of convergence of a power series, 59
Reflection of a point with respect to a circle, 12
Regular curve, 17, 18, 33
Regular function, 19
Regular part of the Laurent series, 66
Regular point, 61
Removable singularity, 41
Representation
    Fourier series, 112
    Mittag–Leffler, 77
Residue of an analytic function, 43
Residue theorem, 45
Riemann sphere, 8
Riemann theorem, 110
Robin's constant, 126
Rouché's theorem, 54

Schwarzian derivative, 98
Schwarz–Christoffel formulas 100
Schwarz–Christoffel integral 105
Series
    power 58
    Taylor's 60
Sequence
    Fibonacci 62
    of functions almost uniformly convergent 57
Single-valued branch of a function 95
Singular part of the Laurent series 66, 67
Singular point 61

Singularity
    essential isolated 41
    isolated 41
    removable 41
Sink of intensity 122
Solomon's seal 103
Source of intensity 122
Sphere of Riemann, 8
Spherical derivative of an analytic function, 75
Spherical distance between two points, 9
Spherical image of a complex number, 8
Starshaped image curve, 94
Stationary field, 121
Stationary two-dimensional flow, 121
Steiner symmetrization, 134
Stereographic projection, 8
Stieltjes integral, 127
Stieltjes–Osgood theorem, 75
Stolz angle, 65
Subordinate function, 92
Symmetric points w.r.t. a circle, 12
Symmetrization
    circular, 134
    Pólya's, 134
    Steiner's, 134
Symmetrization principle, 136
Szegö, 134

Taylor coefficients, 61
Taylor's formula, 42
Taylor's series, 60
Theorem
    Abel's limit, 65
    area, 131
    Cauchy's, for a rectangle, 38
    Cauchy's, for simply connected domains, 38
    Cauchy's, homological version, 38
    Hadamard's three circles, 90
    Herglotz's, 127
    Gauss', 125
    Gauss–Lucas, 6
    Hurwitz's, 76
    Jordan's, 36
    Koebe's one quarter, 136

Theorem
    Liouville's, 40
    on decomposition of an entire function Weierstrass, 84
    Osgood–Taylor–Carathéodory, 110
    Phragmén–Lindelöf, 115
    residue, 45
    Riemann's, 110
    Rouché's, 54
    Stieltjes–Osgood, 75
    Toeplitz's, 7
    two constants, 115
    uniqueness, for the velocity potential, 121
    Weierstrass' mean value, 35
Toeplitz's theorem, 7
Toeplitz's transform, 65
Two constants theorem, 115
Two-dimensional electrostatic field, 124

Uniqueness theorem for the velocity potential, 121
Unoriented line integral, 34

Velocity potential, 121
Vortex, 122

Weierstrass ζ-function, 106
Weierstrass $\wp$-function, 106
Weierstrass mean value theorem, 35
Weierstrass M-test, 57
Weierstrass primary factors, 84
Weierstrass product, 84
Weierstrass σ-function, 106
Weierstrass theorem on decomposition of an entire function, 84
Winding number, 36